Anatomical Forms

ALEMBICS: PENN STUDIES IN LITERATURE AND SCIENCE

Mary Thomas Crane and Henry S. Turner, Series Editors

A complete list of books in the series is available from the publisher

Anatomical Forms

The Science of the Body in Early Modern Women's Poetry

Whitney Sperrazza

PENN

UNIVERSITY OF PENNSYLVANIA PRESS

PHILADELPHIA

Published by
University of Pennsylvania Press
Philadelphia, Pennsylvania 19104–4112
www.pennpress.org
Printed in the United States of America on acid-free paper
10 9 8 7 6 5 4 3 2 1
Library of Congress Cataloging-in-Publication Data
Names: Sperrazza, Whitney, author.
Title: Anatomical forms : the science of the body in early modern women's poetry / Whitney Sperrazza.
Description: Philadelphia : University of Pennsylvania Press, 2025. | Series: Alembics: Penn studies in literature and science | Includes bibliographical references and index.
Identifiers: LCCN 2024050259 | ISBN 9781512827590 (hardback)
Subjects: LCSH: English poetry—Early modern, 1500-1700—History and criticism. | English poetry—Women authors—History and criticism. | Human anatomy in literature. | Human body in literature. | Literature and science. | Poetics.
Classification: LCC PR545.W6 S64 2025 | DDC 821.009/9287—dc23/eng/20241118
LC record available at https://lccn.loc.gov/2024050259
Hardback ISBN 978-1-5128-2759-0
eBook ISBN 978-1-5128-2760-6

CONTENTS

NOTE ON TRANSCRIPTIONS AND CITATIONS

In quoting from early modern texts, I have followed original spelling and capitalization, while silently modernizing *u/v*, *i/j*, and *vv/w*. Titles of early modern books also maintain original spelling, but I have capitalized titles in the body of the text. In the endnotes, I lowercase all early modern title words that were originally rendered in uppercase letters.

The endnotes refer to the *Oxford English Dictionary* in abbreviated form (*OED*), and all such citations are from the *Oxford English Dictionary* (OED Online), Oxford University Press, https://www.oed.com.

A full bibliography of sources cited in the endnotes is available online at https://tinyurl.com/2p8xaf22.

INTRODUCTION

We find a curious note in one reader's praise of Jane Barker's seventeenth-century poetry collection. In Barker's "purest Verse," the reader claims, we can learn "the art of Physick, and Anatomie."[1] The note is curious, first, because poetry seems an odd place to go for an anatomy lesson, certainly with our modern understanding of discrete disciplinary expertise. Second, Barker is a poet best known for her politics as an outspoken Jacobite after the 1688 Glorious Revolution.[2] Scholars do not go to Barker's poetry for science, let alone specific details about anatomy or "physick" (an early word for medicine). And yet this contemporary reader (identified only as "A Table") is adamant that both "Medicine" and "Poetrie" are "at once improv'd" by Barker's pen.[3]

Following this reader's cue, we might turn to the one poem in which Barker references anatomy and physic in the title—a poem called "She begining to study phisick, takes her leave of poetry, so falls into a long degression on anatomy."[4] As she says "fare well" to poetry in order to make "new acquaintance" with anatomy and physic (1–3), Barker imagines anatomists leading her through rooms of a grand palace, each of which correspond to a body part (the ventricles are the kitchen; the liver, the drawing room; the stomach, the dining room). Readers of early modern literature will recognize this structure, perhaps from Edmund Spenser's famous poetic anatomical allegory, the House of Alma. But if allegory is Barker's aim, she is surprisingly literal about it. Spenser's errant knights remain unaware of the anatomical structuring of Alma's castle, and Spenser's thick allegorical descriptions leave little room for explicit anatomical language. Conversely, Barker's poet establishes her somatic setting right from the outset with references to bones, skin, "fretwork muscles," arteries, veins, nerves, and members (19–22). "We began our journey with the blood," Barker's poet recounts, fusing her poem's structure to William Harvey's theory on the circulation of blood, proposed in his 1628 *De Motu Cordis* and still hotly debated when Barker was writing (66).[5] As the poet follows Harvey, her Virgil-esque guide, through the anatomical body, she focuses on the actions of each organ. The liver is actively "sanguificating" and "severing" (52–53); the heart (like a great "furnace" at the center) is "boiling over" and "runing out" (60). And the poet herself starts "meander[ing]" with the "purple flood" of blood (67). The poem takes the form of a compendium of anatomical knowledge, moving through historical foundations ("now Bartholine the first of all this row" [13]) and into contemporary debates ("Harvey that hipotheses deny'd" [82]), and the poet's journey into anatomical knowledge becomes just as active as the body Barker represents: "then passed we" (47); "then came we" (48); "from thence we went" (56); "we began again" (86); "follow me" (65).[6] This 138-line poem begins with 101 continuous lines, primarily in rhyming couplet, followed by three separately numbered stanzas (i, ii, iii) in which the meter and rhyme are more irregular.[7] The simple, sustained pattern of rhyming couplets grounds the anatomical knowledge represented in the first part of the poem, while the enclosed, circular rhymes of the later stanzas (the last, for instance, ending in a quintain rhyming *abbab*) formally reinforces Harvey's theory of circulation.

We are getting closer, perhaps, to understanding what Barker's reader meant by medicine and poetry being "at once improv'd" in her "purest Verse." The form of Barker's poem emulates, even *seems to enact*, the seventeenth-century body's internal, organic, and dynamic form, powered by the "circles of the circulating blood" (91). It is this specific confluence of poetic and scientific forms that interests me in this book. In fact, it is not just a confluence (which implies a merging or intermingling of once separate things), but rather an inextricability of poetic and anatomical forms in early modern poetry that lies at the heart of my investigation. If "poetry is capable of thinking thoughts that are particular to its techniques and that cannot be thought elsewhere and by other means," as Colleen Ruth Rosenfeld argues, then the anatomical knowledge on offer in Barker's poem is importantly different from the knowledge on offer in Harvey's *De Motu Cordis*, even as she champions and reiterates Harvey's theory.[8] Barker says "fare well fare well, kind poetry my friend" as she crafts a poem about the function of different body parts, the relationship of parts to one another, and the heart as both "fountain" and "furnace" of the whole structure. The very form of Barker's anatomical study resists her farewell act, as if, in the same parting breath, she realized that poetry is good to think with—that is, good to think about anatomy and physic with.

Anatomical Forms is a book about early modern women's scientific work, told through an archive of poetry. Looking backward from Barker's poem to trace forms of women's poetic-scientific work, I argue that poetry was a scientific instrument, wielded by women writers to study, explore, shape, and sometimes challenge sixteenth- and seventeenth-century developments in anatomy. In each chapter I pair Englishwomen's writing with transnational anatomical texts, seeking to understand the inscriptive technologies that simultaneously animate the bodies represented within their pages and engineer innovative relations between these texts and the bodies that craft and read them. Some of the poets I consider engage directly with contemporary scientific discourse and debate (Margaret Cavendish and Hester Pulter); for others, the scientific connection is more oblique (Mary Sidney Herbert, Mary Wroth, and Aemilia Lanyer). In both cases, my aim is to excavate the shared formal and material practices of anatomical study and poetic craft, the titular *anatomical forms* of this book. In the pages that follow, *form* is always both textual (entangled in the representational and material conditions of the text) and somatic (of the body, with a shape as well as a feel). When coupled with the adjective *anatomical*, form orients my reading of poetic and scientific work around early modern experiments with the relationship between bodies and texts—experiments that were pervasive and potent in what historians of science have called "the century of anatomy."[9] This coupling is also a persistent reminder that the work of form contributed substantively to the epistemological project of early anatomical science. The concept of *anatomical forms* insists that poetry is more somatic than we have understood and that anatomy is more textual than we have understood.

Central to my inquiries is the relationship between poetic language and poetic form. I do attend to *what* poems say, but much more often I am interested in *how* poems say, and my close readings are attuned to the specific materials and techniques through which that *how* manifests. I am especially curious about the particular kinds of knowledge poetry produces as poets leverage the material conditions that make something poetry rather than prose. For early modern poets cued to the textures and ecosystems of their writing materials, embedded in "a media landscape in which the imprecise boundaries between literary and textual craft were profoundly generative," poetry was a mode of inscription that refused to obscure the material conditions of writing and reading.[10] It enlivened the surface on which it was inscribed as it took form through and with that surface.[11] In the sixteenth and seventeenth centuries, anatomy was, likewise, a practice of enlivening surfaces. The process of dissection—reclaimed as the foundational method for anatomical study in the period—penetrated the body's surface in order to expose, study, and know the body's insides. The subsequent, and still very new, process of representing and circulating anatomical knowledge through text relied on animating the surface of the textual page in ways that recaptured some of the dimensionality, texture, and tactility of dissection as the privileged mode of knowledge making. Coupling poetry's unique relationship to writing matter and anatomy's complex theorization of surfaces, this book focuses on the processes of knowledge making as they are jointly bound to bodies and materials.

The Archive of Women's Writing

It is through women's poetry that we can most thoroughly excavate the complex tension between creativity and vulnerability generated by early modern experiments with translating fleshly body to flat page. Women's writing serves as this book's primary site of investigation for two important reasons. First, as foundational feminist social histories have demonstrated, early modern women's lived experience was fundamentally different from men's within the gendered politics of early modernity.[12] If, following Elizabeth D. Harvey's contention, we can look to early modern England as a historical site where it is "possible to witness an intensive articulation of the power and danger associated with tactility,"[13] then we must also recognize that those tactile relations were deeply implicated in and determined by the period's social and political power structures, particularly those surrounding sex and gender. Women's bodies were subject (as in subjected, overpowered, controlled, objectified) to forms of touch, handling, and penetration that were different from men's bodies—including, as we will see, as objects of anatomical study. "In a culture that construed women's bodies as inherently open and improperly bounded," as Laura Gowing argues, "touch and intimacy were highly charged for women," a politics of touch that would have been even more heightened for black and brown women, poor women, dependent serving women, and enslaved women.[14] As touch increasingly became the privileged mode of anatomical knowing in the period, women would have recognized the implications of the hand's newfound epistemological power in relation to the systematic study of bodies.

Second, early modern women had extensive experience with—and, therefore, expertise in—a diverse range of media forms. Overwhelming extant evidence reveals that women's expressive work in domestic and public spaces encompassed textual, textile, and culinary activities.[15] In Whitney Trettien's study of the "radical, creative bookwork" practiced by women at the Little Gidding religious house, for instance, she illuminates the women's attentiveness to "polyphonic and variable" textual forms.[16] When we "attend to the particularities of making, and the sensuousness of books as the products of human hands," we must also consider, as Joshua Calhoun posits, that early modern women possessed "what we might call advanced literacy in the reading of textual media" because of their extensive work with textiles and paper as everyday technologies.[17] As the rich body of critical work on early modern women writers continues to establish, women's media expertise was anchored in touch—in the highly tactile practices prompted by their daily work and the activities in which they felt most authorized to participate.

Women writers, we might hypothesize (as, indeed, this book does), understood the hand as an instrument of form and forming. If anatomical science was increasingly harnessing the epistemological power of the hand, women writers were primed to engage with the processes of formation underlying that epistemological project. These twinned somatic ways of being—a heightened attention to the politics of touch and penetrative possibilities of surface, along with a highly tactile media literacy—facilitated women writers' participation and innovation in the capacious reimaging of bodies and texts that we see in the wake of Renaissance science. By centering women poets in my exploration of handwork as a complex site of knowledge making, I take up calls by scholars such as Pamela H. Smith and Wendy Wall in their arguments to expand the ranks of early modern scientific communities beyond the period's formal institutions of science. Smith urges us to attend to "artisanal literacy," or the process of gaining knowledge through "experience and labor," which relies on an intimate relation to "material objects and the manipulation of material" rather than the "corpus of knowledge contained in texts."[18] Turning to domestic communities of practice, Wall excavates the exceptional "economies of knowledge enmeshed in the intertwined acts of making, writing, and reading," particularly the practices of recipe making and preserving.[19] My focus on women poets builds on the foundational work of these scholars by, first, turning our attention quite explicitly to form as both ontological and epistemological category. Second, I wield the analytical frameworks that feminist theory and gender studies make possible. Applying these frameworks to early scientific work on bodies allows us to consider the immense implications such histories still have for how female, queer, trans, and nonbinary bodies are treated today within official scientific institutions.

Anchored in the specificity—and heterogeneity—of women's embodied experiences in early modern England, *Anatomical Forms* turns to women's poetry in order to revise and relocate the literary and cultural sites where we search for the genealogies of scientific discourse and practice. Scholars often describe the critical history of early modern literature and science studies in terms of first- and second-wave thinking. The first wave of criticism read literature *through* science and laid the foundation for interdisciplinary analysis by, problematically, ceding authority to science as the driving force behind the knowledge making at work (a critical impulse scholars have referred to as "scientism").[20] More recently, the second wave of critical work in the field, led by Liza Blake, Mary Thomas Crane, Jenny C. Mann, Carla Mazzio, Debapriya Sarkar, and Henry S. Turner, insists on a dialogic relationship between early modern literary and scientific spheres, one that illuminates the shared discourses and "modes of knowing" (the procedures through which we come to know) circulating in the period's "pre-disciplinary organization of knowledge."[21] Representative of this new critical turn, the essays in Mann and Sarkar's special issue of *Philological Quarterly*, "Imagining Early Modern Scientific Forms," bring the widely hailed "return to form" in literary studies into direct contact with early modern science's "imaginative engagements with the problem of form."[22] Mann and Sarkar urge us to approach "literary form as an engine of knowledge production" replete with complex "philosophical texture[s]" that have much to offer historians of science.[23] "How might the specificity of *early modern* imaginative forms," Mann and Sarkar ask, "provide new avenues of connection and differentiation between humanistic and scientific methods?"[24]

Even as such interdisciplinary work continues to productively complicate questions of literary form and expand the bounds of what counts as the history of science, the categories of both "form" and "science" have remained surprisingly stable. The rough contours of both what we read as *form* and what we read as *science* continue to be established by a limited (and limiting) archive of texts, dominated by the writings, ideas, and case studies of men—primarily white, educated, upper-class men.[25] Our inquiries, in other words, remain bounded by the terms and paradigms set by a very narrow group of early modern writers and practitioners.[26] As we navigate the promising shift Mann and Sarkar identify as a "new methodological paradigm in literature/science studies," we need to bring new writers into the conversation and learn from them how to deconstruct our entrenched categories of *form* and *science*.[27]

The writers central to my argument, though, might not seem "new" to many readers. Margaret Cavendish is already a well-established figure in both literary and history of science studies and, for many, she remains a singular, shining example of women's direct participation in early modern scientific work. Correspondingly, Mary Sidney Herbert, Mary Wroth, and Aemilia Lanyer are increasingly canonical figures in the field of early modern literary studies. Danielle Clarke has argued that Lanyer's *Salve Deus Rex Judaeorum* (1611), which I take up in Chapter 4, is "now one of the most widely read religious poems of the seventeenth century."[28] And, while Hester Pulter may be the most recently recovered writer (her manuscript only rediscovered in 1996), she is already a familiar name on early modern syllabi, largely due to the robust, open-access digital edition of her poetry led by Leah Knight and Wendy Wall.[29] For my purposes, the stature of these writers is precisely the point. This is not a book about recovery; it is a book about recognizing expertise when we see it.[30] All of the women poets I discuss have been well recovered and, yet, only Cavendish consistently earns a place in histories of science (although, as we will see, Pembroke and Pulter are frequently cited in the importantly distinct history of *women's* science). *Anatomical Forms* situates the writings of all of these poets at the center of early modern anatomical inquiry, and simultaneously brings into focus a feminist history of poetic form centered on material practice. Following Margaret J. M. Ezell's important call to "re-vision" feminist literary history, this book does not position any of these writers as extraordinary or anomalous.[31] Instead, I use these poets' acute attention to form and materiality to construct "a new framework of understanding," one with significant implications for the fields of literature and science studies, poetics, book history, and feminist literary studies.[32]

From the outset, it is important to recognize the advantages the poets of my archive would have had in early modern England—advantages that enabled them to undertake the deep engagement with contemporary science that I find in their writing, and advantages that have afforded them a place of privilege in later studies of their work. These are white, educated, primarily upper-class women, and their national, racial, and class-based privileges gave them access to literacy training and the protection of influential men as they navigated the tricky gender politics of early modern England's patriarchal system. My analysis of their writing is anchored in foundational premodern critical race studies work that reminds us to remain alert, in Ania Loomba's words, "to the difference that the color of patriarchy (and the color of feminism?)" makes to our reading of historical texts.[33] I position these writers firmly within the history of anatomical science, always mindful that theirs is a science by white women (Mary Wroth's innovative theorization of surface anatomy is partly constructed through a comparison between her own white skin and the black skin of "Indians") and, predominantly, for white readers (Lanyer's gynecological thinking is premised on gathering a group of elite, white mothers as her text's primary audience).[34] Decades of intersectional feminist work has also thoroughly demonstrated the importance of always taking "women" as a heterogenous category, and the women writers of this book are not a homogeneous group of white women writers. Each chapter draws out the experimental heuristics that come from different life experiences and, as a result, each writer offers a different way into anatomical inquiry.

Science of the Body

By turning to women poets, *Anatomical Forms* moves margins to center and, in the process, models a more dynamic and inclusive approach to the study of early modern literatures and sciences. There is, I demonstrate in this book, a critical intervention in early modern literature and science studies that can only be made by turning to women's writing, an archive that immediately defamiliarizes some of the field's central categories of inquiry. Anatomy, for instance, is most often treated as a science of spectacle. Anatomical science progressed rapidly in the sixteenth and seventeenth centuries, so significantly that science and art historian Domenico Laurenza dubs the sixteenth century the "century of anatomy," arguing "never before or since was anatomy so important."[35] Following Jonathan Sawday's influential argument on the Renaissance "culture of dissection"—his term for the "network of practices, social structures, and rituals surrounding [the] production of fragmented bodies" in early modern England—the messy act of dissection, as well as the anatomy theater as its most spectacular site, continue to be central to our understanding of what was, in practice, a much more expansive, theologically and aesthetically informed emerging science of the body.[36] As we will see in Chapters 2 and 4, there is a long history of women's autopsy and embalming work within religious and familial contexts, and women dominated the practices of midwifery and herbal medicine throughout the sixteenth and early seventeenth centuries, practices that helped to shape and were increasingly shaped by developing anatomical knowledge.[37] The history of anatomy is also a history of women's bodies as a highly charged site of investigation—the female form as a privileged *object* of inquiry. As Katharine Park's research on gender and the history of human dissection has shown, "understanding the secrets of women" was one of the "principle goals" of Renaissance medical writers.[38] If we limit our studies to a "culture of dissection" read within and through the work of male writers, how can we provide informed and nuanced accounts of the gender ideologies undergirding early anatomical study? What do we miss by focusing on dissection as the defining feature of Renaissance anatomy?

These are some of the questions driving *Anatomical Forms*, and I approach them not by expanding what counts as science, but rather by shifting where and how we look for science in the archives.[39] Early modern anatomy will look different at the end of this book—its central keywords will not be *dissection*, *fragmentation*, and *spectacle*, but rather *surface*, *texture*, *handwork*, and *form*—because when we read women's poetry as scientific work, a different science of the body (and a different *approach* to the science of the body) emerges. Detouring, then, from histories of literature and science intent on tracking networks of influence between key figures (Milton's reading of William Harvey; Helkiah Crooke's reading of Shakespeare; Cavendish's reading of Jan Baptista van Helmont), what I offer instead is a web of scientific engagement that threads through early modern poetry without clear or explicit corresponding social networks. Additionally, the question of direct influence between early modern anatomical and poetic texts is not my subject here. Readers who are interested in doing so will be able to trace some evidence for that kind of influence (we know, for instance, that the Sidney family library at Penshurst Place included a copy of Helkiah Crooke's *Mikrokosmographia*), but the connections I trace in this book are conceptual and methodological.[40] By studying connections within the work of writers and practitioners participating in speculation about the body and its representation in texts, we illuminate an early modern culture of knowledge and practice circulating around questions of form, both textual and somatic. This is a wide, rather than deep, archive that testifies to the concord between poetic craft and the developing science of the body. Understanding the theories and methods of poetic form and anatomical science as overlapping spheres of cultural practice allows us to locate science in sites where we have not yet looked for it. What emerges from the web I trace is an early modern scientific poetics that forces us to understand literary texts as scientific instruments, just as useful and generative to the writers and thinkers of the period as the scalpel and the microscope.

Guided by the tenets of feminist science studies, my interdisciplinary approach aims to draw out the convergences between literary and scientific inquiries and to trace how and why women employed poetics for scientific thinking. Feminist scholar and evolutionary biologist Banu Subramaniam identifies three "sites or levels of analysis" in the genealogy of feminist science studies, a critical tradition with a long history of radically disrupting established epistemological narratives.[41] All three sites of analysis inform my close readings and interventions in the chapters to come. The first is documenting the histories of women scientists and then using those histories to confront ongoing problems and assumptions surrounding underrepresentation. In many ways, this is the underlying warrant driving *Anatomical Forms*, and it is my hope that by the end of this book, readers will be able to confidently situate five new writers within the history of early modern anatomy. Because I rely on canonical literary writers, though, I complicate the question of underrepresentation (and recovery) and crack open space to think critically about method. How do we do the work of literature and science studies? What counts as evidence of scientific work? How can we return to the archive attuned to the epistemological power of the fleeting allusion, the loose accord, the break in formal pattern, the disturbing of formal convention?

In addressing these questions, we can also use feminist science studies concepts to press on the status of women's writing (and questions of underrepresentation and recovery) within the field of early modern literary studies. Standpoint epistemology theory, for instance, is especially good at deconstructing how the production of knowledge is shaped by and helps to shape surrounding politics, a project that directly overlaps with keen methodological interventions articulated by scholars of women's writing.[42] As Margo Hendricks has argued, "in order to practice a feminist historiography that acknowledges not only the diversity of women in Renaissance and early modern England but also the politics of patriarchal historiography when dealing with these women," we need to step back into the archives with new parameters and expectations for "what constitutes 'factual evidence' and 'truths' about Renaissance societies and their cultural practices."[43] Holding Hendricks's argument together with feminist science studies, this book persistently asks: what are the critical paradigms through which we evaluate, read, and situate women's writing?[44] My aim in this book is to model one way we might think otherwise about the work of women writers.

Subramaniam's second site of analysis in feminist science studies is challenging the persistent tendency in the history of science to think about science as a "culture of no culture."[45] The historical specificity of critical work in early modern literary and science studies already mounts a strong rebuttal to the "culture of no culture" approach (or, science as a "view from nowhere"). Including women writers in these histories expands and nuances our understanding of the specific times, places, and bodies within which science was practiced in early modern England. And, as we will see, women writers consistently frame their scientific work through their embodied, specific, lived experiences. They are situated knowledge practitioners and thinkers, learning in their bodies, all too aware of the limitations of their positions.[46] Throughout this book, while I resist framing my study of these writers' works through their autobiographies, I do consider the different ways these poets use their experiences as a foundation from which to theorize and experiment with modes of knowing. Pulter's poetry, for instance, which I consider closely in Chapter 5, is so often anchored by a first-person, semi-autobiographical voice. We could (and should) read Pulter's poetic "I" within the rich tradition of scholarship on poetic subjectivity in Renaissance lyric.[47] But when we also approach Pulter's poetic voice as a model of situated knowledge-making, the "particular and specific embodiment" of her poetics becomes a significant response to the increasing trend toward objectivity in the seventeenth-century study of bodies.[48]

Finally, Subramaniam outlines a third site of feminist science studies work that interrogates the development and production of scientific knowledge through the critical and political frameworks of sex and gender.[49] The history of anatomy is a history most often told through the archives of learned institutions and the writings of the men associated with them, creating, as Sawday describes it, "a patrilineal diagram of understanding, with medical knowledge passed from father to son in a line of descent."[50] By locating a history of anatomy—a deep engagement with the field's methods, as well as its knowledges—in women's poetics, this book goes some way in tracing an alternative diagram of understanding. It is crucial to remember that the field of early modern anatomy as it was developed within colleges, universities, and learned institutions was a science shaped not only through the bodies of marginalized people (most often white bodies codified as criminal, poor, and working class), but also through women's bodies as a kind of absence. Female corpses were especially difficult to procure for dissection, and while, in part, these circumstances resulted in the cult of curiosity surrounding the "secrets of women" that Park documents, it also resulted in a science of the body developed by, through, and for male bodies.[51] Patriarchal ideologies are baked into the very fabric of modern-day Science (capital "S") surrounding the human body—the privileged methods, the resulting knowledges, and the institutional structures. The writers of this book will illuminate for us paths not taken in that history, even as they sometimes think with and reinforce some of those same ideologies. Their work, to put it simply, gives us new tools with which to think critically about how scientific knowledge has been shaped.

Anatomical Forms

I offer the concept of *anatomical forms* as one of those new critical tools, a particularly powerful one that has implications for both literary and history of science studies. One of my central goals is to craft my argument *with* rather than *about* these writers and, as a result, the interanimation of scientific and poetic forms at the heart of this book emerges, in large part, from the writers I study.[52] We find the most explicit invitation to consider something like "anatomical forms" in Margaret Cavendish's *Poems, and Fancies*, the poetic-scientific treatise I take up in Chapter 1. In one of the book's prefaces, Cavendish draws on anatomical language to distinguish between "fancy" and "words": "Fancy the Form is, Flesh, Blood, Skin, and Bone; / Words are but Shadows; Substance they have none."[53] In this comparison, Cavendish outlines a poetic theory that values "fancy" (her preferred term for the complex workings of the imagination) over language. Equated directly with form, "fancy" here is imbued with both shape and substance, first through the categorization of words as "shadows" and then in Cavendish's use of anatomical language. This is not simply a materialization of fancy as form, but an anatomization—as in, the *rendering anatomical*—of form. We will see in Chapter 1 just how tangible and dimensional the concept of "fancy" is for Cavendish, but with this introductory glimpse at her poetic theory I want to draw attention to how she conceives of form as somatic parts. Poetic forms are fleshly components ("Flesh, Blood, Skin, and Bone") that the poet stitches together to construct the body of the poetic text. Cavendish's forms are not flat, static rhetorical figures but rather the matter of imagination—from the Latin *māteria*, material of which a thing is made. Metaphors, similes, and poetic conceits are figured as materials, as pieces of matter with heft, dimension, and, for Cavendish, *fleshliness*.

Cavendish's anatomizing act prompts us to deconstruct some of the usual paradigms through which we read the intersections of anatomy and poetics. Scholars have tended to focus on anatomical language—the language of bodily partitioning, in particular—as evidence of writers' engagements with the Renaissance "culture of dissection" Sawday identifies. Yet here Cavendish invites us to see anatomy in form, to consider how poetic texts are corporeal entities with surface and depth, porous membranes, animating fluids, and an internal structuring apparatus. If early modern anatomy was "a science that entailed the description of forms,"[54] as Domenico Laurenza asserts, what kinds of textual physicality become possible when poetic and somatic forms collide? When poetry's formal features are understood as fleshly? When poems or poetic collections are presented as bodies by a poet who conceives of the act of writing as the stitching and weaving together of parts? Each poet in this book actively takes up such questions in their work, although not all of them will be so explicit in their assertions of anatomical form as Cavendish. In tracing their different responses to these questions, we find an early modern science of the body that emerges at the level of poetic form and is, in fact, profoundly dependent on form.

What does it mean, though, to understand anatomy as a science of and about forms? It is, first, as Ellen Rooney insists, to "acknowledge[ ] the work that form does in every imaginable text."[55] It is also to recognize that Renaissance anatomists were particularly invested in defining and theorizing the concept of form as it applied to what they called the "human frame." In 1638 English anatomist Alexander Read defined anatomy as the "artificiall separation of the parts of the body by section, practised to attaine to the knowledge of the frame of it, and the use of each part."[56] In practice, Renaissance anatomy was a much more expansive field than Read's definition suggests, and I use the phrase "science of the body" in this book to refer to the range of methods and modes of inquiry—surgery, physic, dissection, postmortem care, anatomical illustration, midwifery, and microscopy—that encompassed the developing field of anatomy.[57] But Read's concise definition betrays an important feature of this burgeoning science: the field's emphasis on form. Emphasizing the "knowledge of the frame" of the human body, along with the "use of each part," Read describes the aims of anatomical practice in terms that resonate with scholars of poetics.

"We may say that 'formalism,'" as Henry S. Turner observes, "begins in the very act of defining a 'significant' part and continues in attempts to determine the relation between part and whole."[58] Formalist readings, like forms themselves, are active and dissective in Turner's definition. Drawing on Bruno Latour's definition of form as "a displacement from ideal to material," Turner urges us toward "a notion of *function*" (the "use of each part") in relation to form: "form *does* things, it doesn't simply mean things."[59] Throughout this book I engage with forms as active entities that are both textual and somatic. Defining form in relation to textuality, I draw on ongoing work in book history and material poetics that takes as its focus "the ostentatious materiality of sixteenth-century literary writing—the fact that it is born of and bound in matter."[60] I think with Turner, Juliet Fleming, Whitney Trettien, Dianne Mitchell, Megan Heffernan, William H. Sherman, and Adam Smyth, holding at arm's length contemporary theories of form as figured language in order to instead foreground the formal structures of textual phenomena.[61] Exploring form in relation to bodies—to *soma*—my close readings gain traction from intersecting histories of art and science that have been vital to understanding the dialogic pages of Renaissance anatomical texts. Anatomy books relied on the interplay of word and image, and I follow Laurenza, Sachiko Kusukawa, James Elkins, and Mechthild Fend in considering the "visual argument" of such texts—the ways in which anatomists "write explicitly about the usefulness of pictures in forming knowledge about nature" *and* "actually make their pictures integral to that knowledge."[62]

I bring these strands together in the book's central concept—"anatomical forms"—and the two words I marshal for this concept work in constant dialogue with one another. *Form* modifies *anatomical* just as often as *anatomical* modifies *form*. This dynamic bodies forth in the central forms around which I shape each chapter: the flap (Chapter 1), the part (Chapter 2), the blank (Chapter 3), the fold (Chapter 4), and the surface (Chapter 5). As we will see, each of these forms apply to bodies and poems. Collectively, this constellation of forms reveals a kind of early modern anatomical thinking that was shaped by and helped to shape the period's distinctly material poetics.

In my close readings of anatomical texts throughout this book, I advance the formal investments of early modern anatomy as a kind of poetic practice and, in the process, demonstrate how the anatomical archive proves a rich seam of evidence for tracing the conceptual history of form. As Read's definition of anatomy makes clear, anatomists understood bodies as structural forms—or frames—and the "art of anatomy" was the study of how individual "parts" or "sections" of that form functioned in relation to the whole.[63] Helkiah Crooke echoes Read's language and adds the word "structure" to describe the "constitution or composition" of each body part within the "frame of man" in the preface to his influential *Mikrokosmographia, a Description of the Body of Man* (1615).[64] Grounded in this fundamental philosophy, anatomical science simultaneously bolstered the microcosm-macrocosm theory of the universe and participated in its gradual fracturing throughout the early modern period. Some anatomists vehemently justified their work as an act of "divine complicity," understanding dissection to be "the highest form of religious devotion."[65] If "the image of the divine nature is most lively imprinted" in the "frame of man," then, as English surgeon John Banister claimed, anatomists performed and revealed "the handy worke of the incomprehensible Creator" anytime they stepped up to the dissection table.[66] For others, anatomy was a technical, mechanical project, justified by its use value for surgeons, physicians, and "those who begin to addict their minds to this study."[67] The third edition of Read's popular dissection manual, for instance, opens with only one short preface by the anatomist, assuring that "all things are set downe more fully, methodically, and correctly, than in the former" edition, but that he has still "endeavored to set down all things . . . plainly and shortly."[68] Crooke's *Mikrokosmographia* is a masterful example of the tensions at work here—a text that simultaneously "looks back to medieval visions of the body and forward to Harveian mechanics."[69] Early modern anatomy's emphasis on form exists within a similarly complex and generative tension between, on the one hand, form as the work of divine formation (and anatomy as a science that reveals God's *forming* work as "chief architect") and, on the other hand, form as an intricate, aesthetic event.[70]

Questions of form also held traction for Renaissance anatomists because of a dramatic shift in the methodology underlying anatomical study, a shift that brought the work of dissection into direct contact with the form of the book. Through the fifteenth century, anatomists relied on classical texts for knowledge of the body—primarily Galen and Aristotle's writings—and the sporadic, infrequent task of human dissection was relegated to barber-surgeons.[71] In late fifteenth- and sixteenth-century Italy, however, anatomists began challenging these methods of study, arguing persuasively for the importance of hands-on dissection as the "firmest foundation of the entire medical profession and its first constituent part."[72] As anatomists started dissecting human bodies, they gained new information that challenged the knowledge presented in classical texts. In the preface to Andreas Vesalius's influential *De Humani Corporis Fabrica* (1543), he reminds his readers that Galen's anatomical knowledge relied wholly on animal dissections, primarily his study of simian anatomy. "It is just now known to us from the reborn art of dissection, from the careful reading of Galen's books, and from the welcome restoration of many portions thereof," Vesalius reports, "that he himself never dissected a human body, but in fact was deceived by his monkeys."[73] Driven by scholastic, political, and financial motivations for publishing this new information, anatomical science became, in part, a study of textual representation—a systematic experiment in the somatic capacity of the page—as anatomists explored new ways to evoke the dimensionality of the body (and their hands-on methods) within the pages of printed texts.[74] The production of anatomical knowledge in the early modern period was, fundamentally, a book-making project.

Anatomy, then, was not only a science of and about forms, but also a site where the epistemological and ontological statuses of form were active questions. Consequently, it is a rich historical site in which to take up Turner's call to examine form "as a plurality" and "as an ongoing process."[75] Exploring new ways to suture anatomical and poetic practices in early modern England, *Anatomical Forms* builds on critical formalist approaches that aim to theorize form as both epistemological tool and ontological category.[76] As epistemological tool, form functions as an instrument for and "engine of knowledge production," to return to Mann and Sarkar's language.[77] The act of *forming* is also an act of knowledge making. As ontological category, form (along with the process of formation) is a structuring apparatus for how knowledge is shaped, conceived of, and materialized. The concept of form and the processes of forming themselves come under scrutiny as we consider how certain forms make possible (and limit) certain kinds of knowledges. Studying form within both frameworks, literary scholars increasingly understand form as a process rather than a fixed set of end products, a reimagining of poetic form that has been central to work on women's writing and especially generative for thinking about the intersections of literature and science.[78]

In my close readings of both scientific and poetic archives, my formalist methodology aligns with the modes of knowing I trace in early modern texts. I animate questions of form that resonate for both literary and history of science scholars and, in this act of animation, I follow the models of the poets and practitioners I take up in this book, all of whom understood texts as corporeal forms that could be cut, pierced, unfolded, and stitched to produce knowledge. I frequently use the term "inscription" to describe the acts of forming practiced by the poets and anatomists I study because it is a term with deep material resonances. "Inscription" always brings our attention back to the work of bodies. As a verb, to inscribe is the act of writing on or in something, with emphasis on writing *in* because of the term's connotations of carving, etching, and engraving. As a noun, an inscription refers to the characters, words, and symbols on a surface that function as both a trace of the expressive, inscribing act and a record of the expressed sentiment or piece of knowledge. Inscription carries a material heft that puts pressure on the surface on which it acts and results in a textured record of process that both holds something of the inscriber's body within it (we will see this quite explicitly in Mary Wroth's poetry in Chapter 3) and invites haptic engagement from its readers (most conspicuous in Aemilia Lanyer's poetry in Chapter 4). There is, the poets in my archive insist, a dynamic relationship between embodied poetic practice and the fleshly, material techniques of the science of the body, both as it was practiced in the dissection room and as the resulting knowledge was translated into textual forms. By lingering in the visceral, tactile folds of that relationship, my own work of inscribing shares something intimate with the methods of the writers I study.

An Opening Dissection

Anatomical Forms tracks the scientific work of five women poets. I use the book's organization to reinforce my methodological arguments about where and how we look for science by progressing from writers more familiar in histories of science to writers increasingly less familiar as scientific thinkers. Beginning with Margaret Cavendish, I launch my study on well-trodden early modern literature and science studies ground. Cavendish is consistently upheld as the best (and sometimes only) example of women's direct engagement with scientific thinking and institutions in seventeenth-century England. She published several treatises on natural philosophy and experimental science during her lifetime, and was the first woman invited to attend a meeting of the Royal Society. I begin with this familiar figure and the conventional critical methodologies that position her squarely within the history of science in order to introduce how my book will proceed differently. Rather than take up Cavendish's explicit interventions in natural philosophy and experimental science, I consider her persistent attention to parts and wholes in her poetic collection *Poems, and Fancies* (1653). Cavendish uses this collection, I argue, to experiment with texts as animated "paper bodies" crafted in concert with the dynamic, ongoing textual experiments of early modern anatomists.[79] I read *Poems, and Fancies* alongside anatomical flap figures, which featured superimposed flaps that could be lifted to simulate the act of dissection, enabling the reader to participate in the forming of anatomical knowledge. Cavendish's poetry, I demonstrate, similarly invites its readers into the process of formation, so poetic meaning is shaped through the interactions among writer, reader, and dimensional forms—the "Flesh, Blood, Skin, and Bone" of her poetics.[80] Defamiliarizing Cavendish's poetry through my attention to form over content, this opening chapter situates her as an anatomical thinker with a complex and highly somatic theorization of poetic form that becomes the foundation for my exploration of anatomical forms in the book's subsequent chapters.

Following the paradigm set by Cavendish's explicit (and copious) scientific interventions, scholars have mined the archives for similar evidence of other early modern women's scientific work, and Mary Sidney Herbert, Countess of Pembroke, has been of particular interest to those recovery efforts. Pembroke was well connected to contemporary scientific thinkers of her day, and these biographical circumstances have anchored claims about her active participation in chymical, alchemical, and entomological experiments. Foregrounding Pembroke's writings rather than her social network, I trace a sustained, systematized study of death—or thanatology, as the field is known today. Women were often responsible for caring for the dying and the dead in early modern England, and in this chapter I consider how those routine, familiar practices might be reanimated by women's poetic archives. I take embalming and other forms of postmortem care as a new framework for reading Pembroke's psalm translations (ca. 1586–1599), her poetic tribute to her deceased brother, Philip Sidney. Pembroke's psalms traditionally have been read as an exemplary model of early modern women's devotional translations, but I draw out how postmortem processes infuse the collection. In her poetic forms—famously varied and complex—Pembroke leverages her own experience with death, dying, and loss in order to theorize how experience can function as a mode of knowing. As she investigates experiential knowledge in her psalm translations, I argue, her poetic work becomes a record of postmortem practices, a complementary study to anatomists' ongoing experiments with corpse preservation. Studying her poetic translations as history of science texts, I pay special attention to the highly collaborative way she frames her psalm collection. Pembroke's poems are always a collaboration between the living (Pembroke herself) and the dead (her brother Sidney and the other writers who influence her translations), and that collaborative act becomes not only a basis for her knowledge-making work, but also the underlying mechanism for her postmortem poetics. Reading her poetry through this anatomical framework, I extend familiar accounts of her posthumous collaboration with Sidney—often read as a smart, but ultimately deferential, act of self-fashioning to maintain the poetic and political influence of the Sidney Circle—and establish how Pembroke uses not only her brother's reputation but also his dead body to contribute to the period's developing science of postmortem care.

In Chapters 3 and 4, I turn to two writers who are almost completely absent from critical work on literature and science—Mary Wroth and Aemilia Lanyer—in order to demonstrate the promise of an interdisciplinarity that attends closely to form. Chapter 3 destabilizes a key relationship in early modern literature and science studies: the connection between the blazon (the poetic technique of fragmenting and cataloging the physical attributes of a subject, usually a woman) and dissection (the scientific technique of cutting open and partitioning the human body for study). Scholars often cite the blazon tradition as *the* representative example of the conceptual links between poetic and scientific cultures in early modern England. I upend this critical tendency through my reading of Mary Wroth's manuscript poetry (ca. 1614–1621) and the glaring lack of a blazon we find in her sonnets. In Wroth's poetry, this chapter argues, we find an alternative mode of crafting knowledge about women's bodies that simultaneously confronts the violent tendencies of the blazon tradition and generates striking new interconnections between the poet's body and the materials of writing. Throughout her manuscript, Wroth explores paper as an instrument for knowledge making that goes beyond its role as a conduit for content (what I call Wroth's "paper epistemology"). She transforms her poetic page into a conceptual skin, theorizing the concept of surface and the process of poetic inscription in ways that align closely with contemporary experiments in anatomical illustration. I frame my reading of Wroth's material practice through an analysis of copperplate engraving, an increasingly popular anatomical illustration technique in the sixteenth and seventeenth centuries. The actions involved in copperplate engraving—carving, cutting, and cross-hatching—offer a useful lexicon of verbs through which to understand Wroth's persistent attention to the interactions between inscriptive surfaces and the materials and processes that mark them. The key to Wroth's theorization of surface is ultimately *absence*, and by attending closely to the white spaces, line breaks, gaps, and blanks of her manuscript pages, I illuminate her complex engagement with the popular early modern genre of anatomical poetics.

My fourth chapter installs the writing of Aemilia Lanyer as the most remarkable and complex instance of anatomical form. I analyze Lanyer's *Salve Deus Rex Judæorum* (1611), a poetic adaptation of the biblical story of Christ's crucifixion, alongside contemporary anatomical representations of the womb as a mysterious, sacred vessel. In her telling of Christ's Passion, Lanyer includes eleven prefaces addressing influential English noblewomen, a defense of Eve's original sin as the birth of human knowledge, and an extended meditation on the paradox of Mary's virginal womb. Taking seriously what scholars have described as the "wombs within wombs" structure of *Salve Deus*,[81] I draw out how the womb functions as an anatomical-textual form within which Lanyer gathers the many women represented in her text. For Lanyer, the womb is a powerful site of incarnational possibility (the location of Christ's transformation from word to flesh in Christian theology) and, by animating her text through the form of the womb, *Salve Deus* becomes a similarly significant incarnational site. *Salve Deus* is not an anatomy treatise, but it does mount the argument that women's reproductive anatomy (as distinctly different from men's in the binary Lanyer constructs) allows them privileged access to intimate devotional knowledge. Pairing poetic and anatomical wombs, I show how Lanyer repurposes anatomy's investment in women's reproductive bodies to craft a formal gynotheology that relies on women's embodied acts of creation and the particular devotional capacity enabled by those acts.

I begin the book's final chapter, which functions as an extended epilogue to *Anatomical Forms*, with the story of two lost manuscripts, both recovered in the decade from 1996 to 2006. The first (1996) was a volume of poetry and an unfinished prose romance by Hester Pulter, the second (2006) a volume of Royal Society meeting notes by Robert Hooke. In the years since their recovery, Pulter's manuscript has been read predominantly as a literary text, while Hooke's manuscript has been added to the Royal Society archives, filling an important gap in the record of seventeenth-century scientific work. I insist in this chapter, however, that we overlook an important, nuanced lesson in early modern poetic-scientific forms as a result of this stark disciplinary divide. I structure the chapter as a methodological thought experiment, first tracing Hooke and Pulter's different approaches to natural history. My methodological approach in the first part of the chapter aligns with conventional approaches to literature and science studies and, with this critical framework in hand, I demonstrate Pulter's keen theorization of the act of observation and use her poetry to reveal Hooke's potential as a situated knowledge practitioner. In the chapter's second half, I experiment with how the pairing of these early modern thinkers could be reimagined through the methodological framework of *Anatomical Forms*. Pulter and Hooke both experimented with the epistemological capacity of texts (how texts function as formal conduits for knowledge) and the multidimensional capacity of textual surfaces (how acts of representation can exist on multiple planes within a text). These experiments, I argue in this chapter, are the result of anatomical thinking. Such experiments with writing, reading, and textual bodies are only possible, to borrow Gillian Beer's language, within "a society whose determining circumstances are the direct outcome of applied science"—anatomical science, in this case.[82] As a performance of my book's methodological stakes, this last chapter is as much about rereading Pulter and Hooke's work as it is about how we do literature and science studies. Writing alongside the early work of the Royal Society, Pulter shows us a path not taken in the history of science, the final node in a variegated web of women's anatomical work that has deep implications for our understanding of early modern poetic and scientific forms.

CHAPTER 1

Paper Bodies

Anatomical Flapbooks and Margaret Cavendish's Dimensional Forms

In a published letter to an imaginary interlocutor, Margaret Cavendish, Duchess of Newcastle (1623–1673), confides that she thinks of her writings as bodies. Citing a rumor that a manuscript of her plays was lost at sea while being transported to London for publishing, Cavendish declares that she would have accounted the loss of her twenty plays "as the Loss of Twenty Lives."[1] Fortunately, she explains in the letter, "I alwayes keep the Copies of them safely with me, until they are Printed, and then I Commit the Originals to the Fire, like Parents which are willing to Die, whenas they are sure of their Childrens Lives."[2] Cavendish then ends the letter with a short description of her burning ritual: "howsoever their Paper Bodies are consumed . . . they make a great Blazing Light when they Burn."[3] Silvia Bowerbank and Sara Mendelson take this evocative phrase—"paper bodies"—as the title of their anthology of Cavendish's writing, prompting their readers to consider "what sorts of embodiments are texts?"[4] For Bowerbank and Mendelson, this question concerns the relationship between historical bodies—the "actual lived context of the past"—and the extant materials of the early modern archive. "How do we—three centuries later—go about restoring or recovering the 'body' out of the textual remains?"[5] I want to first ask a simpler (but, of course, not so simple) question based on Cavendish's letter: what is a paper body?

The history of anatomy has this question at its center. When Andreas Vesalius published *De Humani Corporis Fabrica* (1543), his prefatory material was consumed by this question as he tried to reconcile his argument for hands-on dissection with the circumstances of the printed book. The illustrations in particular, Vesalius insists, "will especially satisfy those who do not always have the opportunity to dissect a human body."[6] Likewise, the title page of Alexander Read's *Manuall of the Anatomy* includes the enticing promise that the book "contain[s] the enumeration, and description of the parts of the same, which usually are shewed in the publick Anatomicall exercises."[7] Renaissance anatomy was as much a science of representing bodies in texts as it was a science of dissection, a point I will continue to unfold throughout this book's chapters, and one neatly encapsulated by Cavendish's phrase "paper bodies." In *Anatomical Forms,* we will see poetic texts perform embalming work, take shape as wombs, and engraft the poet's own skin, which makes the phrase "paper bodies" a particularly rich starting point for this book's formal explorations, always both textual and somatic. With this phrase, we touch the place where paper and body meet. In the break between these two nouns, one functioning like an adjective, we encounter the instant at which and the processes through which body is transformed into paper and paper is transformed into body. The text itself becomes a corporeal form, and the body is absorbed into the textual form. This phrase marks Cavendish's participation in the wide-ranging early modern experiments on how to translate fleshly bodies to texts, and it provides important insight into her poetic practice.

In this chapter, I put Cavendish's poetry in conversation with the most literal version of paper bodies we find circulating in early modern England—the anatomical fugitive sheet, or flap figure. These printed illustrations were constructed with a series of hinged flaps, allowing the reader to study the body's internal organs by successively lifting flaps and simulating various stages of dissection. If Renaissance anatomists were particularly interested in involving the reader in the material, hands-on work of dissection *through* their texts, anatomical fugitive sheets were the fullest expression of this objective. Formed by and through the act of reading, flap figures relied on the intrinsically interactive properties of text as matter. It is this emphasis on reader interaction, along with the exceptional dimensionality of flap figures, that holds my interest here because Cavendish was also particularly interested in textual dimensionality. We glimpse this interest in her concept of "paper bodies," and notice it again when she describes the tiny creatures ("Fairy Wits") that "dance upon" poetic lines, "pass through" the "small Crevice[s], and Crannie[s]" of texts, and "pinch . . . black, and blue" any sections they find to be "dull, and sleepy."[8] As we will see throughout this chapter, Cavendish imagines her text as a three-dimensional place wherein writing, reading, and paper bodies meet.

Cavendish's poetry, I will argue, constructs a dialogic relationship between the science of the body (or, more precisely for Cavendish, the science of *bodies*) and a poetic theory keenly interested in how writing and reading bodies act on and with poetic matter. I need to be more precise about what I mean by "poetry" in relation to Cavendish's writing, because many of her texts—fourteen published in her lifetime, most in multiple editions—include poetry in some form.[9] Poetry is consistently a mode of knowing for Cavendish, a form of writing that helps her make sense of things (if she were to transform her atomic philosophy "out of verse into prose," she notes at one point, her "brain would be like a river that is turned from its natural course").[10] Here, though, I will focus on the collection of writing that she herself titles as poetry, the folio volume *Poems, and Fancies*, her first published work. *Poems, and Fancies* (hereafter, *Poems*) consists of five discrete "Parts," all containing different poetic genres and employing different formal structures, along with four transitional sections that Cavendish calls "Claspes," a prose allegory titled "The Animall Parliament," and extensive prefatory and concluding material. Cavendish first published the collection in 1653, then heavily revised the text for a second edition in 1664 ("The Second Impression, much Altered and Corrected"), followed by a more lightly revised third and final edition in 1668 (newly titled, *Poems, or, Several Fancies in Verse: With The Animall Parliament, in Prose*). For Cavendish, *Poems* was a living, breathing thing, an organic, dynamic paper body, and the extended life of this text is part of my interest in it. Emulating and repurposing the collaborative knowledge-making model of anatomical fugitive sheets—with anatomist, reader, and text working together to make meaning—Cavendish invites her reader to take an active part in forming the dimensional text that is *Poems*. We will see this in the book's prefatory material when she instructs us to "read this part of my Book very slow" and "observe very strictly every word," and when she inserts a poetic masque into the middle of the text and prompts us to visualize a "dumb show" at its opening.[11] And we will encounter it most forcefully in a collection of recipe poems wherein the *stuff* of poetic matter is churned, baked, roasted, boiled, and served. Through the many parts of this collection, Cavendish constructs a poetic-scientific treatise that relies on multi-sensory, collaborative acts of formation.

Most critical work on *Poems* up to this point either treats the text's science with little regard for its poetics, or its poetics with only cursory treatment of its science.[12] This is, in large part, because *Poems* lays the foundation for Cavendish's natural philosophy, which is most explicit in the book's first section of "atom poems" detailing Cavendish's vitalist theory of matter.[13] Engaging with the contemporary theories of Thomas Hobbes, René Descartes, and Pierre Gassendi (all of whom we know Cavendish met and conversed with), the atom poems take up questions about the formation of matter, the motion of atoms, and the agency driving atomic motion—all topics of central import to seventeenth-century natural philosophy debates. The atom poems, however, are only one section of this multi-part volume, and it is only when we put the different sections in dialogue with one another that we can understand Cavendish's sustained exploration of forms and formation. The project of *Anatomical Forms* is to explore less obvious modes of scientific engagement, to trace material rather than conceptual links, links that have less to do with content and more to do with process, practice, and ways of knowing. Through such an approach to literature and science studies, Cavendish's *Poems* asks to be read very differently. Reading *Poems, and Fancies* as a body constantly forming, we will find a deep interest in figuration and formation that applies to both textual and somatic forms, returning us by the end of this chapter to the evocative phrase "paper bodies" and the promise it holds for reconsidering histories of science.

Tactile Craft (or, Making Poetry with the Hands)

First, a short detour into the site of Cavendish's "paper bodies" reference—*CCXI Sociable Letters* (1664)—that will help orient us to the opulent materiality (and, we will increasingly see, corporeality) of *Poems*. *Sociable Letters* itself is not a poetic text, but one letter in particular details an imaginary feast of poets that Cavendish relates to her imagined interlocutor (named only "Madam" throughout the text). "Nature sent the Muses to Invite all the Poets to a Banquet of Wit," Cavendish relates, and "Invited also me a Poetess, or rather Poetastress."[14] The poets were seated around a "strange Table," one that was "made of all the Famous Old Poets Sculs" with a tablecloth "made of their Brains."[15] The brains, Cavendish explains to Madam, were "Spun by the Muses, for they are Spinsters of Men's Brains," and they first spun the brains into "Cobweb Threads, as Soft and Thin as Air."[16] Then "Old Time" wove the threads into a "Web like Damask or Diaper, in Works and Figures of Golden Numbers."[17] Also laid out on this banquet table of anatomical-textile delights are napkins of "Pure Fine White Paper, all over-wrought with Black Letters," plates made of the "Films or Drums of Sensible Ears," knives made from "Orators Tongues," "Writing pens" as forks, and saltcellars "made of the Crystalline part of Observing Eyes."[18] With the table set and all the guests seated, Cavendish goes on to detail the various dishes that were served to the poets (a "Hash of Anagrams, Letters, and Names," a "Dish of Funeral Elegies," and a "Bag-pudding of Sciences, made of Mathematical Cream, Logistical Eggs, and Astronomical Spices").[19] Textiles, body parts, and food take center stage as Cavendish imbues a material heft, a *stuff-ness*, to the world of poetry. Writing materials and body parts become culinary utensils; poetic forms and styles become dishes to be eaten. Cavendish's lavish "Banquet of Wit" is not so much allegory as it is *fancy*—her complex and flexible term for the workings of the imagination—and, in this letter, we peek into the workings of Cavendish's imagination as she thinks about the animated life of poetic work.

Bracketing for now the anatomical and gustatory elements of Cavendish's poetic banquet, I want to first take up her textile metaphors—brains spun into "Cobwebs Threads"—because here we find an important link to the way Cavendish describes poetic making in *Poems, and Fancies*. Cavendish frequently draws on textile craft metaphors to describe both the book of *Poems* and her "plain" writing style. In her prefatory material, she refers to the book as a "Garment of Memory," spun "to lap up [her] Name, that it might grow to after Ages,"[20] later shifting the metaphor slightly to compare the text's "plain" poetic style to a "garment": "though her Garments are plain, and unusual, yet they are clean, and decent."[21] She employs this conceit again in one of the book's concluding poems, eventually titled "Of the Style of this Book" in the 1664 edition. Acknowledging that her poetry might be "more courted" if it could "make a better show" (with ornamental trappings like "Silver Lace" or a "rich Hood"), she again describes her poetics as "plain" and "clean."[22] "I Language want, to dress my Fancies in," Cavendish laments, but instead, "the Hair's uncurled, the Garments loose, and thin."[23] In references to *Poems* as a "coarse piece" and her poetics as a kind of "fashion," Cavendish draws on familiar etymological connections between "text" and "textile," both from the common Latin root *texere* (to weave). For Cavendish, though, it is not the etymological connections but the material connections that are important—connections between cloth and paper, needle and pen, and, as we will see, threads and fancies.[24]

These material connections ground Cavendish's more complex and sustained interest in the writing process itself as a form of textile craft. Cavendish opens *Poems* with an account of poetic practice that rivals Philip Sidney's famous description of the poet as "maker" (one who, Sidney claims, "make[s] things either better than nature bringeth forth or, quite anew, forms such as never were in nature").[25] For Cavendish, the poet is also a maker, but a more precise kind of maker—specifically, a spinner. "True it is," she writes in the dedicatory epistle to *Poems*, "Spinning with the Fingers is more proper to our Sex, than studying or writing Poetry, which is the Spinning with the Braine."[26] This gendered framing returns again in the "Epistle to Mistress Toppe," where Cavendish declares that women have too much "waste Time," which forces their thoughts to "run wildly about." For Cavendish, "wild thoughts," with "nothing to fix them upon," result in "snarls" of "Thread."[27] Much better, then, to order and "spin" those threads into something. "All brains work naturally, and incessantly, in some kind or other," she reflects in another preface, but it is women's brains that seem most suited to the "spinning" work poetry requires. In the preface "To All Noble, and Worthy Ladies," Cavendish speculates that women can "claim" poetry "as a work belonging most properly to themselves," not only because "their Braines work usually in a Fantasticall motion," but also because of their expertise in "Wrought Works" and "divers sorts of Stitches."[28] She reiterates the spinning analogy one final time at the end of the collection in a wonderfully dramatic poem about spiders, books, and dust. "Since Men doe spin their Writings from the Braine," she posits, when books are "laid aside" and "buried" under "a heap of dust," spiders, "for honour of that Art," raise "thin, high Altars" and "offer Flyes, as sacrifice of praise."[29] Employing textile metaphors to talk about the mental activity of writing, Cavendish imbues not only the text itself but the process of *making poetry* with tactility and texture.

Crucially, Cavendish's poetic matter is recycled, repurposed, rearranged. Spinning, to borrow Raymond Williams's language on form, is the act of "making indeterminate material into a determinate or specific being or thing."[30] Rather than, in Sidney's formulation, "grow[ing]" things "quite anew," Cavendish's poet-spinner relies on matter that already exists—and this distinction between making new and repurposing existing matter is central to Stephen Hequembourg's account of Cavendish's "poetics of materialism."[31] For Hequembourg, this concept reconciles Cavendish's natural philosophy with her poetics: "if all that exists is matter in its various and shifting aggregations, how does the poetic mind create a world that previously did not exist?"[32] We will return to Cavendish's natural philosophy later in this chapter, but here I bring Hequembourg's argument into the discussion because he offers a generative concept about Cavendish's material poetics while falling into a familiar critical trap. Much of the history of Cavendish scholarship is a history of trying to reconcile generic tension—the "tension between philosophy and fiction," as Hequembourg articulates it.[33] When we start from a place of tension, however, we are applying our own contemporary disciplinary frameworks to early modern texts and, in the process, neglecting the promise of Cavendish's highly interdisciplinary thinking. In the specific case of Hequembourg's "poetics of materialism," what we miss by starting from a place of tension is the rich, complex materiality through which Cavendish articulates her writing process—a materiality that has much to do with the answer to my opening question (*what is a paper body?*). We must get closer, that is, to the *materials* of Cavendish's materialism, the material stuff that fills her poetic pages and the different ways she figures her poetic text as a made, textured, tactile thing.

In the prefatory material addressed directly "To Poets," Cavendish takes her contemporaries to task for not being inventive enough. She critiques poets that act simply as "Messengers, that only make Report," comparing them to "those that . . . do win Inventions Fort."[34] "Fancies, in the Brain that Nature wrought, / are best; what Imitation makes, are naught," Cavendish asserts at the poem's opening.[35] Her distinction here between "imitation" and "invention" initially seems to echo Sidney's ranking of "three general kinds" of poets.[36] For Sidney, the first and lowest rank are those poets who simply "imitate," while the highest ranking poets (the "right poets") "borrow nothing of what is, hath been or shall be."[37] The way Cavendish defines "invention" in her poem, though, is very specific. In the above line, she glosses "invention" as "Fancies, in the Braine that Nature wrought," so the inventive emphasis here is on "fancies," mental *matter* that nature has "wrought."[38] As a verb and adjective (as in the "wrought works" women perform with their needles), "wrought" connotes acts of intricate making and subtle craft. In the early modern period, the term was often used in relation to the textile arts—"to make (a fabric, garment, etc.); to weave, spin, sew, knit"[39]—and Cavendish signals that connotation in this poem by again incorporating a textile metaphor to describe the charge of "imitation" she brings against her fellow poets. "Many there are, that Suits will make to wear," she critiques, "of several Patches stol'n, both here, and there."[40] Instead of stealing patches "here and there" from other writers, Cavendish values working in new ways with the material nature provides. For this reason, Cavendish aligns herself with Penelope, the famous spinner of Greek mythology. Her verses "are like Chast Penelope's work, for I wrote them in my Husbands absence, to delude Melancholy Thoughts, and avoid Idle Time."[41] Cavendish explicitly connects her situation to Penelope's because of her husband's absence, while implicitly invoking Penelope's famously reiterative textile work.[42] Penelope passed her time not just by weaving but by unraveling and reweaving the same threads—forming, deforming, and re-forming. The material was the same (repurposed, recycled, rearranged) and her handicraft was focused more on the sustained process of formation than the final product of that process.

In her material orientation to the process of poetic writing, Cavendish resists any polarization between mind and matter, a synthesis on full display in her imagined "Banquet of Wit" in *Sociable Letters*. Bringing some of the anatomical elements from that lush tableau back into our discussion, I want to turn our attention to how very *tactile* the poetic process is for Cavendish. Understanding poetry as "spinning with the braine" gives her a way to make tangible an invisible, interior process. There are many poems in *Poems, and Fancies* that move beyond textile metaphors in their attempts to make tangible the activities of the brain—to turn *fancy* into something that both reader and writer can grasp. For example, in a poem titled "Similizing Thoughts," Cavendish figures thoughts as a "Pen . . . writ[ing] upon the Braine."[43] Another poem compares the brain to a beehive, with "every Thought just like a bee . . . each by it selfe within a parted Cell."[44] The bees of fancy then fly "into every Garden" and "suck[ ] the Flowers sweet." In one especially noteworthy poem, Cavendish describes not just the motion of her thoughts, but also the physical effects of that motion. "Some Fancies, like small Gnats, buz in the Braine," she explains, and "sting so sore the Poets Head" that the mind becomes "blister'd" and "Thoughts turn'd red."[45] Writing poetry, it turns out, is the "perfect Cure" for this blistering mind: "Nought can take out the burning heat, and paine / but Pen, and Ink, to write on Paper plaine."[46]

Cavendish's "poetics of materialism" is, indeed, materially invested, highly attuned to associations between the various forms of animate matter involved in the poetic process.[47] It is also, though, corporeally invested. As the above metaphors suggest, Cavendish consistently foregrounds the writing body in relation to the text in her attempts to convey the internal, embodied process of writing with and *as* matter. In the prefaces addressed to poets, where she outlines most explicitly her definition of "invention," she concludes her introductory material with a short, argumentative poem, titled "Wherein Poetry chiefly consists" in the 1664 and 1668 editions of *Poems*. This poem ends with the striking collision of anatomy and poetics that I briefly discussed in this book's introduction: "Fancy the Form is, Flesh, Blood, Skin, and Bone; / Words are but Shadows; Substance they have none."[48] This oft-cited line offers a concise articulation of Cavendish's investment in imagination over style. "Words" (ornament, artifice) are "Shadows" without "Substance," while "Fancy" is the substantive "Form" of poetry. Fusing *fancy* and *form* in these lines, Cavendish prompts us away from a kind of New Critical formalism (more on that later) and toward a mode of reading that takes the sensuous *materia* of fancy as a formal feature. She then anatomizes form itself, breaking down the *materia* of fancy into component, somatic parts. In this convergence of anatomy and poetics, poetry becomes an organic creature.

Paper Flaps

What would it mean to think about Cavendish's *Poems, and Fancies* as an anatomical experiment? How might we read it as a book about the science of bodies (corporeal forms and the corporeality of form as "Flesh, Blood, Skin, and Bone") and a book that activates a touch-centric epistemological and aesthetic experience? In order to do so, we have to adjust our understanding of Renaissance anatomy, expanding our sense of the field's history so that a phrase like "anatomical experiment" does not primarily invoke the messy work of the dissection room and the spectacle of the public anatomy theater. Anatomical fugitive sheets are what we might call an anatomical-textual archive, extraordinarily interactive paper objects that have been of great interest to both historians of science and historians of the book. Grounded in this archive, any inquiry into the history of anatomy takes on a distinctly textual register. Anatomical fugitive sheets were constructed using cut-and-paste techniques, with layers of movable flaps that could be lifted to reveal the body's interior. Internal organs and body parts were printed to be cut out separately and then attached in consecutive layers to a base sheet representing either male or female anatomy. These flap figures allowed the reader to simulate the act of dissection through the medium of paper. As their name implies, anatomical fugitive sheets were part of what scholars refer to as "popular print culture," the vibrant "social dialogue" and "mediations of experience" facilitated by the circulation of cheap, printed texts.[49] Fugitive sheets "broke the academic monopoly on the knowledge of anatomy," offering basic (and often incorrect or outdated) anatomical knowledge to a much wider audience of readers than the expensive folios of Andreas Vesalius, Helkiah Crooke, and many other anatomists we will meet over the course of this book.[50] These were texts, in other words, that passed through many hands and invited handling.

Anatomical fugitive sheets are tricky archival materials—inherently fragile, often incomplete (with movable flaps frequently missing), and extant only in small numbers. Andrea Carlino, whose comprehensive archival study of these texts is, coincidentally, titled *Paper Bodies*, notes that the sheets were "probably sold at low prices" when first published, dooming them to "a short, ephemeral life."[51] Of those that have survived, most were preserved because they were pasted or bound into other texts. The ephemerality, however, is a testament to their popularity throughout sixteenth- and seventeenth-century Europe. Evidence suggests that anatomical fugitive sheets were first published in Germany in 1538 by Heinrich Vogtherr the Elder and Jobst de Negker.[52] As would become standard for such texts, Vogtherr and Negker's flap anatomies were printed on cheap paper, sometimes bound into short booklets, and featured one male and one female figure cut from the same woodblock (Figures 1.1 and 1.2). Vogtherr and Negker's figures were widely reproduced throughout the early modern period and became a model for the genre. Suzanne Karr Schmidt cites "roughly 65 known male and female pairs" circulating throughout Europe by the end of the seventeenth century.[53] Perhaps most famously, Andreas Vesalius included uncut anatomical flap figures in both his *De Humani Corporis Fabrica Libri Septem* (1543) and *De Humani Corporis Fabrica Librorum Epitome* (1543), directing readers to reinforce the uncut flap page, cut out each individual flap, and assemble their own fugitive sheet—instructions to which we will return. Flap anatomies lost popularity in the eighteenth century, but found a resurgence during the nineteenth century with works like Edward William Tuson's *Myology* (1828) and Frederick Hollick's mass market reproductive health manuals (late 1800s).[54] As one early reviewer of Tuson's texts notes, "these plates must not be regarded as mere drawings or paintings, but as dissections of drawings, and, in point of accuracy and utility, are second only to actual dissections of the human body." Studying these plates requires "the sense of touch . . . as well as that of sight," the reviewer reports, activating what we would now call a haptic reading experience.[55]

The flap mechanism of the anatomical fugitive sheet is a paper technology that does the work of form, and here *form* is a fusion of structural composition and material conditions. As a "paper technology," the flap functions in a "literal, three-dimensional sense . . . to contest and restructure the very making of knowledge."[56] Structurally, each anatomical flap figure begins with a base layer of paper on which the flaps are attached, and a new three-dimensional structure (a paper body) is constructed through the embodied actions of cutting, folding, composing, and gluing. The resulting representation is made out of paper flaps, so these are anatomical-textual forms wherein the inscriptive work relies on and takes the form of the material out of which it is made. The earliest versions of these sheets usually contain a single, central figure that dominates the page, surrounded by columns of text that offer the names of major body parts and short descriptions of their functions. Each central figure contains six to ten flaps visually representing different internal organs, although more highly designed figures also include the names of body parts printed on each organ. A particularly striking example in the Wellcome Library's collection, for instance, is a female figure printed by Hans Weygel (based on Vogtherr's figures) that includes eight flaps total, all of which fold up from a top center point until the final flap (the colon), which is attached directly to the base sheet (Figure 1.3). Some of the flaps on this figure are very fragile (for example, the thin strips of paper representing the "lacmamil," thought to convert blood to breast milk) and have been carefully glued not only to the supporting page but also to each other to ensure that the organs are positioned correctly within the body cavity. A testament to the concept of *poiesis*—"thinking through making"—anatomical flap figures evince the epistemological work of form as a convergence of textual material and *soma*.[57] The resulting body then becomes a prompt for further embodied interaction with anyone who handles these pages, inviting a mode of reading that is highly attentive to the intrinsically interactive properties of text as a material medium.

Essential to the interactive work of these texts are the reader's own hands. Anatomical fugitive sheets require an incredible amount of handling, whether one is actually constructing the flap figure or "reading" the sheet once it has been constructed. As a result, these texts force us to inhabit an early modern ethos wherein reading required an expanded sensorium. Renaissance readers read "in and through their own bodies," Elizabeth Spiller reminds us, and anatomical flap figures provide a potent example of how reading, as "a distinctly corporeal act," could contribute to the very formation of the text itself.[58] In these texts, the reader rather than the anatomist is central to the work of anatomical form, and the epistemological work of the text relies, in large part, on the reader's interactions with it. This could go very wrong, as evidenced by one extant copy of a French female flap figure in which all of the flaps have been attached (improperly ordered and some inverted) at the top of the sternum (Figure 1.4). Anatomically, this figure leaves something to be desired—the colon, for instance, does not sit between the ribs and the womb does not hang just beneath the solar plexus. But, as an archival object, this sheet holds a record of *use*, a record of (in this case, incorrect) anatomical knowledge making.[59]

As paper bodies, anatomical flap figures were formed through a series of relations among anatomists, illustrators, printers, readers, and the resulting anatomical-textual figures (which, themselves, are both records of past acts of forming and prompts for future readerly interactions). This complex relational network, centered on the representation of bodies in texts, reorients our understanding of Cavendish's reference to her texts as "paper bodies."[60] She wields this term to describe the relationship between writer and text, but we are also starting to see how very central handwork was to Cavendish's representation of poetic craft, not only her own craft but also the capacity for poetic work she attributes to her readers (such as the women whose "Braines work usually in a fantasticall motion"). We are also building the foundation for a web of women's poetics that instrumentalizes texts as laboratory spaces (Mary Sidney Herbert and Hester

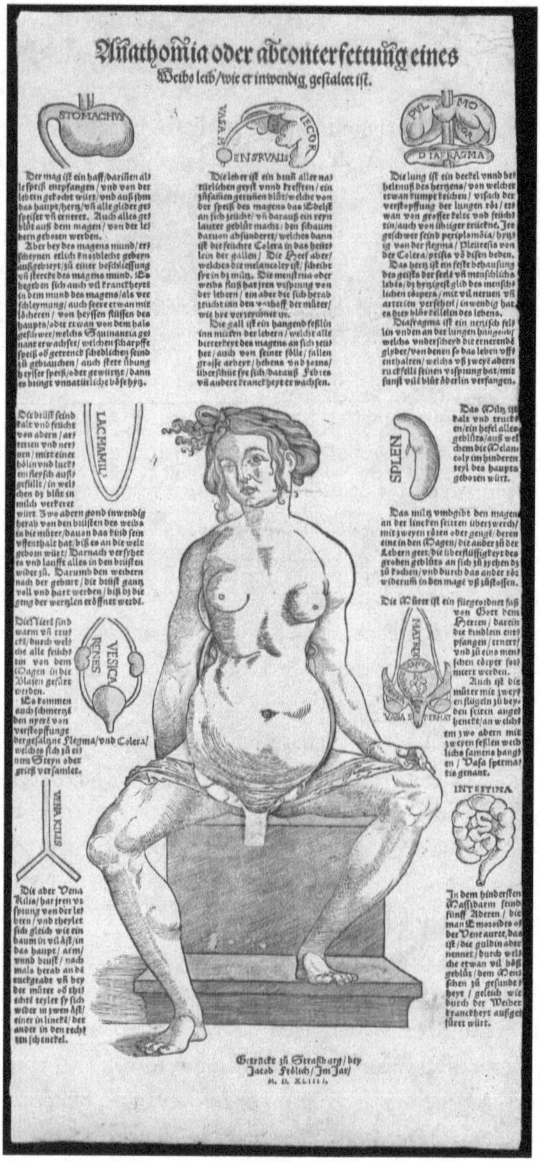

Figure 1.1. Female anatomical fugitive sheet, from German woodcuts (1544). The anatomical flap figures pictured here and in Figure 1.2 are not Heinrich Vogtherr the Elder and Jobst de Negker's sheets, but rather those of Jacob Frölich, who used the same woodcuts to reprint the sheets in Germany in 1544. Jacob Frölich, *Anathomia oder abconterfettung eynes Mans leib* (Strassburg, 1544), Wellcome Collection, EPB/293.11.

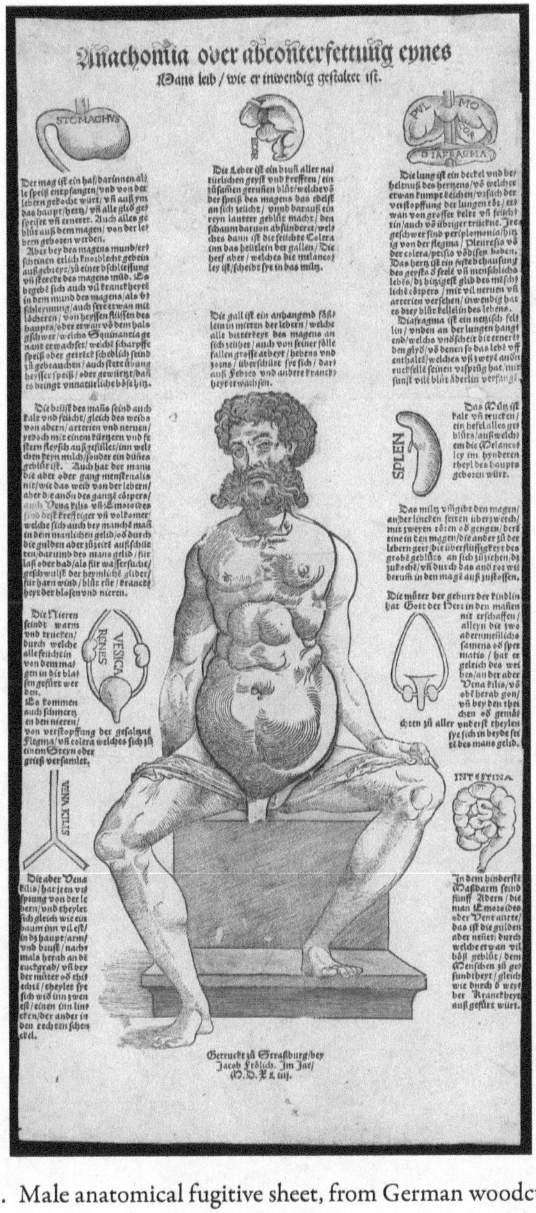

Figure 1.2. Male anatomical fugitive sheet, from German woodcuts (1544). Jacob Frölich, *Anathomia oder abconterfettung eynes Mans leib* (Strassburg, 1544), Wellcome Collection, EPB/293.10.

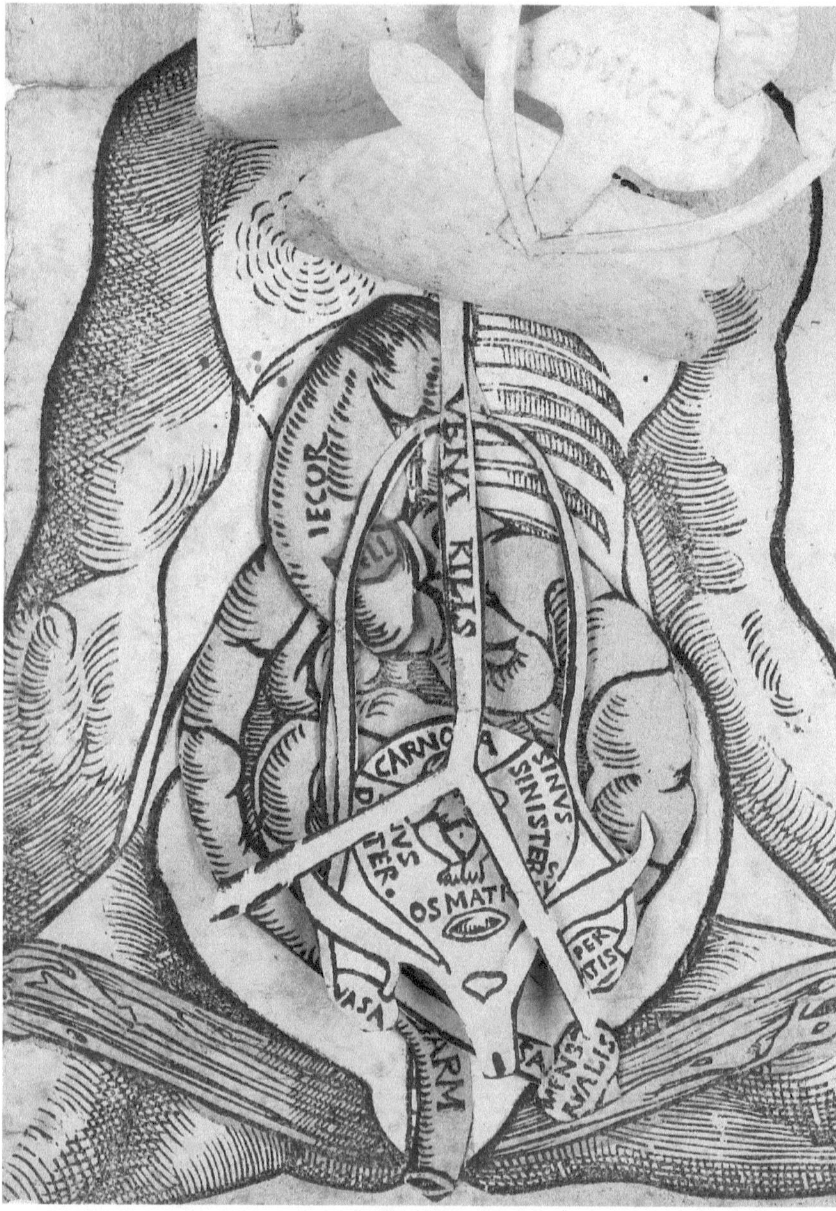

Figure 1.3. Female anatomy flap figure detail. Hans Weygel, *Anathomia oder abcontrofettung eines Weibs leib* (Nürmberg, 1564), Wellcome Collection, EPB/294.12.

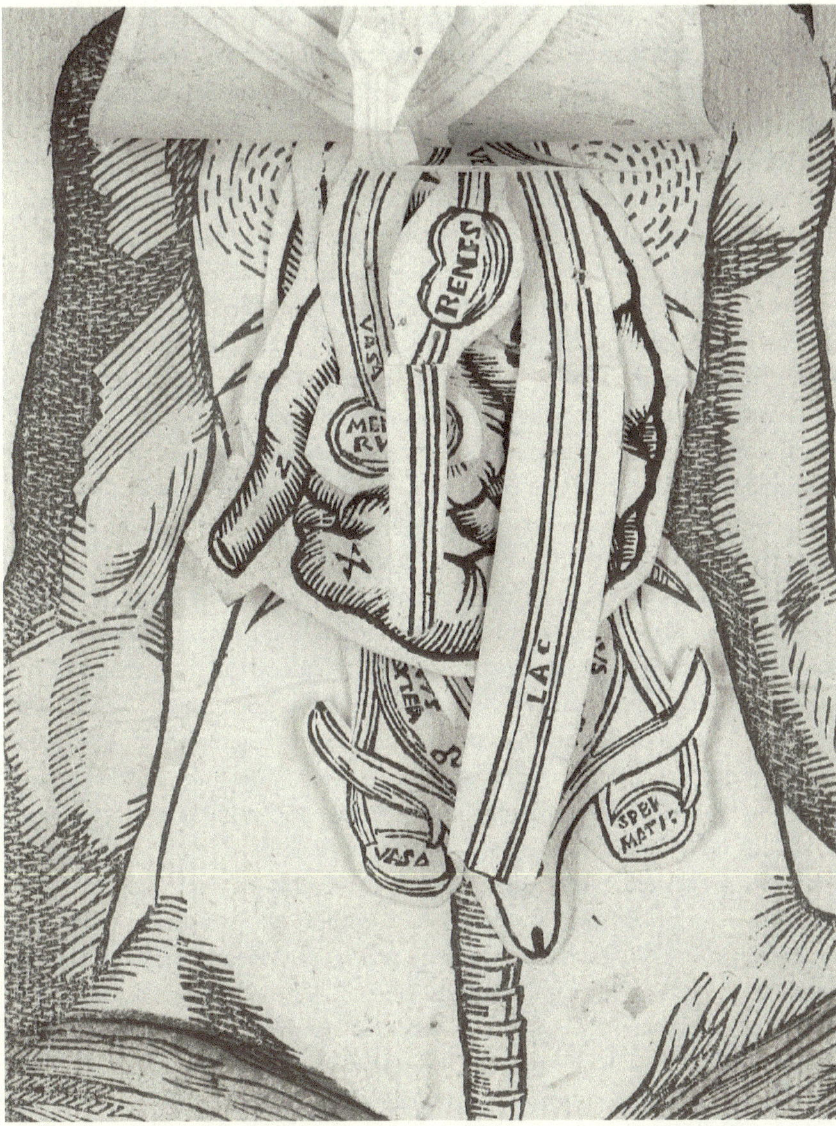

Figure 1.4. Female anatomy flap figure, assembled with all flaps secured at the top of the sternum. *Vif pourtraict des parties interieures du corps humain* (Paris, c. 1560), Wellcome Collection, EPB/292.9.

Pulter), activates a pronounced intimacy between skins and writing surfaces (Mary Wroth), and figures books as generative anatomical parts (Aemilia Lanyer). The poets central to this book, we will see again and again, are exceptionally interested in the processes of anatomical forms and the potential for those processes to reverberate through their poetics and extend into the space of reading. They experimented not only with how their own bodies relate to texts, but also with the complex and dynamic relations between texts and bodies more broadly defined. Yoking Cavendish's *Poems* to the somatic materiality of the anatomical flap figure, her corporeal understanding of form ("Flesh, Blood, Skin, and Bone") takes on new dimensionality.[61] If pen, ink, and paper are the "perfect Cure" for fancies "buzing in the Braine," paper flaps model a textual mode in which fancies keep moving, in which they remain animated through the transformation from brain to page.[62] The makers of anatomical fugitive sheets instrumentalized paper as animated matter that could be formed into a body, then pulled apart, then reformed. The resulting textual forms were constructed, animated, composite figures—the very stuff of Cavendish's natural philosophy.

Before shifting to take a closer look at Cavendish's philosophy of matter, though, I want to consider one more anatomical example: Lucas Kilian and Johann Remmelin's *Catoptrum Microcosmicum*, often cited as the epitome of the Renaissance anatomical flapbook. The *Catoptrum* was first printed (possibly unauthorized) as a series of broadsheets in 1613, and explanatory pamphlets were quickly printed to supplement the flap figures in 1615.[63] The first full edition containing figures, accompanying text, and title page was printed as a short book starting in 1619, with at least six subsequent reprintings in Latin and German, along with translated editions in Dutch, French, and English.[64] Suzanne Karr Schmidt's description of the *Catoptrum* as a "sculptural" print series testifies to just how intricate Kilian and Remmelin's text is compared to the early flap figures modeled on Vogtherr and Negker's sheets.[65] With more than one hundred superimposed flaps, the *Catoptrum*'s pages function like anatomical curiosity cabinets, and as a result this text invites exploration rather than reading, a distinction that will be important for understanding Cavendish's experimental work in *Poems*.

The *Catoptrum*'s plates are slightly different depending on the edition and translation, but the most prominent features remain fairly consistent. The "Visio Prima" features two mostly complete human figures, one male facing forward and one female facing backward, mirrored on the left and right sides of the page (Figure 1.5). Both figures stand on plinths with emblazoned emblems and, in later editions, the bases of the plinths celebrate the makers involved in the project

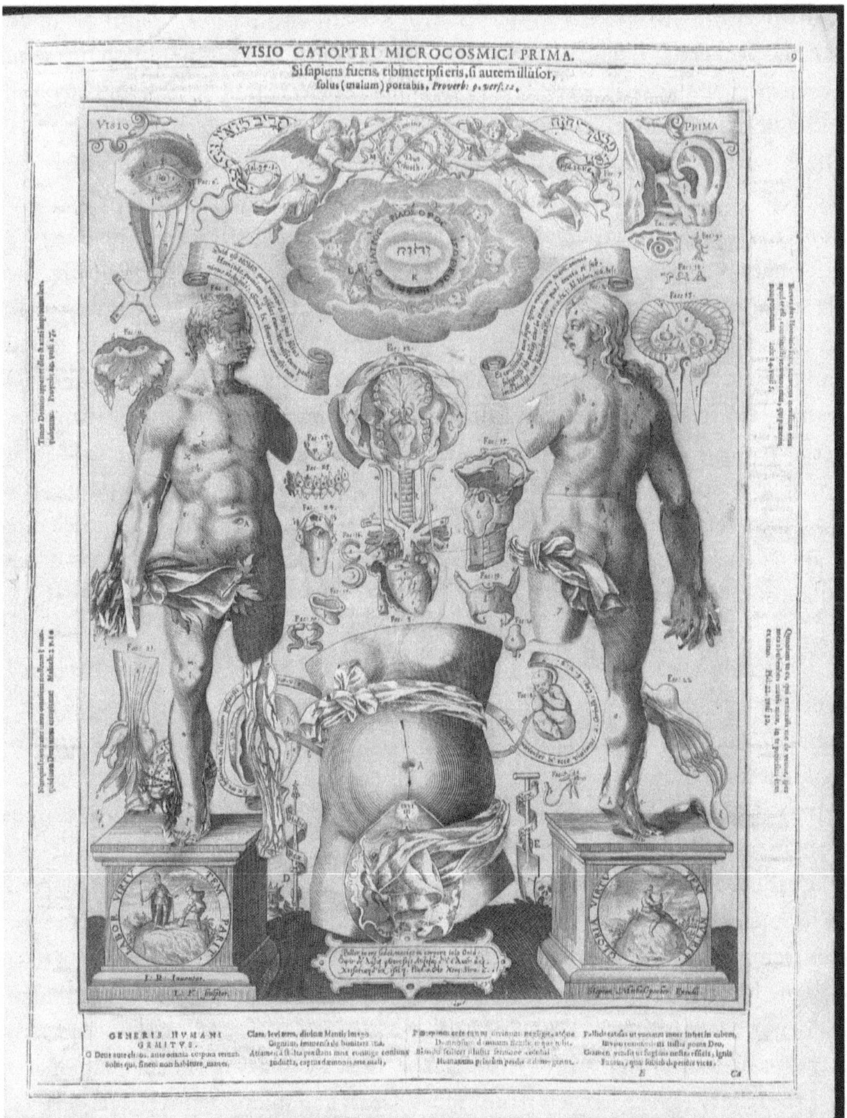

Figure 1.5. Visio Prima. Johann Remmelin, *Catoptrum microcosmicum* (Augustae Vindelicorum [Augsburg], 1619), Wellcome Collection, EPB/F/5418/1.

(I. R. [Remmelin] as "Inventor," L. K. [Kilian] as "Sculptor," and Stephan Michelspacher as "Excudit," or printer). There is also a prominent flap figure of a truncated, pregnant female torso positioned at the center bottom of the page, and in some editions the vulva is covered by a flap portraying the face of a devil. On this crowded page, one also finds a small series of flaps inscribed with the Hebrew word for "Yahweh," surrounded by clouds and cherub faces. In later versions, the cloud is replaced with a cartouche displaying title, author, and printer information. Finally, there are many small anatomical diagrams surrounding the plate's larger figures (including dissections of the eye, ear, mouth, tongue, teeth, and upper circulatory system), some of which also include several superimposed flaps. The book's second and third plates ("Visio Secondo" and "Visio Tertia") feature separate male and female flap figures, similarly accompanied by smaller, detailed anatomical diagrams, eschatological emblems, and Christian iconography (Figure 1.6).[66] In the female plate of the 1639 Latin edition, for instance, the woman's foot rests on a skull through which a snake winds its way, an apple tree branch caught in its mouth, evoking the biblical story of Adam and Eve—a frequent subject in anatomical plates, which I will discuss in more detail in Chapter 4.

Unlike the early German plates, in which all the flaps lift vertically from a central point, the flaps in Kilian and Remmelin's plates lift vertically and horizontally, sometimes in one direction and then another. In both the male and female plates, the outermost pair of flaps open horizontally from the center to expose the chest cavity, mimicking the initial incision of the anatomist's scalpel from breastbone to pubic bone. Some flaps include engravings on both sides, so when the reader lifts the flap they see the underside of the organ in question. And in several extant versions of the *Catoptrum*, small paper organs can be removed completely from the chest cavity (the heart and lungs in the British Library's 1695 edition, for instance), allowing for a particularly close encounter with the body's hidden inner workings.[67] Drawing on Cavendish's use of textile metaphors, this is truly a textual *fabric*, formed through the intricate weaving of paper body parts and the reader's touch as their hands move across and manipulate the text's pages.

Invoking textile language to describe this remarkable anatomical-textual object (and, in doing so, recalling Vesalius's use of the word "fabric" in the title of *De Humani Corporis Fabrica*—the "fabric" of the human body) might help us contemplate an unexpected but significant detail in one extent copy of Kilian and Remmelin's *Catoptrum*: British Library C.107.k.13. In both the male and female plates of this copy, several flaps have been stitched rather than glued together. The paper flaps in the chest cavities are literally woven, with some flaps

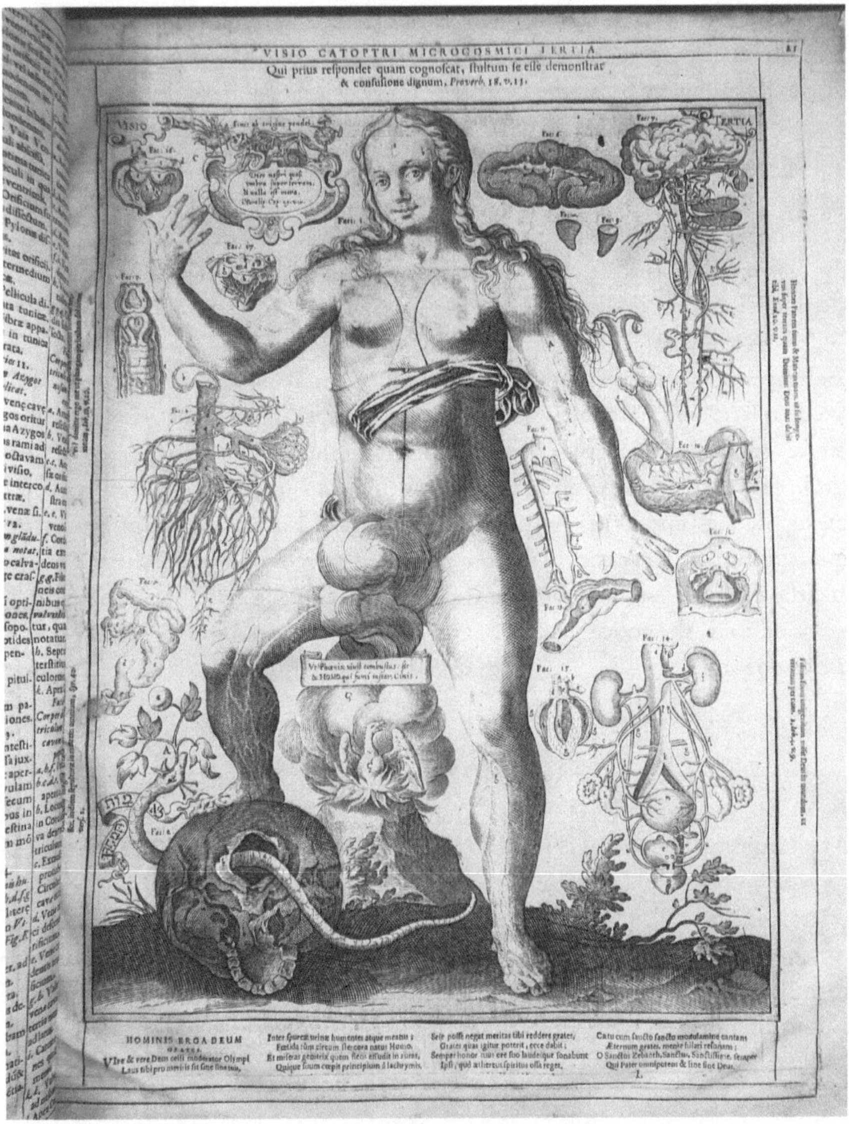

Figure 1.6. Visio Tertia. Johann Remmelin, *Catoptrum microcosmicum* (Ulm: Johannes Görlinus, 1639). From the British Library Collection: Tab.583.f.45.

cut to include small slits through which another flap has been inserted, and others held together with yellowing thread. The female organs have been stitched in three separate places, and the flexibility of the stitched joints allows the organs to unfold, accordion-like, from the figure's chest cavity (Figures 1.7 and 1.8). It is difficult to assess exactly how anatomical flap figures would have been assembled, or who would do the assembling. As we have seen with the Wellcome Collection's 1560 French sheet (Figure 1.4), there is some evidence that readers took on the task and constructed wildly inaccurate paper bodies. Vesalius explicitly invited his readers to do the work as part of the pedagogical project of the *Fabrica*, but given the complexity of the *Catoptrum*'s flaps, scholars speculate that the sheets were preassembled and then sold.[68] One textual maker decided that thread would be the best material for joining scraps of paper into a body. Those stitches, preserved in the paper folds of this text, simultaneously evoke women's textile work, the needlework involved in bookmaking, and anatomists' practice of sewing up tissues and skins post-dissection.[69] The archival page—infused with overlapping tactile practices—becomes sutured membrane and stitched sampler, and the multidimensional forms of Kilian and Remmelin's text hold within their textured folds the convergence of paper, body, and thread so central to Cavendish's poetics.

Cavendish's Science of Bodies

As anatomical-textual experiments, flap figures are both exemplars of and responses to the complex experiments with bodies and texts that were of active interest in the early modern period, which Katharine A. Craik marks as a historical moment of renegotiation "between texts and those who encountered them."[70] This archive of paper bodies also draws heightened attention to the relationship between parts and wholes in its presentation of a whole (the human body) as an assemblage of parts. Kilian and Remmelin's *Catoptrum*, in particular, is a text of parts, a paper experiment with parts and wholes that also models a mode of early modern reading grounded in exploration, motion, and handicraft.[71] Cavendish may or may not have been familiar with this highly dimensional genre of popular print culture (although, she would likely have seen René Descartes's *De Homine* [1662], which contained a flap anatomy of the heart), but as a writer of many different genres, thinking about her own texts as "paper bodies" and armed with a comprehensive theory about bodies as matter, Cavendish has much to contribute to these experiments. In order to understand her innovative contribution, we need to take a closer look at her natural philosophy, or, more precisely, her theory of forms.

In *Poems, and Fancies*, Cavendish outlines her natural philosophy most explicitly in Part I of the book. Known as the "atom poems," this first section of poems (105 total in the 1653 edition) imagines the world to be comprised of four types of atoms (square, round, long, and sharp) and explores how those atoms move around, come together, and break apart as they create and destroy all the matter that makes up the world. In later poems in this section, Cavendish uses that foundational vitalist theory to speculate on different sensations ("What Atoms Make Heat and Cold") and the possibility of different worlds ("Of Many Worlds in this World").[72] The atom poems directly follow Cavendish's opening prefatory material, which includes her reference to the text as a "garment," her assertion that writing poetry is "Spinning with the Braine," and her speculation that women are naturally inclined to poetry as a form of "wrought work." Already, then, Cavendish has oriented her readers to a poetics that is both tactile and dimensional, and in the atom poems she outlines the philosophy underlying that poetics.

The atom poems are about formed matter and how atoms function in the process of forming matter. For Cavendish, atoms are the smallest parts of matter, the "foure principall Figure[s]," "perfect in themselves alone, / not taking any shape, but what's their owne."[73] Atoms themselves do not alter but rather move and combine to make "mixt" forms: "what Forme is else, must still take from each part, / either from Round, or Long, or Square, or Sharp."[74] Cavendish prefers terms like "matter" and "figure" to reference the atoms themselves, usually reserving "forme" for the combinatory shapes atoms take as they move and join.[75] In a poem focused on the "Atomes long" that make air, for instance, Cavendish describes the constancy of the atomic "figure" even as individual atoms combine into "Formes."

> And let it move, join to what Forme it will,
> Yet lies in every Line, that Figure still.
> For Atomes long, their Formes, as Thread, are spun;
> And, like a Cobweb, interwoven run.[76]

Air itself is a *form*, composed as it is of many long atoms, which, when combined, make forms "like a Cobweb" of interwoven threads. Air can also join to another *form* ("join to what Forme it will"), but the principle *figure* of its composition (the long atom) will not alter ("in every Line, that Figure still"). When "fixed"

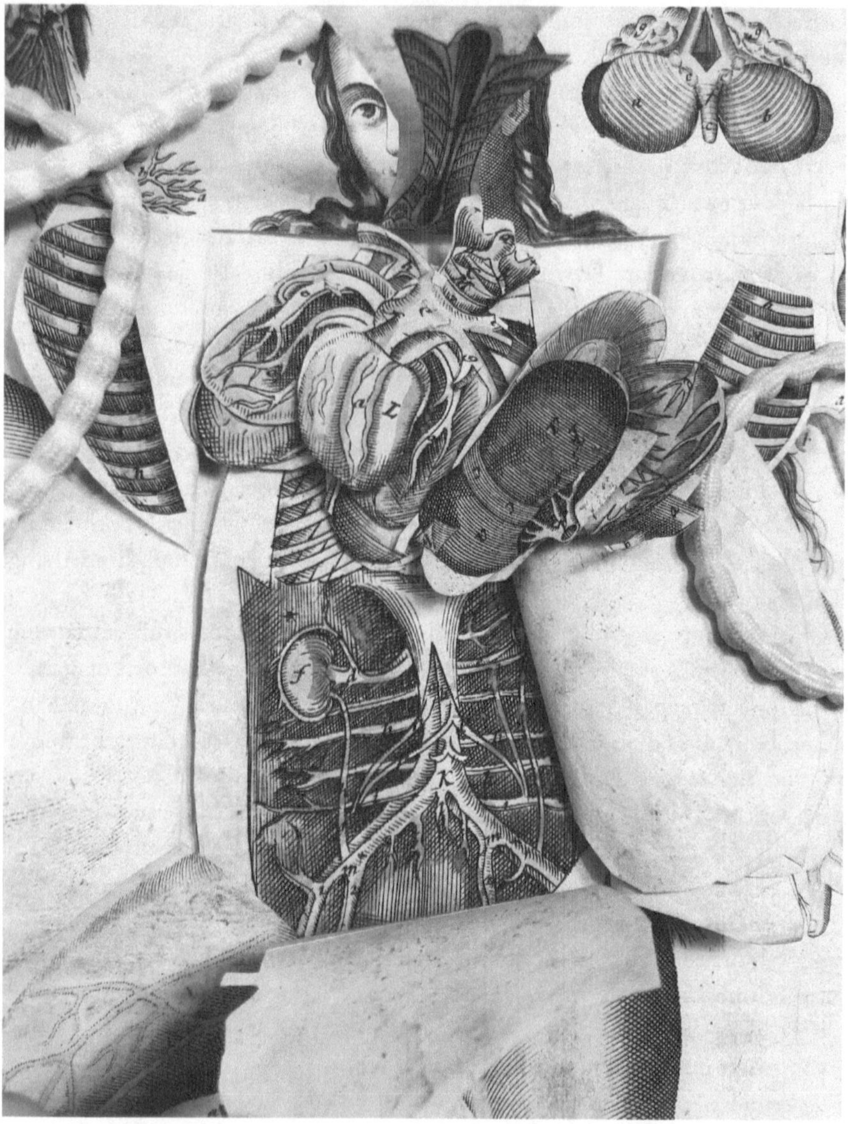

Figure 1.7. Visio Tertia flaps detail. Johann Remmelin and Michael Spaher, *A survey of the microcosme: or, the anatomy of the bodies of man and woman* (London: James Moxon, 1695). From the British Library Collection: C.107.k.13.

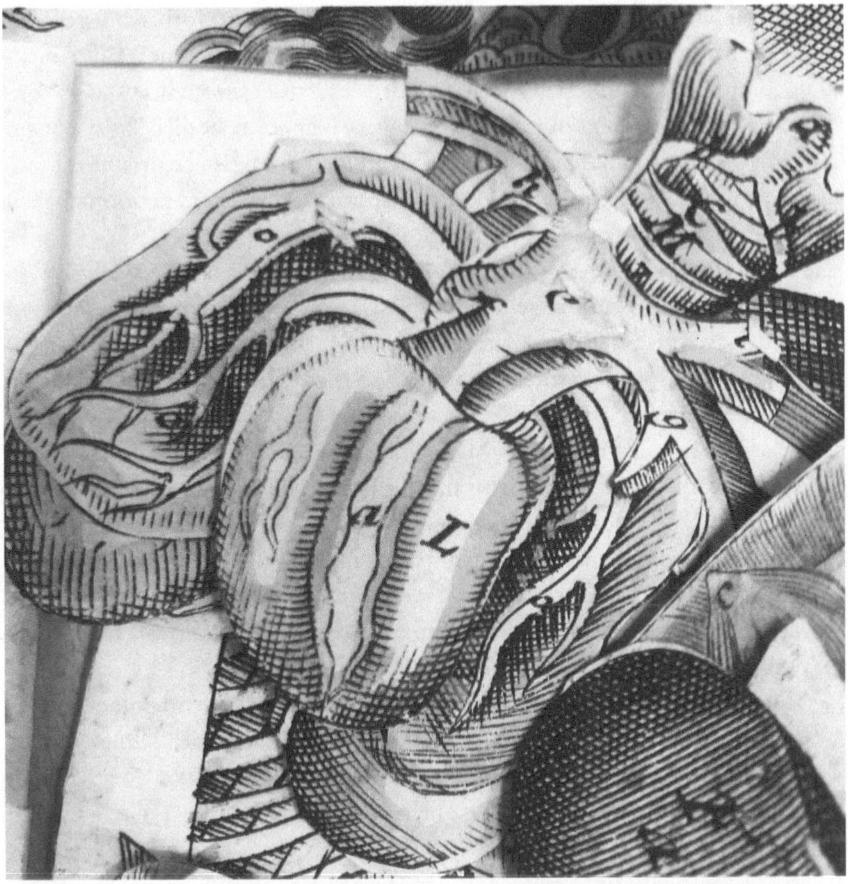

Figure 1.8. Visio Tertia stitched flaps detail. Johann Remmelin and Michael Spaher, *A survey of the microcosme: or, the anatomy of the bodies of man and woman* (London: James Moxon, 1695). From the British Library Collection: C.107.k.13.

atoms (atoms combined into a form) "do dissolve, and new Formes make," Cavendish explains in another poem, "they still to their first Principles do take."[77] For Cavendish, form is a combinatory system that can be "dissolved," returning its essential matter (atoms) to their "first Principles."

The theory of form and matter Cavendish outlines in the atom poems is both natural philosophy and poetic theory. On this point, I think with Liza Blake's important argument that we must refuse "to separate [Cavendish's] poetry from her natural philosophy."[78] Close attention to "the particular forms and poetics" of Cavendish's *Poems*, Blake argues, reveals a concept of form that is fundamentally "combinatory and flexible."[79] Cavendish's theory of form urges us to consider "the process of combination, of joining, of composite bodies."[80] In a poem that is very explicitly about the "joyning" of atoms, Cavendish returns to the language of textiles to describe the result of atoms' combinatory motion. When "severall figured atomes well agreeing" do "joyn[ ] severall waies," she explains, "the Fabrick of each several Creature raise."[81] In her use of the word "fabrick" to describe the forms atoms make, we glimpse the important material connection between Cavendish's theory of form and her theory of poetic making—the process, recall, of spinning threads (fancies) "in the Braine that Nature wrought" into new and creative "joynings." Foregrounding Cavendish's understanding of poetry as "Spinning with the Braine," what are the implications of this theory of "combinatory and flexible" forms in relation to the heightened materiality of her poetics?[82] How does a formalist theory centered on combining, joining, and arranging affect our understanding of poetic matter *as matter*? And how, ultimately, does this formalist theory—Cavendish's science of forms—animate the poetic text, transforming *Poems* itself into a complex, collaboratively formed paper body?

There is a persistent thread through the atom poems that might help us answer these questions. Cavendish associates "sharp atomes" with "fancy"—that important term that we have already seen Cavendish define as the "Flesh, Blood, Skin, and Bone" of poetry.[83] In one of the earliest poems in Part I, Cavendish applies her vitalist philosophy directly to the human brain, composed of the four different types of atoms.[84] Here is that four-line poem in full:

> In every Braine loose Atomes there do lye,
> Those which are Sharpe, from them do Fancies flye.
> Those that are long, and Aiery, nimble be,
> But Atomes Round, and Square, are dull, and sleepie.[85]

Ascribing "Fancies" to sharp atoms, Cavendish gives them first billing in this hierarchy of the brain's "loose Atomes." This short poem is light on the details, but when we "joyn" this poem to others in the collection, a more complex representation of fancy emerges through Cavendish's theories on sharp atoms. Consistently throughout the atom poems, Cavendish describes sharp atoms as the most "lively" of the four figures, "active" and "light" compared to the "heavy, dull, and slow" flat atoms.[86] Sharp atoms compose the element of fire, so they are responsible for heat (they "keepe the Body hot") and even sustain life: "Life is Atomes sharpe, which we call Fire, / when those are stopt, or quench'd, Life doth expire."[87] They are also capable of strong productive and destructive actions. They "eat . . . Bodies of lesse force" and "all things . . . dissolve, or bow, or breake," "keeping [their] strength, by making others weake."[88] Through its association with sharp atoms, fancy—the essential *stuff* of Cavendish's poetics—gains a material heft. Fancy is a powerful force ("sharpest Atomes have most power thereon"), with stinging, piercing, needle-like properties ("Atomes sharpe . . . so far do peirce into our Skyns"), that moves higher and faster than all other figures ("all pointed Atomes . . . they so light become: / above the rest do flye").[89] The complex materiality of Cavendish's fancy emerges when we apply her theories of form and formation to the atom poems themselves, combining and joining poems across this section into a composite body.

Theorizing fancy in such a way throughout the atom poems, Cavendish insists that poetry, too, is a vital form, a material body, an animated creature. Or, as she states more explicitly in her preface "To Naturall Philosophers," an *atom*. Toward the middle of that preface, Cavendish offers an explanation for why her natural philosophy is written in poetry rather than prose.

> And the Reason why I write it in Verse, is, because I thought Errours might better passe there, then in Prose; since Poets write most Fiction, and Fiction is not given for Truth, but Pastime; and I fear my Atomes will be as small Pastime, as themselves: for nothing can be less then an Atome.[90]

Cavendish begins this preface by apologizing for her "ignorance," first because she has not read an "English Booke" on natural philosophy and then noting that, while she has "heard of Atomes, and Figures, and Motion, and Matter," she has not had a chance to "throughly reason[ ] on" these topics.[91] Consequently, when she asserts here that she wrote about these subjects "in verse" rather than prose because "Errours might better passe" in verse, we might well gloss "errours" as "mistakes" or "incorrect knowledge," an extension of the "ignorance" she cites earlier in the preface. This long sentence, though, also establishes an oblique connection between "errours" and "pastime" (diversion or recreation, "a practice commonly indulged in") that draws on a different early modern connotation of "error" as the "action of roaming or wandering."[92] Might we gloss "errour," then, as a kind of speculative thinking—an epistemological mode that leaves room for error? This alternative reading finds support in the preface's opening, where

Cavendish cites "Thought, and Speculation" as different "waies in Observation," different modes of knowing for the natural philosopher.[93] In her later *Observations upon Experimental Philosophy* (1666), Cavendish claims that "experimental and speculative philosophy do give the surest informations, when they are joined or united together"—revisiting her interest in "joyning" from the atom poems and affirming "speculative philosophy" as a fundamental epistemological mode.[94] If we read "errours" in the *Poems* preface as "speculation," poetry becomes the form best suited for "thoughts" to "run wildly about," a very deliberate generic choice for Cavendish's material-textual experiments in form.[95]

This preface's justification for poetry performs another important function: it creates a rhetorical link between *atoms* and *verse*. That is, it invites us to think *atoms* and *verse* together. Through the many clauses of her characteristically labyrinthine explanation for writing "in Verse" rather than prose, Cavendish weaves a series of keywords that all refer in some way to her writing's epistemological work. If we focus on just those keywords, we unravel a thread that looks something like this: reason, verse, errors, prose, poets, fiction, truth, pastime, atoms. Her reference to "Atomes"—"I fear my Atomes will be as small Pastime"—gets absorbed into the discursive fabric of the forms and functions of writing surrounding it. "My Atomes" stands in for "my poems," reinforced when Cavendish then adds, "my desire that they should please the Readers, is as big as the World they make," with "my Atomes" as the referent for the third-person plural "they."[96] Her "verse" becomes "Atomes" presented to "readers," and poetry is the combinatory, material form that "makes" the world of her text. Cavendish's sustained exploration of forms and formation in the atom poems has as much to do with poetic form and the construction of her poetry collection as it does with her vitalist theory of matter. We must read *Poems*, then, as vital matter, as a combination of atoms—imbued with all the material heft, dimensionality, and motion such a formulation holds within Cavendish's vitalist philosophy.

Cavendish's natural philosophy rests not just on a theory of forms, but on a theory of *formation* with motion at its center. She invites us, as Jen Boyle argues, to "reimagine form as in process and as a mutable physical and conceptual determinate in meaning."[97] If poems are atoms, who or what is the force of motion that forms, deforms, and reforms poetic material? The first, perhaps obvious, answer to this question is: the poet. And, indeed, Cavendish does offer this as one possible answer in a poem titled "Motion makes Atomes a Bawd for Figure." This poem centers on "wild Motion," who forces atoms to "run from place to place" and make "new Formes" in a constant cycle of forming, deforming, and re-forming.[98]

Did not wild Motion with his subtle wit,
Make Atomes as his Bawd, new Formes to get.
They still would constant be in one Figure,
And as they place themselves, would last for ever.[99]

Cavendish begins with an if/then structure: if motion "did not . . . Make Atomes as his Bawd," then they "would last for ever" in "one Figure" ("Figure," here, used interchangeably with form to designate the combinatory shapes atoms take). This structure initially frames the poem as a thought experiment, an exploration of the relationship among motion, atoms, and form, all of which are referenced in the poem's first two lines. Motion, though, is a precisely defined force in Cavendish's atomic imaginary, and the next cluster of lines shift the poem's tone from exploratory to explanatory.

But Motion she perswades new Formes to make,
For Motion doth in Change great pleasure take.
And makes all Atomes run from place to place;
That Figures young he might have to imbrace.[100]

Cavendish's hermaphroditic motion (alternatively "she" and "he" throughout the poem) takes "great pleasure" in "Change" and, consequently, forces atoms to be "his Bawd," motion's go-between, the facilitating agent for her desire for "Figures young."[101] "And thus are most things in the World undone," Cavendish finally declares—perhaps even poetry.

In the poem's opening line, we might notice, Cavendish attributes a "subtle wit" to motion, a characteristic that she repeatedly uses to describe poets. Recall that the poets in *Sociable Letters* are invited to a "Banquet of Wit" and, in Part II of *Poems*, a personified Fame declares to a group of poets that "Wit, in Honors Court, doth greatness make."[102] In the prefatory poem, "Wherein Poetry chiefly consists," which ends with Cavendish's assertion that "Fancy the Form is, Flesh, Blood, Skin, and Bone," Cavendish critiques poets for considering "not the Fancy but the Phrase: / as if fine Words, were Wit."[103] If poetry "chiefly consists" in "fancy" and "wit," then motion's "subtle wit" aligns this vital force with the poet.[104] As if to reinforce—even *enact*—this association, Cavendish changed the poem's opening lines for the 1664 edition of *Poems*, revising them to, "Motion makes Atomes, by his subtle Skill / his Bawds, to get new Formes him to his will" (she changed the end of this couplet again for the 1668 edition: "to get new Formes unto his will").[105] Cavendish's own poems are constantly in formation and, wit(h) these revisions, the poet truly is Motion, working to "undo" previous forms and reshaping poetic matter into "Figures young."

The 1664 revision from "subtle wit" to "subtle skill," however, has another important implication. "Wit" may be one of Cavendish's favorite terms for poetry, but "skill" is one of her favorite terms for textile craft, central to the textile metaphors through which we saw her frame her poetic work. "Having no *skill*" in the art of "Spinning with the Fingers," she "delights" in "Spinning with the Braine."[106] In a later poem in the collection, the poet describes a spring landscape as a "wrought Carpet," detailing how "Spring doth spin fine grasse green silk" and how "the ground was wrought like threads drawne from the Sun."[107] "Skill" enters at the poem's conclusion, where Cavendish reasserts her argument on invention and imitation: "this peice the patterne is of Artfull skill; / Art, Imitator is of Nature still."[108] Nature's landscape becomes a "pattern" for "wrought works" ("Fancies, in the Braine that Nature wrought / are best"), just as Cavendish's poems are hailed as a "Pattern" from which future poets will "take their Sample" and "measure their Conceits and Fancies."[109] The shift from "wit" to "skill" in "Motion makes Atomes a Bawd for Figure" returns us to Cavendish's prefaces and the way she draws her readers into the book. It returns us to the highly tactile process of forming the "garment" that is *Poems, and Fancies* and to the "natural . . . fantasticall motion" of women's brains. For Cavendish, Motion is also the reader, handling the pages of her text and weaving their own garment of poetic matter.

Three-Dimensional Forms (or, How to Read Dimensionally)

It is one thing for a reader to handle a text, flipping the pages of a book, perhaps reading poems out of order, perhaps adding a marginal note here and there. It is quite another thing to understand the reader as *motion* as it is theorized in Cavendish's philosophy, literally an active agent of form (material formation and reformation) in relation to the text. But that is exactly the relationship between reader and text when the text in question is an anatomical fugitive sheet. Returning briefly to that archive, Andreas Vesalius's uncut flap anatomy sheets offer a particularly illuminating example of the reader's formative role in the text's epistemological and aesthetic work. Throughout his prefaces, which I will take a closer look at in Chapter 4, Vesalius insists that his texts are meant to put bodies in front of his readers and engage them tactilely in a way that simulates some of the work of the dissection room. Urging readers to instrumentalize his book through very active reading, he notes in the preface to Book II, for example, that the musclemen figures should not be studied in isolation ("it will not suffice to look at a muscle in a single picture even if I have marked it with a character").[110] Rather, readers should "investigate" individual muscles as they are represented in every figure. "To see where each stands out or lies beneath another," Vesalius insists, "it must be investigated everywhere, and viewed especially where it presents itself to view as a whole."[111] In both the *Fabrica* and the abridged, student-oriented *Epitome*, these instructions for active reading culminate in the anatomical flap figures.

About halfway through Vesalius's *Fabrica*, the reader encounters a page with eight small woodcuts, each representing a different body part. A note at the top of the page explains that this "chart combining several small figures" has been prepared for the reader to construct their own anatomical flap figure. These "small figures," Vesalius specifies, "should be attached to the figure seen on the page marked m3," a large fold-out diagram of the human body showing the systems of veins and arteries.[112] In one 1543 copy of the *Fabrica* held at King's College London, both the uncut flap sheet and the full body diagram are bound as fold-outs at the book's center, ready for easy extraction to complete the flap figure assembly process (Figure 1.9).[113] Vesalius hails the reader as an active former of the anatomical knowledge contained in the book's pages, and the work of forming here is simultaneously textual and somatic—the formation of a paper body. Vesalius directs the reader to first "attach a layer to the back" of the sheet with the uncut flap figures. Presumably, one would also have to remove the page from the book to do this, thereby changing the composition of their copy of the *Fabrica* (momentarily deforming the text in order to re-form it in a new way).[114] Then, "cut out each of the figures from the surrounding paper," some of which are incredibly delicate parts (the second ["secunda"] woodcut, for instance, is actually two flaps of thin arteries and veins, one that attaches "beneath the pectoral bone to the upper part of the abdomen," and another that "entwine[s] the lower part of the abdomen").[115] Finally, the reader can assemble the anatomical flap figure, taking care to attach the organs in the proper order and, in some cases, performing extra tasks to bolster the dimensionality of the fully constructed figure. The sixth figure, for instance, "should first be attached under the seventh" (the sixth being the "lower surface of the penis" and the seventh being the "front of the bladder and penis"), so the two flaps are "combined" before being secured to the full body diagram.[116] To conclude this step, the reader should then "fold[ ]" the combined figure "into an S like a penis." Many readers did this work, as

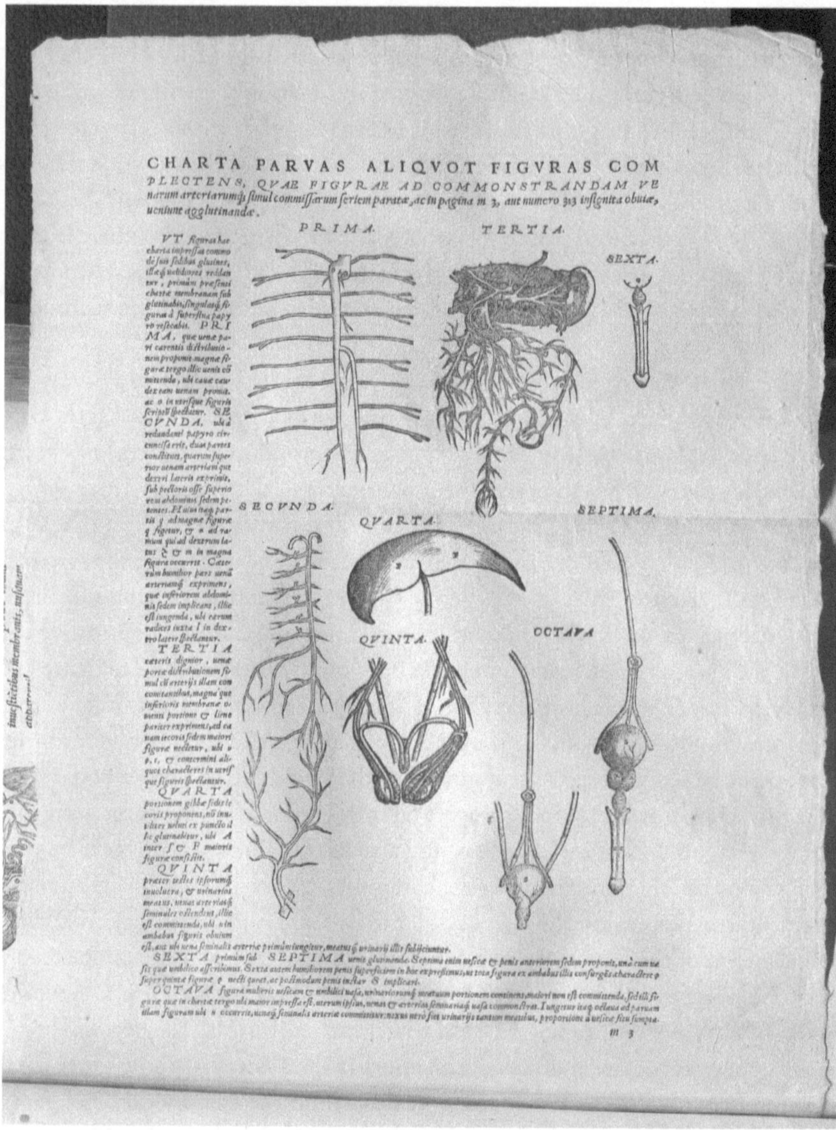

Figure 1.9. Uncut flap anatomy sheet, sig. m3. Andreas Vesalius, *De humani corporis fabrica libri septem* (Basel: Johann Oporinus, 1543), King's College London, Foyle Special Collections Library, FOL. QM21 VES(2).

several extant copies of the *Fabrica* with fully assembled flap figures testify (Figure 1.10).[117] As we engage with these archival materials, we encounter paper figures and anatomical knowledge formed by readers' hands and, even where we find uncut, untouched flap pages, Vesalius's instructions invite collaborative knowledge making that brings together anatomist, text, and reader.

Cavendish is similarly instructive throughout *Poems* as she consistently reminds her readers: this is not the poetry collection you are used to. She includes three prefaces addressed directly "To the Reader," each at the start of a different part of the book (Parts I, III, and IV), in addition to the many prefaces addressed to individual communities of readers (such as, "To Naturall Philosophers," "To Poets," and "To all Writing Ladies"). In one extant copy of the 1653 edition, Cavendish has handwritten an additional note after the opening prefatory material. "Reader," she begins, "let me intreat you to consider only the fancyes in this my book: of poems."[118] Her instructions for reading are particularly explicit in the reader's preface to Part III, where she directs, "read this part of my Book very slow" and "observe very strictly every word."[119] Being a good reader of Cavendish's *Poems* requires work. If one does not understand her fancies, they must find a poet and ask "where the Conceit lies"—not a "Rhimer," Cavendish specifies, but a "Judge . . . learned in the Lawes of Poetry" and free from "Bribes of Envie."[120] This part of the book is "thick of Fancies," she continues, "and therefore requires the more Study."[121] Any reader who does not carefully "mark" each fancy, or who "skip[s]" through with "too hasty reading," will "intangle the Sense of the whole Copy."[122]

We think of books as objects with "thickness"—Cavendish's folios are themselves quite thick—but we must pause over the curious description of "fancies" as "thick." What does it mean for poetry to be "thick of fancies"? And why does such a quality call for a different mode of attention on the part of Cavendish's reader? Following Amy Scott-Douglass's important reconsideration of Cavendish's paratexts, this preface becomes one of many "deliberately crafted devices for establishing [Cavendish's] public authority," and it feeds into the "pro-laureate" stance Scott-Douglass gleans from Cavendish's writing.[123] The most important part of Cavendish's self-fashioning, Scott-Douglass argues, is her persistent claim that she is "the first English poet who ever wrote this way."[124] This claim, however, is not simply a rhetorical self-fashioning device. As I have been aiming to demonstrate, Cavendish understood her text as an experiment, and striking out new poetic ground required her to recalibrate readers' expectations for their engagement with her book.

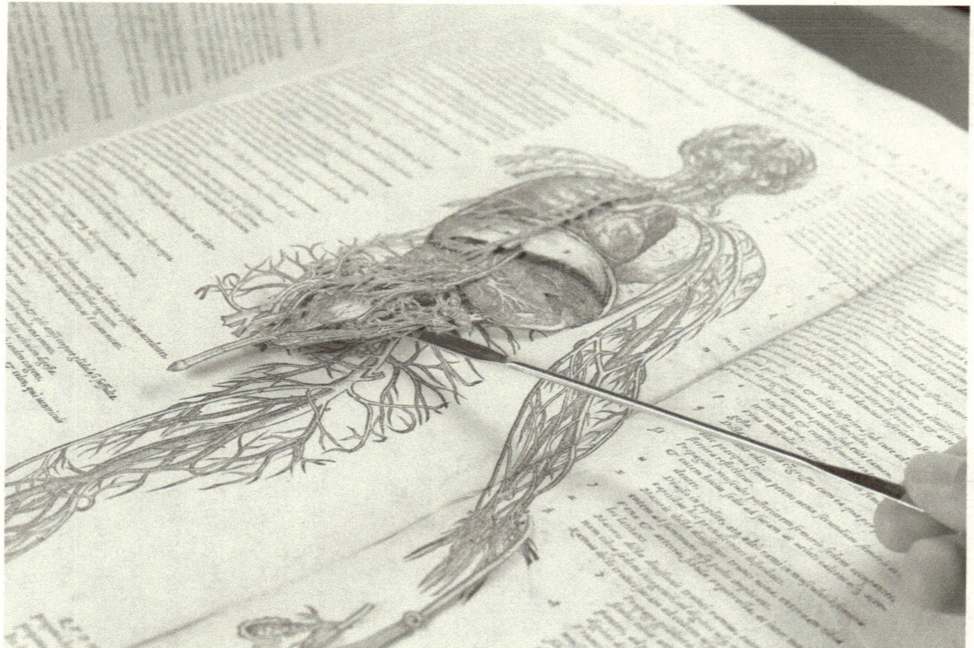

Figure 1.10. Assembled anatomical flap figure. Andreas Vesalius, *De humani corporis fabrica libri septem* (Basel: Johann Oporinus, 1555), University College London, Special Collections, FOL. 1555 V 27.

Poems, and Fancies is a manifesto not only for an alternative practice of poetic making, but also for an alternative practice of poetic *reading*. Cavendish's theory of forms—her philosophy of how matter forms, deforms, and reforms—centers on the relations between bodies and, accordingly, she is keenly interested in how writing and reading bodies engage with the poetic matter of *Poems*. This is most fully evident in Part III, the part that is "thick of fancies." In fact, Cavendish titles this section of the book "Fancies," inviting the reader, even with that simple gesture, to adjust how they are reading as they move from the atom poems of Part I, to the dialogue poems and "Moral Discourses" of Part II, to the cluster of forty-eight "fancies" contained in Part III. If we read this section of the book through the framework of the atom poems, with the association between fancy and "Atomes sharpe," we are primed to make this adjustment. Fancy, we have learned, is an animated force with material heft, susceptible to motion and forced to "run from place to place" in an infinite cycle of forming, deforming, and re-forming, and Part III is about this cycle of formation.[125] It is a section about "the very idea of a poem," as Liza Blake argues—a section that "imaginatively explore[s] the process of conceit-making."[126] It is also the section of the book with the most body parts, and the one framed directly with the prefatory poem rendering poetic form anatomical: "Fancy the Form is, Flesh, Blood, Skin, and Bone."[127] And it begins with a flap anatomy.

The first poem of Part III is titled "The severall Keyes of Nature, which unlock her severall Cabinets"—although, typographically, this "title" reads more like a section header than the title of a poem. In the 1653 edition of *Poems*, this descriptor is positioned just below the heading "FANCIES," printed in large font, all capital letters. Heading and subheading are then followed by a line in the white space between the section heading and the first poem, printed with an illuminated "A." Graphically, then, this first poem is structured as a kind of opening argument for the "FANCIES" section of *Poems*, and here it is transcribed in full.

A Bunch of Keys did hang by Nature's side,
Which She to open her, five Boxes tri'de:
The first was Wit, which Key unlock'd the Ear,
Open'd the Brain, to see what things were there.
The next was Beauty's Key, unlock'd the Eyes,
Open'd the Heart, to see what therein lyes.
The third was Appetite, which quick did go,
Opening the Stomack, to put Meat into.
The Key of Scent unlock'd the Brain, though hard;

For, of a Stink, the Nose is much afear'd.
The Key of Pain did open Touch, but slow;
For, Nature's loath any Disease to show.[128]

This poem imagines a different kind of flap mechanism, structured as locked boxes ("the five senses," Cavendish glosses in the marginal note) opened with different keys. The keys correspond to poetic themes and effects: wit unlocks the ear, beauty the eyes, appetite the mouth, scent the nose, and pain unlocks touch. Each sensory system also "open[s]" a corresponding, internal body part—the brain, the heart, and the stomach (and, perhaps, also "Touch," which is both sensory system and organ of the body). "Observ[ing] strictly every word" of this poem, we are invited to match allegorical keys with our own embodied sensory organs. As Nature unlocks her "five Boxes," Cavendish unlocks her reader's five senses, calling attention to each of them in turn and lifting the flaps of corresponding internal body parts. She creates her own kind of poetic paper body, a flap anatomy figure formed through poetry whose flaps open in response to poetic matter.

This poetic flap anatomy prepares readers for the poems that follow in Part III, all of which target different bodily senses as they figure mental activities as garments to touch ("Gloves of Remembrance" and "Colour'd Ribbons of Fancies"), age as ingredients to taste ("Juice of Youth" and "Childish Macaroon[s]"), and affects as herbs to smell ("Ginger of Revenge" and "Heart-burning Parsley").[129] Nature is the primary protagonist and agent in many of the poems. The first half of this section contains poems detailing various elements of Nature's cabinet and table (with poems like, "Natures Cook," "A Posset for Natures Breakfast," and "Natures Wardrope"). The second half contains "simile" poems, which compare different parts of the body to other forms in the world ("Comparing the Tongue to a Wheele" and "Similizing the Head of Man to a Hive of Bees"). The cluster of eight poems known as the "recipe poems" have been of particular interest to scholars, offering, as Mihoko Suzuki describes, "a *tour de force* mixing of the materiality of the body and of food."[130] These poems in particular force the reader into multi-sensory modes of engagement. "A Bisk for Natures Table," for instance, plays in the spaces between looking and touching ("a Forehead high, broad, smooth, and very sleek"), hearing and smelling ("a sharp, and quick, and ready, pleasing Tongue" joins "a Breath of Musk, and Amber" in the pot), and touching and tasting ("two Cherry Lips, whereon the Dew lies wet").[131] In a particularly visceral instance of this interplay, Cavendish describes "a Hand that's fat, smooth, and very white, / the inside moist, and red like Rubies bright" as one of the ingredients in Nature's bisk.[132] Such a description activates visual and tactile sensory systems (a quality like "moist" is perceptible through vision and touch) and recalls the haptic technology of the anatomical flap figure as we move with the line break from the exterior ("smooth, and very white") to interior of the hand ("moist, and red").

Extending Suzuki's argument on the "materiality of the body and of food," we must also consider Cavendish's experiments with the materiality of poetry itself in this section of *Poems*. These poems transform body parts (the many ingredients of Cavendish's recipes) into the material stuff of the poetic text, and Cavendish wields highly dimensional paper technologies—first, the anatomical flap (the successive "opening" of different organs), and then the recipe—to animate body and text together within the pages of *Poems*. As instructions for making a thing, forming parts into a whole, Cavendish's recipe poems are about making poetry as a collaborative act between writing bodies and material fancies. The individual ingredients that make up the recipe poems are excessive in their materiality, calling heightened attention to these "ingredients" as building blocks of poetic matter. Each line in Cavendish's catalog of ingredients transforms a poetic fancy—the stuff of imagination that we have heard Cavendish describe as threads "in the Braine that Nature wrought"—into something the reader can grasp, hold, and manipulate.[133]

This is particularly evident in a twelve-line poem titled "A Tart," which takes the conventional blazon mode as its conceit.

> Life took some Floure made of Complexions white,
> Churnd Butter, by Nourishment; as cleane as might:
> And kneads it well, then on a Board it laies,
> And roules it oft, and so a Pye did raise.
> Then did she take some Cherry Lips that's red,
> And Sloe-black Eyes from a Faire Virgins Head.
> And Strawbery Teats from high Banks of white Breast,
> And Juice from Raspes Fingers ends did presse.
> These put into a Pye, which soone did bake,
> Within a Heart, which she strait hot did make;
> Then drew it out with Reasons Peele, and sends
> It up to Nature, she it much commends.[134]

In this poem we encounter a striking mix of culinary and dissection imagery that Elizabeth Scott-Baumann argues "wittily undermines and reworks the blazon tradition."[135] The key ingredients of the titular tart are the fragmented pieces of the female body so familiar from early modern blazon poems: complexion, lips, eyes, breasts, hands (with requisite dashes of whiteness and virginity thrown in). But the complexity of this poem lies in its mixed metaphors—the constant blurring of body and food—that Cavendish enhances with her syntax. The first line designates "Floure" as the ingredient and then includes the body part ("Complexions white") in the description of how the flour is "made." Changing this construction in line 5, Cavendish lists "Lips" as the ingredient and only includes the food item ("Cherry") as an adjective. Consequently, the poem does not offer a strictly ordered poetic conceit—body parts as ingredients in this tart for Nature. Instead, Cavendish forces a constant shuttling between the tenors and vehicles of these metaphors, repeatedly blurring the line between anatomical part and foodstuff.[136]

The poem's muddled arrangement of metaphors does two important things. First, it draws our attention to the making of the poem. We start to notice how the poem itself is formed, even as the poem's language details a process of formation, of making parts into whole. The process of making on display here, importantly, models the theory of formation Cavendish outlines in the atom poems. Our poet, as motion, acts on poetic matter that would have been very familiar to her readers, deforming the blazon conventions by extracting them from the familiar context of love poetry, and then re-forming them into new poetic matter—a poem about Nature's relationship to bodies, structured as a recipe. Conceptually, the process of forming parts (ingredients) into a whole makes a "tart" for Nature; formally, the process of transforming the material stuff of poetic matter makes a new aesthetic object.

Second, the metaphors themselves, along with their muddled arrangement, activate a multi-sensory shuttle of perception for Cavendish's readers, or what Harry Berger Jr. calls an "observer shuttle." These poems "construct[ ] a viewer who is invited to move up close, touch the surface, run fingers over it, make manual contact with the work" of the poet's hand and then "move back and recompose the texture within the image."[137] In "A Bisk for Nature's Table," for instance, we are invited into intimate contact with the "moist, and red" inside of the hand, and then shuttled back out to partake in the completed dish at the end of the recipe (served in a "Dish of Admiration").[138] In "A Tart," our attention is drawn to an even more granular level—the processes of making individual ingredients, violent processes that veer into the realm of the grotesque. Cavendish is not explicit about these processes; rather, we have to "observe strictly every word" of this poem in order to glean how certain ingredients are made. Revisiting the first line, for example, Cavendish describes a seemingly mundane action to begin the recipe ("Life took some Floure made of Complexions white"), but this action contains within itself the grisly process through which this flour was made by grinding "Complexions white." Similarly, line 8 ("and Juice from Raspes Fingers ends did presse") is poised between acts of production—an ingredient added to make the tart—and destruction. "Raspes" is an early modern word for raspberries and also a term that describes the process of grating, filing, or scraping something (here, "Fingers ends") through a rasp (a kind of cheese grater).[139] Attending closely to these fancies reveals how each line—each ingredient, each piece of poetic matter—has the work of deforming and forming packed into it.

The recipe poems offer a glimpse into the workings of this entire section of *Poems*, a textual anatomy animated by its perpetual construction. Befitting its opening argument, Part III of *Poems* functions according to the formal mechanic of the paper flap, but what is revealed in the cutting, assembling, and handling of Cavendish's flaps is not human anatomy. It is an anatomy of poetics and an argument for reading this poetic text as a body in parts. Each poem in Part III takes up the familiar parts of seventeenth-century English poetry (later allegorical poems in this section compare the tongue to a spinning wheel and the lips to a loom; the heart to a harp and the passions to "strings on which the mind doth play") and exposes those parts *as* matter. Cavendish forms her poems with dimensional, tangible poetic material; they are not just "thick of fancies," but *thick fancies*. Her readers are then invited to dissect and explore around, through, and under the three-dimensional poetic objects she forms with those material parts. Cavendish shows us how to manipulate the stuff of poetic matter, and poetic formations become something we make and unmake alongside the poet. In this section of poems animated by formation rather than form, Cavendish seeks to adjust the limits of what counts as poetic craft and models an alternative way for readers to experience her material book. For Cavendish, the force of motion that forms, deforms, and reforms poetic material encompasses all of the bodies that collide in the act of poetic making. Writing, reading, and paper bodies together form poetic matter, and it is the relations among these bodies that generate the epistemological work of poetry.

The Paper Body of *Poems, and Fancies*

Throughout this chapter, I have studied pieces of *Poems* (*Poems* in parts) in order to demonstrate the dialogic relationship Cavendish constructs between a science of bodies and a theory of poetic formation with writing and reading bodies at its center. I want to turn, finally, to *Poems* as a whole because this is still, after all, a *book* of poetry and, as Jeffrey Masten posits, Cavendish's books are best understood "as performances—as performing."[140] We might begin (as many readers do) with epistemological performance, for, as we have seen, Cavendish uses *Poems* to explore her vitalist, monist philosophy, and each section of the book functions as a complementary epistemological experiment, offering readers many different ways into Cavendish's natural philosophy.[141] We might proceed (and many readers do) with domestic performance, reading *Poems* as a book that persistently aligns acceptable, appropriate feminine labor with the act of writing.[142] We might finally consider (again, as many readers do) Cavendish's self-fashioning and read *Poems* as a tour de force authorial performance. This final approach to *Poems* would put us in good company with some of the most insightful literature and science studies work on Cavendish, much of which challenges Catherine Gallagher's influential argument on Cavendish's absolutist female subjectivity to instead consider how her radical scientific theories deconstruct subject-object binaries.[143]

Poems, and Fancies is, in part, all of these things. What I have been trying to show in this chapter is that there is another kind of project—an anatomical project—that also frames this work. *Poems* is about the complex relationship between parts and wholes that constitutes bodies—a theory of formation and a science of bodies knit together across the pages of Cavendish's book. As a combinatory, textual form (as a paper body), *Poems* offers itself as a poetic-anatomical experiment, infused with Cavendish's corporeal, physical, and material way of understanding the world. Like the combinatory forms described in Part I's atom poems and the culinary olios prepared in Part III's recipe poems, *Poems* calls attention to the materiality and tangibility of its component parts even as they work together to make a whole. To understand *Poems* as a whole—that is, to understand its status as a paper body—we need to look, briefly, at its "claspes," the transitions between each part of the book marked by the recurring title "THE CLASPE." This repeated word, as Randall Ingram argues, is a "perpetual reminder[ ]" of the book's physical and conceptual "constructedness." Physically, clasps bind loose leaves into a book and, conceptually, Cavendish's clasps "yoke topics that are sometimes widely disparate."[144]

The "claspes" are stitches, sutures, connective tissues that expose the formation of *Poems* itself, constantly reminding us "that what has been clasped may be unclasped."[145] The first clasp, between Parts I and II, contains poems that critique attempts to "order" the "circle of the brain." "The Braine is round," Cavendish declares, but since "Archimedes, or Euclid's time," "Men of Arts" have "stretched" the brain "upon a Line," trying to understand it using "Division, and Subtraction," "Lines drawne out, and Points exactly layd." "But yet None can demonstrate it plaine," Cavendish concludes.[146] This first clasp insists on the brain's natural "fantasticall motion" ("Easie, Free, / not to be bound"), which then corresponds with the "Free" and "wild" poetic style Cavendish extols at the start of the second clasp between Parts II and III.[147] The second clasp contains the oft-anthologized "The Hunting of the Hare" and "The Hunting of the Stag" poems, in addition to two poems about a fantastical island paradise ruined through pride and ingratitude, often read as allegories on events of the English Civil Wars. As L. E. Semler convincingly demonstrates, though, "the second clasp . . . is, as with the other clasps, primarily about poetry."[148] The hunt and island become "poetic surrogates" in response to the actual, lived conditions of the Civil Wars, and the movements of the hunted hare and stag embody the "free" and "wild" motions of Cavendish's fancies.[149] Fancies and bodies literally take center stage in the book's third clasp, titled "Phantasmes Masque" (to which we will return momentarily), and then become refigured as "fairies" and "animal spirits" in the fourth and final clasp between Parts IV and V. Opening the fourth clasp with a thought experiment ("Who knowes, but in the Braine may dwel / little small Fairies; who can tell?"), Cavendish imagines fairies as the animating figures of fancy and sensation, moving around a very anatomical brain—a "thick scull" surrounded by "Dura Mater" and "Pia Mater."[150] The poems in this final clasp recall the first clasp's framing of the brain as a circle, along with the poet's "fairy wits" in the prefaces to Part III (which "dance upon" poetic lines and "round every fancy").[151] The "animal spirit" poems then prime the reader for "The Animall Parliament," a political allegory of the anatomical body as a "parliament" of Upper (head) and Lower (heart) Houses, included after the poems in Part V.

As the connective tissue between the various pieces of *Poems, and Fancies*, these clasps do an extraordinary amount of infrastructural work. They embody quite fully the knitting together of Cavendish's theory of formation, her science of bodies, and the materiality of *Poems* itself as a paper body, most evident in the third clasp, "Phantasmes Masque." In this clasp, Cavendish stages a performance within the pages of her poetry collection ("books as performance—as performing," indeed).[152] "The Scene is Poetry," she begins this clasp, and "the Stage is the Braine, whereon it is Acted."[153] Cavendish takes the familiar spatial contours of a performance "scene" and applies them to poetry. "Poetry," this masque insists, has a topology within and through which bodies move. After describing a "dumb show" featuring "a young Lady in a Ship" (anticipating, perhaps, the famous opening of her *Description of a New World Called a Blazing World* [1666]),

Cavendish announces the "six Masquers" who have each been "dressed by" a different allegory. We encounter a "Lady drest by Youth," whose "bashfull Eye-lids" hang down "as Lace"; a "Souldier arm'd by Mars," wearing a "breast-plate made of Courage" and holding a sword made from "pure bright Honour"; and a "Lady drest by Love," who wears "Strings of threaded Teares" and her lover's sighs "as Pendants . . . at either Eare."[154] These are not the allegorical characters Youth, War, and Love, but rather bodies "dressed by" youth, war, and love. These are "paper bodies," dripping with the material stuff of fancy and formed out of familiar allegorical language, imagery, and tropes—combinatory forms that take on dimension within the "scene" of Cavendish's poetic performance.

Scholars often note that there is no *body* at the center of Cavendish's poetry collection, no first-person poetic "I" or clear addressee (the "you" of so many early modern English sonnet sequences).[155] When we do encounter bodies in the text, they are allegorized, similized, or consumed in ways that defy our expectations for the poetic representation of bodies. They are bodies conspicuously made of fancies, like the masquers' bodies, or body parts figured as fancies, like in the recipe poems of Part III. To return to the question with which I opened this chapter—*what is a paper body?*—*Poems* responds with: a paper body is material and dimensional textual matter, perpetually forming through the motions of other bodies. Texts themselves, like all matter, are "combinatory and flexible" corporeal forms with weight, dimension, texture, and *liveliness*.[156] In "Phantasmes Masque," texts are performing, dancing, costumed bodies, choreographed by the materials of their making. The nine muses "dance a measure in foure and twenty Figures," which, as Cavendish glosses in a marginal note, "are the 24 letters in the Alphabet."[157] The "scene" of poetry—or, perhaps, the "scene" of *Poems*—is filled with the essential stuff of poetic language in motion. A musical measure is one part of a whole, but also the indicator of how the whole has been broken into parts. It signals to other musicians how to play the piece and signals to the audience how to interact with the piece. Cavendish's opening measure in "alphabetic time" is a rhythmic mark of collaboratory making that invites creativity even as it sets a foundational structure. Poetry ceases to be an inevitable, immutable form. Instead, Cavendish offers us a masque (another form that is simultaneously structured and flexible) in which even letters of the alphabet, the "principall figures" of text, become fancies.[158] By transforming letters from elements of language and graphic marks on the page into a rhythmic measure, Cavendish puts poetry in motion at its most fundamental, granular level. The scene of poetry, she argues, is a collection of material parts made into a whole through the collaborative forming acts of the bodies with which it comes into contact.

When we read *Poems* as a paper body, simultaneously dissected and stitched together for and by its readers, then we begin to understand Cavendish's active participation and intervention in the vibrant anatomical-textual experiments of seventeenth-century England. Through this, her first published text, she calls out to fellow scientists and poets to say: look at what is possible if we understand textual forms as dimensional, tactile—*fleshly*. Famously, Cavendish was the first woman invited to a meeting of the Royal Society, founded in 1660 and soon after granted a royal charter by King Charles II.[159] She was also in regular correspondence with prominent Society members, such as Kenelm Digby, Walter Charleton, and Joseph Glanville. Her *Philosophical Letters* (1664), *Observations upon Experimental Philosophy* (1666, 1668), and *Grounds of Natural Philosophy* (1668) contain thorough and thoughtful engagements with the writings of Thomas Hobbes, René Descartes, Henry More, and Jan Baptista Van Helmont. She read and praised John Gerarde's *Herball, or Generall Historie of Plantes* (1597), and her own writings were read and annotated by Nehemiah Grew, English physiologist and plant anatomist.[160] Consequently, of all the poets I take up in this book, she has certainly garnered the most interest from historians of science and philosophy and scholars of literature and science. But she has never been granted the status of anatomist. Never positioned as an interdisciplinary thinker with the science of the body at the center of her inquiries. And never studied as a writer with a significant contribution to anatomical science as it came to bear on early modern textual forms.

This is what the methodological apparatus of *Anatomical Forms* allows us to see. I have attended closely to not only *what* Cavendish's poems (and *Poems*) say, but *how* they say. In considering the *how* of *Poems*, I built my formal analysis with Cavendish, allowing her theories to guide my study of form as it manifests in the interdisciplinary space of poetic and anatomical thinking. As a result, we have not only excavated the shared formal and material practices of anatomical study and poetic craft that shape Cavendish's work, but we have also started to sketch new contours for histories of science—contours that I will sharpen in this book's remaining chapters. *Poems, and Fancies* joins the ranks of archival objects that refuse to obscure the messy materiality of their making, demanding the reader's "interaction and skill as a *bricoleur*,"[161] not because the book object itself (like anatomical fugitive sheets) is formed of small scraps of paper that require certain kinds of handling, but because in the folds of the book's pages Cavendish exposes the warp and weft of *Poems*. She draws our fingers to the stitches that form this paper body and then invites us to make something of it.

CHAPTER 2

Postmortem Poetics

Mary Sidney Herbert's Studies in Death, Dying, and Loss

Mary Sidney Herbert, the Countess of Pembroke (1561–1621), has been of particular interest to scholarship on early modern women and the history of science, not because of any overwhelming evidence of her scientific work, but rather because of a significant *lack* of evidence. Hers is a tantalizing narrative because so much about her connection to contemporary science is based on tenuous biographical information and scattered manuscript references across the early modern English archive. It is a story, in other words, that allows for imaginations to run wild through extant scraps and fragments of knowledge. According to antiquarian John Aubrey, Pembroke had a private experimental "laborator[y]" at Wilton House, the Pembroke family estate.[1] For Hilary Rose, Pembroke's laboratory is one example of how "domestic technological resources" in early modern England fed developing knowledge in "chemistry and biology and medicine."[2] We have some evidence that Pembroke did indeed participate in the domestic scientific economy Rose links her to, like a distillation recipe titled "tragea Comitisse Pembrokiae," bound in a manuscript that includes an almost identical recipe for "Adrian Gilbert's Cordiall Water."[3] Half brother to Sir Walter Raleigh, Adrian Gilbert (ca. 1541–1628) was a chemist who, according to Aubrey, was often in residence studying at Wilton House. Gilbert was also connected with the more famous Renaissance chemist and astrologer John Dee (1527–1608), who tutored Sir Philip Sidney in chemistry.[4] Aubrey records that Pembroke herself "was a great chymist and spent yearly a great deal in that study," and, while Aubrey's account is never the most reliable, extensive scholarship on English noblewomen's learning suggests that Pembroke likely had access to some of the same education as her brother.[5]

Scholars have also located snatches of scientific engagement in Pembroke's writings, drawing on autobiographical information to link Pembroke's lived experience with the knowledge on display in her poetry. Margaret P. Hannay pinpoints the embryological knowledge evident in the third stanza of Psalm 51, where the poet's sin is "conceave[d]" in her mother's womb and "cherished" by her mother's "living heat" (Ps 51.16–17).[6] This stanza, along with similarly vivid childbirth imagery in Psalms 48, 58, and 139, Hannay suggests, not only reflects Pembroke's own experience with pregnancy and childbirth, but also "demonstrate[s] a real interest in embryology."[7] Such interests, scholars then speculate, might have informed her relationships with the prominent physicians that served the Pembroke family. Thomas Moffet (1553–1604) was a naturalist and member of the College of Physicians in London. He studied entomology (even dedicating a treatise on silkworms to Pembroke, where he addressed her as the "nurse of learning") and was a well-known supporter of Paracelsian medicine, a system motivated by religious and theological questions that relied on chymical principles as the grounds of natural philosophy.[8] After Moffet's death, Matthew Lister became the family's physician, also a member of the Royal College of Physicians and interested in Paracelsian "chymical medicine."[9] The true nature of Pembroke's relationship to Lister is unknown, but Lady Mary Wroth allegorized their relationship in her closet drama *Love's Victory* (the love plot between Simeana and Lissius), and Aubrey speculates that the two were secretly married.[10]

Biographical snippets, enticing archival traces, and the dubious account of a seventeenth-century antiquarian: these are the pieces of evidence most often leveraged to make the case for Pembroke as a "visionary scientist."[11] The story outlined above relies on science looking like we expect it to look, as if Pembroke's place in the history of science could only be secured through a connection to established, well-known early modern scientific thinkers, or evidence that she was actively experimenting in a laboratory setting, documenting recipes and chemical trials for circulation and testing. It is also a story that relies on "experience" as an incontestable, undertheorized category of evidence, a methodological tendency Joan W. Scott cautions against in her landmark essay "The Evidence of Experience." "When experience is taken as the origin of knowledge," Scott argues, "questions about the constructed nature of experience, about how subjects are constituted as different in the first place, about how one's vision is structured—about language (or discourse) and history—are left aside."[12] When we use Pembroke's experience with alchemical study, for example, to explain the alchemical knowledge represented in her psalm translations, which then "bear witness to the fact that she regarded the intensive study of nature and chemistry in the manner of Paracelsus as a spiritually improving pursuit," experience becomes both "a starting point and a conclusive kind of explanation," and we lose sight of the fact that experience is a discursive production.[13] "It is not individuals who have experience," Scott reminds us, "but subjects who are constituted through experience."[14] Some of the assumptions Scott questions are outdated (particular to the predominantly white feminist recovery work of the early 1990s), and her argument presupposes presence in language as a default condition for historically marginalized figures, a presupposition that has been scrupulously challenged by intersectional feminist work that takes absence as its starting point. Her argument, though, remains all too relevant for the study of early modern women's writing because the assumptions she critiques continue to shape critical work in the field.[15] As we will see, Pembroke's poetry is an ideal site for thinking critically about the category of "experience" as it has been leveraged in scholarship on early modern women's writing because it is precisely the question of experience that Pembroke theorizes most thoroughly in her writing.

Following Scott's imperative to historicize experience and "to historicize the identities it produces,"[16] the story I tell in this chapter looks quite different from the story outlined above, even as it builds on the important fragments and tantalizing traces Pembroke scholars have recovered. Focusing on Pembroke's formal experiments, I consider the sustained—distinctly corporeal—study of death and dying that extends across the literary, philosophical, and theological inquires that propel her writings.[17] Pembroke translated three texts associated with the *ars moriendi* tradition—literature meant to offer spiritual guidance in the face of death and instruct the reader on the art of dying well. In 1592, she published a translation of Philippe Duplessis-Mornay's *Excellent Discours de la Vie et de la Mort* (*A Discourse of Life and Death*) bound with a translation of Robert Garnier's play *Marc Antoine* (*Antonius*). Several years later, Pembroke translated sections of Petrarch's *Trionfo della Morte* (*The Triumph of Death*), maintaining the original *terza rima* stanza form to tell the story of Death coming to Petrarch's Laura.[18] In addition to these ars moriendi texts, Pembroke's extant original poems deal directly with the untimely death of Philip Sidney—first, an elegy to Sidney titled "To the Angell spirit of the most excellent Sir Phillip Sidney," and, second, a prefatory dedication to Queen Elizabeth that eulogizes Sidney and offers the siblings' collaborative psalm translations as an act of praise and mourning. Pembroke was, indisputably, a student of the strange and unsettling circumstances of death, an early practitioner of thanatology, the interdisciplinary field that aims "to construct a scientific comprehension of death, its rites, and its meanings."[19] Thanatology is an anachronistic term, to be sure, but it neatly encapsulates Pembroke's poetic explorations of morality and grief, as well as her formal experiments with the strange felt experience of dying—an extensive epistemological project that I call her *postmortem poetics*.

It is in the psalm translations that we find Pembroke's most robust engagements with early modernity's widespread cultural fascination not just with death but with the *body* in death. Undoubtedly the most well-known of Pembroke's writings, and the focus of much of this chapter, her and Sidney's collaborative translation of the one hundred fifty biblical psalms (often referred to as the Sidney Psalter) are currently extant in eighteen different manuscript versions, none written in the siblings' hands, some missing substantial sections, and all containing variant versions of certain psalms.[20] From these manuscripts, scholars gather that Pembroke revised Sidney's translations of Psalms 1–42 and completed the collection with her own translations of Psalms 43–150.[21] The psalms were a potent site for early modern engagements with the topics of death, dying, and loss during a time, as Michael Neill documents, when people were searching for ways to "put a shape upon the confusion of death."[22] The formation of knowledge surrounding death and mortality was motivated by ontological questions that informed both the period's developing scientific knowledge and the period's devotional poetry: questions about how God formed and unformed matter; about how and where death (and the dead) existed in the material world of the living; and about the status of souls and care of bodies in the aftermath of death.

In Pembroke's psalm translations, the dominant formal mode is repetition, which functions as an act of remembering, accustoming, preserving, and theorizing. If, as Hannibal Hamlin argues, translations of the biblical psalms are "a testament to the author's efforts to make the strange familiar,"[23] then we might begin our investigation with the tentative hypothesis that Pembroke employs repetition in order to grapple with the strange, the uncertain, the unknown of death. The effect of repetition in Pembroke's psalms, as we will see, is a kind of partitioning that invites the reader to engage closely with the relationship between parts and wholes. Her collection of psalm translations, she insists in the dedicatory poem, is "peec't" (110.24), its structure reminiscent of both the fitful, fragmented process of mourning and the prevailing approaches to postmortem care in early modern England—partitioning the body in order to preserve the body.[24] The anatomical forms of Pembroke's poetics might surprise us, because they reveal a poet whose spiritual and devotional work acts in service of a sustained investigation of the science of bodies.

Women's Work with the Dead

Let us begin with that tricky category of "experience." Pembroke would have been particularly interested in the postmortem handling of bodies around the turn of the sixteenth century. Philip Sidney died of a gangrenous battle wound fighting in the Queen's army in October 1586 and was buried five months later in London. By then, Pembroke had already lost her infant daughter (1584) and her parents (1586), and she would lose her husband in 1601.[25] In 1603, just four years after Pembroke penned her politically laced dedication to Queen Elizabeth, the queen's death marked an occasion of universal mourning.[26] Sidney's death is often cited as the most influential of these losses, evidenced by Pembroke's written expressions of mourning and her active management of Sidney's textual corpus after his death.[27] Abraham Fraunce hailed her as "morientis imago Philippi" ("image of the dying Philip") in his dedication to *The Third Part of the Countesse of Pembrokes Yuychurche* (1592), following the example of many other contemporary poets who represented her as "a memento of Sidney's death."[28] Elizabeth Harris Sagaser uses the term "poet-mourner" to describe the persona Pembroke constructs in her poetry, and Marion Wynne-Davies argues that Pembroke "chose to use text as her primary form of mourning."[29] In fact, Pembroke's role as Sidney's mourner is so well established in critical work that Jonathan Goldberg claims this point as a rare instance of unanimous scholarly accord: "as everyone who has written about the Countess of Pembroke agrees," the voice of the mourner "impels all her writing."[30]

What definition of "mourner" are we working with when we apply that status to Pembroke? To be a mourner in early modern England—particularly, a woman mourner—was to be intimately involved in what Patricia Phillippy calls "death's materiality."[31] Women's mourning work (and it was *work*) was part of the wide range of "cleaning, healing, and caring labors" surrounding the early modern body that Kathleen Brown and Mary E. Fissell refer to as "bodywork."[32] Investigating "the relationship between the work we consider medicine and the broader category of attending to the human body," Fissell argues, we recover the "work that was often too ordinary or too unremarkable to be recorded in writing, but was usually performed by women."[33] Snatches of this labor can be glimpsed in written records, particularly when a very important person was involved. We have an account, for instance, of Queen Elizabeth's final days from Elizabeth Southwell, one of the queen's maids of honor. In the account, written or dictated by Southwell and preserved with the papers of the Jesuit priest Robert Persons, Southwell documents the many weeks that she and other ladies-in-waiting sat and cared for the queen by her bedside. As the queen "daily decreased of her rest and feeding," Southwell describes attempts to get the queen to eat ("my lord Admiral one time persuaded her to drink some broth"), the effort required to get the queen out of bed ("one day being pulled up by force stood on her feet 15 hours"), and, finally, the treatment of Elizabeth's body after death.[34] The queen's privy council left her body "with charge not to be opened such being her desire," but Southwell recounts that Robert Cecil had given "a secret warrant to the surgeons" who opened and embalmed the queen's body. As one of Queen Elizabeth's mourners, Southwell was tasked with sitting and watching the coffin ("myself that night there watching as one of them being all about the body"), and it is during the watching period that she records the particularly striking detail that the queen's body cracked open in the coffin: "her body and head break with such a crack that splitted the wood lead and cere cloth." The next day, the queen's body had to be "new trimmed up" and rewrapped for burial.[35]

As Southwell's account makes clear, early modern women were no strangers to the "matter of death" in early modern England—not just to death and grief, that is, but also to the physical processes and rituals necessary for postmortem care.[36] Women were often responsible, Phillippy catalogs, for "washing and winding the corpse, watching the body during its period of laying-out, conducting ritual lamentations within the home, serving as almsfolk and mourners for funerals, and donning mourning garments."[37] There are also records of women assisting even more directly with the embalming process. At the death of James Montegue, Bishop of Winchester, for instance, "10*s* was paid to 'a woman to attend the surgeons with water, mops, cloths, and other things.'"[38] Such records of bodywork take on heightened significance when we situate them within a longer history of women's expertise in embalming and autopsy, a crucial corollary to histories of anatomy that focus solely on academic and professional settings dominated by male anatomists and male cadavers.[39] In Katharine Park's research on the "*ad hoc* cultural practices that had nothing to do with medical instruction," she traces histories of "dissectionlike practices"—embalmings, autopsies, and fetal excisions—dominated by female practitioners.[40] Park recounts, for example, the story of Chiara of Montefalco, fourteenth-century abbess of a small Umbrian monastery, whose body was dissected and embalmed by her fellow nuns. Sister Francesa of Foligno, the account details, "cut [Chiara's body] open from the back with her own hand . . . and [the nuns] took out the viscera and put the heart away in a box, and they buried the viscera in the oratory that evening."[41] Unlike the university and public anatomy theater dissections most often foregrounded in histories of Renaissance anatomy, Park locates "dissectionlike practices" in domestic spaces and in the context of family care and religious ritual, suggesting that such embodied practices were the grounds for women's unique authority on the topics of death, dying, and loss.

Building on these studies of early modern women's bodywork, it is the contention of this chapter that when we talk about mourning in the early modern period, we are talking about the science of the body.[42] As "traditional caretakers of the flesh," women's relationship to the dead and dying was just as, if not more, fleshly than it was spiritual.[43] The physical and material specificity of their work with the dead situated women's experiences of death at the intersection of anatomy and theology—between the developing knowledge of the flesh-and-blood body and the ongoing, ever-present explications of faith, specifically those surrounding the mortality or immortality of the soul in death. How does reframing mourning as a knowledge practice change our understanding of the substantial archive of women's poetry on death and dying?[44] Or, to pose the question differently, what does women's mourning poetry, as a discursive presentation of their experience with death, contribute to the history of science?

Immediately, the category of "women's experience" presents itself as something that warrants further attention and, here, Pembroke's poetry is quite useful in its persistent coupling of death and experience. In Psalm 119D, for instance, Pembroke's contemplation of death takes the form of a conditional statement: "Dead as if I were, / my soul to dust doth cleave" (1–2). The conditional "if" is a "form with strong epistemological and rhetorical associations" in early modern parlance.[45] As Jonathan P. Lamb has demonstrated, "if" functioned at the intersection of religious and scientific discourses, marking appeals to belief and certainty of knowledge fundamental to both spheres.[46] Pembroke's "dead as if I were" sits at this very intersection. It becomes the basis for the poet's prayer (tellingly, a plea for knowledge: "make that I may know, / and thoroughly understand" [Ps 119D.11–12]) and simultaneously marks the poem as a thought experiment based on a kind of experiential inquiry (this is an "as if" statement, rather than simply an "if" statement). Here are the first lines of the poem's opening stanza in full:

Dead as if I were,
my soul to dust doth cleave:
lord keep thy word, and do not leave
me here:
but quicken me anew. (Ps 119D.1–5)

Pembroke gives space to the silence after "here" in line 4 before concluding the sentence in line 5. Using the enjambment of lines 3 and 4 to enhance the effect, Pembroke poetically conjures a charged, still space for death in the poem's white space. The experiential mode of inquiry marked by the poem's opening becomes an invitation to consider how the poem's content and form work together to construct the "as if" of death.

The biblical psalms are full of references to death ("thousands here, / ten thousands there decay" [Ps 91.21–22]), dying ("I fade and fail as shade with falling sun" [Ps 109.65]), and loss ("griefs on griefs on me still heaping lay" [Ps 55.8]). These thematic threads, and much of the language Pembroke uses to foreground them, come directly from her source texts—the 1560 edition of the Geneva Bible, Miles Coverdale's psalm translations circulating in the Book of Common Prayer, and the 1562 French Genevan metrical psalter by Clément Marot and Théodore de Bèze *Les Psaumes de David mis en rime Francoise*.[47] As we will continue to see throughout this chapter, though, Pembroke's translations and formal experiments harness the epistemological possibilities of the original language. The formal structure of Psalm 119D, along with Pembroke's "dead as if I were" framing, recall Philippe Ariès's reminder that, in the seventeenth century, "death and the dead body themselves constitute[d] objects of scientific investigation, quite apart from the cause of death."[48] This is, Ariès goes on to emphasize, "an attitude alien to modern medicine, in which death is inseparable from the disease."[49] The variant extant versions of Pembroke's psalm translations show her compressing and reworking her poems as she crafts precise language to describe death and the dead body. In a variation of Psalm 49, she condenses the line "in shade of death where life shall never show" to simply "endless dark" (Ps 49.39), changes "grave" to "tomb" (16), and swaps out "grave discourse" for "riddled speech" (6). "Grave discourse" instead became her choice phrase for the opening of Psalm 78. "A grave discourse to utter I intend" (Ps 78.1), she announces, playing with the dual connotation of "grave" as both adjective (weighty, important) and noun (a place of burial). With both connotations modifying discourse, Pembroke simultaneously signals the weight of her poetic task and invokes language's memorializing function at the start of a lengthy poem cataloging God's "memorable deed[s]" (Ps 78.13).

In other poems, the poet speaks with authority about the practices of care surrounding death. Women's work with the dying and dead, we might speculate, informs Pembroke's translation of Psalm 51, "one of the most popular texts of the dying,"[50] and the psalm that Anne Locke chose for her twenty-six-sonnet psalmic meditation, *A Meditation of a Penitent Sinner* (1560).[51] The cleansing and purification rituals described in the psalm bring to mind the labors involved in washing bodies before burial. The poet calls on God to "wipe," "cleanse," and "wash" her foul iniquity (Ps 51.4–5), to "purge" her body with "hyssop" (23), and to scrub her heart to make it "pure, clean, spotless" (31). If we read this poem as a record of physical practice, Pembroke's word choice is particularly noteworthy because each of her synonyms for *clean* marks one part of a detailed, sequential process. Pembroke includes a string of these verbs in the first stanza, outlining a series of actions that will continue throughout the poem: "And wipe, O Lord, my sins from sinful me. / Oh cleanse, oh, wash, my foul iniquity" (Ps 51.4–5). The action of *wiping* first clears the body's surface of dust and dirt; *cleansing* performs a deeper cleaning to "free from dirt or filth" and evokes the more figurative sense of "making morally or spiritually clean"; and, finally, *washing* describes the action of immersing the body in water, evoking the sacrament of baptism and the poet's spiritual rebirth. In the poem's fourth stanza, God's mercy becomes a rinse that "flow[s]" over the poet (Ps 51.25) and Pembroke introduces another verb—*purge*—linked to herbal purification rituals: "with hyssop, Lord, thy hyssop, purge me so" (23). Hyssop has a long history in death and burial practices as a common ingredient in embalming powders and an herb used to cleanse the body after contact with a corpse. Pembroke's use of the verb "purge" also draws on the herb's medicinal properties as both mild purgative and expectorant, an aid for the body to purge itself from the inside out.[52]

Such moments in Psalm 51 reveal ways of knowing that work alongside and within the poem's devotional context, and Pembroke's representation of spiritual health seems informed by physical medicinal practice. As editors of her work have noted, Pembroke's invocation, "O God, God of my health" (Ps 51.40) is quoted directly from Locke's translation of Psalm 51. Perhaps Pembroke, picking up on Locke's work, understands this poem as an act of cleansing and preservation that prepares the poet's body for closer connection with God after death. In Philbert Guibert's instructions on embalming, first published in his 1625 *Le Médicin Charitable* (*The Charitable Physitian*, 1639 English translation), he instructs the practitioner to *wash* ("you must *wash* them with warm oxycrat, made with four parts of water, and one of vinegar compounded") and *cleanse* ("the head, breast, and belly inferior being also emptied and *cleansed*, you shall begin to embalm them") the body and its organs in preparation for the embalming process.[53] According to Guibert, embalming required bone-deep cleansing (the same "bruised bones" referenced in Psalm 51) as each cavity of the body was cleaned and packed with preservative powder. "You shall make an Incision," Guibert instructs his reader, "from the shoulder to the wrist in the arme, piercing it with a bodkin to the bone all about."[54] Psalm 51 seems to be about the poet's spiritual health, but the specificity of the cleansing ritual on display relies on the medicinal and postmortem knowledge of its translator.

Pembroke uses her poetry as a site of exploration, especially interested in testing how experiential knowledge ("as if") can inform poetic work—what we might call her poetry's experiential mode of knowing. But my intention here is not to read Pembroke's poetry as evidence of direct experience caring for the dead and dying. Such an approach, like those I outlined in this chapter's opening, would use the details of Pembroke's autobiography as the grounds for analyzing specific thematic threads in her poetry. Rather, I invite us to *start* with her poetry and use it to trace forms of engagement with early modern science that might not align with our usual evidence for scientific work. We are starting to see how the practical, material, and conceptual interconnections surrounding death, dying, and loss impel Pembroke's poetics. For many early modern English women, death came to be known through hands-on work and embodied participation in rituals of care and mourning. Pembroke's poetry calls attention to that experiential knowledge and then asks us to consider how poetic work can not only be uniquely informed by such experience but can also be a rich site for theorizing experience as a mode of knowing—a shift from individual experience to the experiential mode. This shift in approach will allow us to see Pembroke's poetry for what it is: a significant site of thanatological practice.

Preservation and Elegy

In order to understand how poetic acts of mourning have the potential to function as scientific work, we can look to perhaps the most famous early modern elegy written by a woman, Pembroke's "To the Angell spirit of the most excellent Sir Phillip Sidney." This is a thirteen-stanza elegy in which Pembroke mourns the loss of her brother, dedicates her psalm translations to him, and introduces the text as the siblings' collaboration—a "coupled worke" (110.2). The poem is also, as many have noted, an emphatic display of grief, frequently punctuated by exclamations and interjections that seem to erupt into Pembroke's more measured language of memorialization. "Angell spirit" often anchors critical accounts of mourning as a gendered practice in early modern England. The status of mourner authorized women's interventions in a range of social and political spheres and, as Elizabeth Hodgson argues, Pembroke's poem contributes to a canon of women's writing that leverages "mourning's social and literary uses."[55] If "sorrow was not just what a person felt, but also a defining term in . . . what made a man or a woman of influence," elegies were not grief for grief's sake, but rather grief wielded for a specific purpose.[56] Such arguments provide an important foundation for reading early modern women's elegies as more than mourning, but they also tend to elide the physicality of these poems in favor of excavating their political and literary interventions. Celeste M. Schenck associates what she calls the "masculine elegiac" mode with many of the political and literary aims that scholars have attributed to "Angell spirit."[57] The masculine elegy, Schenck argues, is "a gesture of aspiring careerism," and its "most important convention" is "the deification of the dead one in a process that lifts him out of nature, out of the poem, and out of the successor's way."[58] Alternatively, for Schenck, the female elegiac mode works as an act of preservation.[59] Studying women's poetic elegies from Anne Bradstreet to Marilyn Hacker, Schenck demonstrates that the female elegist "protests final separation" and works in the space of "connective tissue" and "fleshly bonds" that "preserv[e] . . . relations in a kind of continuous present."[60] I want to linger with Schenck's description of "connective tissue[s]" and "fleshly bonds" because it is especially useful in my investigation of poetics at the intersection of textual and somatic matter. How might we bring the physical, fleshly capacity of the early modern elegy to the fore, and what would such an excavation reveal?

Pembroke's "Angell spirit" does much of that work for us. It certainly is, as Claire Kinney argues, "a highly nuanced Defense of Poesy" and, as Danielle Clarke claims, an "assertion of ownership" over Philip Sidney's legacy, "political *and* poetic."[61] It is also a striking representation of the "fleshly bonds" that exist between Pembroke's writing body and her dead brother. The poem's oft-cited description of Sidney's bleeding body commingles Pembroke's "hart teares" with the fresh blood of Sidney's wounds: "Deepe wounds enlarg'd, long festred in their gall / fresh bleeding smart; not eie but hart teares fall" (110.19–20). In a second, variant version of the poem, mistakenly attributed to Samuel Daniel and published with his works in 1623, Pembroke's interjection on Sidney's wounds is even more pronounced.[62] "But ah!," she begins, "wide festred wounds that never shall / nor must be clos'd, unto fresh bleeding fall" (113.19–20). If in the first version of these lines (known as the Tixall manuscript version) the poet calls to mind Sidney's wounds, in the Daniel Variant the wounds instead seem to force themselves to mind. The medial caesura created by Pembroke's opening interjection ("But ah!") enacts the disruptive thought of Sidney's "fresh bleeding" body in the disrupted rhythm of the poetic line.

These lines on Sidney's wounds initiate the preservative capacity of Pembroke's elegy. The act of preservation aims to keep something in its original or existing state. It is simultaneously an act of protection and maintenance. Both impulses underlie the early modern practice of embalming, which we have already had a brief introduction to in Elizabeth Southwell's account of Queen Elizabeth's death and Philbert Guibert's popular *The Charitable Physitian*. Embalming, or corpse preservation, was a common funerary practice across Europe from at least the twelfth century, and while it would eventually become closely linked to the practice of dissection, the earliest printed texts circulating on the subject discuss embalming as a customary part of death care rather than a topic of anatomical interest.[63] Peter Forestus, a prominent physician of the Dutch Republic, outlined his embalming process in the appendix to his 1605 treatise on wound treatment. After detailing the herbs and preparations needed to make an embalming powder, Forestus instructs the reader to "lay open the trunk of the body, remove all the viscera, afterwards take such sponges which were previously immersed in cold fresh water, afterwards dipped in aqua vita, and wash out the interior of the body by hand with the sponge."[64]

The process of embalming that physicians like Guibert and Forestus describe (reminiscent, again, of the cleansing ritual performed in Psalm 51) was often a logistic necessity because it allowed bodies to be preserved for weeks—sometimes even months—between death and burial.[65] According to archaeologist Sarah Tarlow, such a delay was especially common when "somebody died far from the place they were to be buried,"[66] famously the case for Philip Sidney, who died in the Netherlands on October 17, 1586, and was buried in London on February 16, 1587. In such cases, "the most corruptible organs" were often removed and buried at the place of death, and the body was then preserved for travel to the burial site, a literal scattering of body parts that perhaps informs the material concerns surrounding death in John Donne's "The Relic," wherein the poet imagines himself and his beloved meeting as their souls reunite with their disarticulated body parts on Judgment Day.[67] Lady Anne Clifford references this custom quite directly in her accounts of family members' deaths and burials. After her father's death in October 1605, Clifford records, "his dead body was opened, and his bowels and inward parts was buried in the Chappell in the Savoy, but his dead body was buried a little after in the vault in Sykpton Church in Craven the 13th of March following."[68] Likewise, when her brother died at North Hall, his "inward parts were buried" at the local church before his "corpse" was transported to Buckinghamshire.[69]

Such practices fragmented the funerary process, preserving both the dead and their mourners in a state of suspended ending. Without the physical corpse or the familiar burial rituals, the mourner's grief was unmoored, uncoupled from death's physical and material touchstones. This is perhaps the source of the "impatient griefe" expressed by the poet of "The Dolefull Lay of Clorinda," one of a series of elegies to Philip Sidney compiled and printed by Edmund Spenser in 1595 (133.2). The authorship of "Dolefull Lay" remains contested but, as Kinney notes, regardless of its true authorship, the poem presents "another avatar of the Countess [of Pembroke] in mourning."[70] "Where shall I unfold my inward paine" (133.3), the poet entreats before apostrophizing death to demand, "what is become of him?" (134.57). The poem articulates an important question: where and how could Pembroke (here represented in her pastoral guise, Clorinda) have mourned for Sidney? The fragmented, extended experience of grief, I want to suggest, incited Pembroke—a writer very interested in death, dying, and loss—to explore alternative possibilities for caring for and preserving her brother's dead body.[71]

"Angell spirit," written about thirteen years after Sidney's death, is a long-term preservation project, a study of how the physical practices of corpse preservation might function poetically. Pembroke reveals her preservation work when she interrupts the work of lament with the physicality of Sidney's corpse. Her impassioned outburst about Sidney's "fresh bleeding" wounds in both versions of "Angell spirit" functions not only to emphasize the poet's grief, but also to conjure Sidney's dying and dead body. As Wendy Wall notes, Pembroke's descriptions of Sidney's wounded body are, importantly, echoed in her references to the text as a "halfe maim'd peece" (110.18). In her "wounding lynes" and "sadd Characters," Wall argues, Pembroke "creates a set of imprecise grammatical associations" that blend Sidney's "anatomy with the very text of the *Psalms* because both are wounded, bereft, and incomplete."[72] Pembroke adds her own physicality to the mix as "hart teares fall" and her body becomes freshly wounded as she continues to mourn. But Pembroke is not simply mourning. She is also writing—her mourning takes shape as a language event—and we must consider the "connective tissue" that binds Pembroke to poet (Sidney) and poetry (the psalms) as caretaker of both corpse and corpus.[73] How does Pembroke construct postmortem care through her poetics? What does elegiac preservation look like if one is working not just to preserve the memory of the dead, but also their body as a conduit for the act of writing?

Each version of "Angell spirit" offers different responses to these questions. The Daniel Variant amplifies the poet's mourning and treats the poem as a substitute for Sidney's funeral rites, aligning Pembroke's work with the *exegi monu-*

mentum tradition, which finds particularly famous expression in Shakespeare's sonnets to the young man ("and thou in this shalt find thy monument / when tyrants' crests and tombs of brass are spent").[74] Pembroke situates her elegy as both active memorializing ritual (Sidney's funeral "rites" [113.26]) and lasting physical structure (a "faire monument[ ]" to his "faire fame" [114.64]), a fitting framework for the Sidney Psalter itself.[75] The poem's memorial function is most explicit in the two stanzas after Pembroke's reference to Sidney's "wide festred wounds." She asserts that her grief will not keep her from doing her "part" (113.25), both to continue her brother's translation work and to memorialize him in that work. The psalms serve as "rites to aright" (113.27), a site where Pembroke can properly treat Sidney's death and his memory. The unusual verb form of "aright" connotes the act of correction or "putting something right," perhaps a suggestion that the poet, through her writing, needs to correct the improper, incomplete, or inadequate funeral rites performed for Sidney's death.[76] The collaborative text of the psalms, Pembroke insists, will "remayn[ ]" and serve "as the tombe of life disceast" (114.32). Later in the poem, Pembroke uses her syntax to blur the line between textual and physical mourning practices when she implores, "receive these Hims, these obsequies receive" (115.85). Bounded by the repeated plea to "receive"—another of the many repetitions we encounter in Pembroke's poetry—the line's semi-chiastic construction substitutes "obsequies," a term reserved for graveside funeral rites, for "hymns," a term that externalizes mourning work in the form of textual and aural rituals. This substitution establishes Pembroke's elegy as a physical mourning object, part of the embodied "rites" surrounding death.

Pembroke's elegiac work in the Daniel Variant of "Angell spirit" becomes more significant if we read it as a prelude to the Tixall manuscript version, in which the elegy takes a distinctly visceral turn. The two stanzas following Sidney's "wide festred wounds" in the Daniel Variant are, importantly, not in the Tixall manuscript version of the poem. Instead, in the Tixall manuscript, Sidney's wounds serve as a fulcrum that shifts the poem beyond the work of elegiac mourning and into the realm of physical postmortem practice.

> Deepe wounds enlarg'd, long festred in their gall
> fresh bleeding smart, not eie but hart teares fall.
> Ah memorie what needs this new arrest? (110.19–21)

In this stanza, raw, opened, "festr[ing]" flesh confronts the poet and, instead of turning to the memorializing language she draws on in the Daniel Variant,

Pembroke begins to transform the poem into an instrument of healing. The referent for "deepe wounds" is unspecific—are the wounds Sidney's, Pembroke's, or those of the "peec't" poem (110.24)? These lines, again, blur the physical boundaries between the living and dead, placing the text between them as an enactment of and physical site for their commingling. Pembroke closes this stanza with the word "arrest," cleverly playing on the term's double connotation, first, in its more specific definition, of keeping the mind fixed on a subject and, second, in its more general meaning, as an act of stopping something in its course.[77] And Sidney's bleeding does, indeed, seem to have stopped by the start of the next stanza.

> Yet here behold, (o wert thou to behold!)
> this finish't now, thy matchless Muse begunne,
> the rest but peec't, as left by thee undone. (110.22–24)

In this turn from stanza 3 to stanza 4, the poem's interstanzaic white space cauterizes the "fresh bleeding" wounds of line 20. The poem performs a kind of macabre medicinal work in the stopping of bleeding *after* death—a shift from elegizing to embalming—as Pembroke describes her labor to "peece" together the poetry that Sidney left "undone" (110.24). If her earlier lines fused Sidney's anatomy with the "halfe maim'd" text (corpse with corpus), here both are healed, sewn up, "finish't now" by Pembroke's skillful, caring hand.

When Peter M. Sacks argued that English elegies should be regarded "as work, both in the commonly accepted meaning of a product and in the more dynamic sense of the working through of an impulse or experience,"[78] he did not have embalming work in mind. But that is precisely what I am suggesting here. This is a tricky claim if the practice of embalming connotes only the messy, physical handling of a corpse. According to the leading textbook on the subject, however, the official definition of embalming is "a means of artificially preserving the dead human body."[79] The key term here is "artificially," which can refer to any system or process performed "by human skill or design, especially in imitation of or as a substitute for a natural process or product."[80] If we think about embalming as a form of artificial preservation and a familiar part of postmortem and mourning rituals, (as, for instance, Elizabeth Southwell and Anne Clifford's accounts suggest) it starts to sound a bit like the poetic elegy, a genre that responds to death with poetic acts of preservation.[81]

For Pembroke, as we are seeing, preservation work always occurs in the space between living and dead, which is also the space of her text. Sidney's wounds, recall, are simultaneously "fresh" and "festred," a seeming paradox, but one that aligns directly with certain early modern conceptions of the cadaver as both "still the body and already the corpse."[82] Philippe Ariès cites early modern reports of *cruentatio* (the bleeding of a corpse in the presence of its murderer) as one piece of evidence for this pervasive belief in "the continuation of a certain kind of life and sensibility in the cadaver, at least as long as the flesh is preserved and the body not reduced to the state of a dried-out skeleton."[83] Pembroke's poetic construction of grief is a physical preservative act (both Sidney and text as simultaneously body and corpse), and perhaps we can better understand the *work* involved with a further comparison between the Tixall and Daniel versions of "Angell spirit." Instead of lingering with her grief and shifting into the conventional language of exegi monumentum in the Tixall manuscript version, Pembroke cuts the two stanzas focused on funeral "rites" and "obsequies," and moves immediately to stop Sidney's "bleeding" wounds. With this edit, that is, Pembroke cuts the language that most directly invokes exegi monumentum conventions. Her elision overwrites and, in the process, merges funeral "rites" with the act of writing—"rites" to "writes."[84]

The penultimate stanza of the Tixall poem is not included in the Daniel Variant, and it begins with the following lines:

> To which theise dearest offrings of my hart
> dissolv'd to Inke, while penns impressions move
> the bleeding veines of never dying love:
> I render here. (112.78–81)

Deborah Uman notes that Pembroke's writing "is very much of and on her body" in these lines, which begin with a familiar elegiac trope (the poem as "offrings" to the dead) only to then fold that familiar gesture into a distinctly somatic postmortem practice.[85] These lines describe two highly physical actions. First, through Pembroke's writing, her blood ("the offrings of [her] hart") is transformed into "Inke." Second, the press of her pen has an effect on "bleeding veines." Again, though, the reference point here is imprecise—whose "bleeding veines" are these? Is the blood a mark of Pembroke's "never dying love" as she mourns for her brother? Or are these Sidney's "bleeding veines" reanimated by the "impression[ ]" of the poet's pen, an echo of his "fresh bleeding" wounds earlier in the poem? As early modern anatomists experimented with improved methods for corpse preservation, one increasingly popular technique was to inject fluid into the body's hollowed cavities or its vascular system, and there were a variety of suggestions for what kind of fluid worked best as a preservative. Jacopo Berengario, who published a particularly influential sixteenth-century anatomical atlas, injected warm water into the veins of his cadavers. Bartholomeo Eustachius, a contemporary of Andreas Vesalius, claimed that warm ink was a particularly effective preservative.[86]

If Sidney's anatomical corpse has been fused with his corpus, the *impression* of Pembroke's pen simultaneously moves through the veins and lines of both—an act that situates the messy handling of Sidney's dead body within the pages of her poetic text. In Pembroke's poetic transformation of blood to ink, the pressure and movement of her "render[ing]" pen, she prompts us to consider how "bleeding veines" might be reanimated by the "penns impressions"—that is, how poetry might do the work of corpse preservation.[87] The work of Pembroke's pen becomes akin to an embalming injection, pressing on the surface of Sidney's body and forcing fluid into his vascular system. In the lines cited above, through the double enjambment of lines 78–79 and 79–80, two long clauses linked by the conjunction "while," Pembroke constructs the physical actions of these lines as simultaneous—Pembroke's blood dissolves to ink which she presses through Sidney's veins. The poem's form creates a constant poetic transfusion in the service of Pembroke's long-term preservation project. Reading beyond the poem's mourning work to consider how "Angell spirit" performs physical postmortem care, we can better understand how Pembroke takes her own experience as a departure point for exploring poetry's epistemological possibilities, and this is only one part of a much wider network of experiments surrounding death, dying, and loss that she takes up in her writing.

The Translation of Grief (or, why Grief is of Anatomical Interest)

Claiming that Pembroke wields her poetry for scientific work, my argument runs into long-standing debates about how to categorize and thereby read Pembroke's psalms given their status as translations. "Angell spirit" is, in many ways, an anomaly in Pembroke's extant archive, one of only two original poems attributed to her amid a much larger collection of poetry and prose translations. Entrenched assumptions about translation—a particularly prominent genre for early modern women writers—as a derivative and menial task have long restricted critical work on Pembroke's writings, with even her most enthusiastic readers at times lamenting her lack of original work.[88] In response to such arguments, Helen

Smith has called for a radical shift in how we think about women's relationships to the materials and technologies of writing, also a central aim of *Anatomical Forms*. How, Smith asks, might we "conceptualize the act of writing as *itself* creative and recreative?"[89] Early modern anatomy texts have much to offer literary scholars on this point because, for anatomists, translation was very explicitly a knowledge-making act. The anatomical knowledge circulating in sixteenth- and seventeenth-century England was rarely original in the way we think about original literary compositions, and many anatomy texts advertised their status as translations on the title page.[90] John Banister's 1578 treatise, for instance, was titled *The History of Man Sucked from the Sappe of the Most Approved Anathomistes, in this Present Age, Compiled in Most Compendious Fourme, and now Published in English, for the Utilitie of all Godly Chirurgians.*[91] Banister's metaphor for translation might be a bit indelicate, but his title reflects the rigorous intellectual processes involved in his translation work: reading and studying anatomy texts in Latin and other learned languages, synthesizing and compiling the information presented in those texts, and conveying the synthesized knowledge in English vernacular "for the utility" of English surgeons.[92]

For Banister and his contemporaries, translation was not only language work but also scientific work. It was one important part of anatomical knowledge development, a perception even more evident when the process of translation was framed as a substitute for hands-on anatomical practice, as Helkiah Crooke claims in his preface to *Mikrokosmographia* (1615). Addressing his fellow surgeons, Crooke notes that he will likely be critiqued for not relying on his own dissections to write his treatise. But "such Guls little know what it is to write an Anatomy," Crooke insists. "Thirtie yea forty yeares practice . . . in dissection, and a hundred bodies more or lesse, small and great every yeare, hath bin thought little enough to make an Anatomist fit to teach this learned age, for which I dare boldly say, no man amongst us that knowes himselfe can esteeme himselfe sufficient, because we want those meanes which other Countries affoord their professors."[93] Crooke defends his translation work, first, by outlining the extensive dissection practice necessary to "write an anatomy," and, second, by critiquing the lack of resources supplied to English anatomists.[94]

The underlying implication here is that Crooke's translation stands in for his lack of hands-on work. I translate, Crooke argues, because I cannot dissect. Thomas Johnson, English translator of the French anatomist Ambrose Paré's writings, takes Crooke's argument a step further by introducing his translation *as* an act of dissection. Johnson spends the first half of his preface outlining Paré's extensive hands-on experience, first trained in the Paris hospital ("an ample theater of wounds and diseases of all kinds"), then "principal surgeon to two or three Kings of France," and, finally, traveling surgeon with the Spanish army.[95] When Johnson then turns to his own translation work in the preface's second half ("Now I come to the translation"), his labor seems a natural extension of Paré's hands-on experience when he asserts that his book "lay[s] open the mysteries of a worthy art."[96] As an act that exposes and reveals the "mysteries" of anatomy, just as Paré "lay[s] open" the body for treatment and dissection, Johnson also understands translation as an act that enables further hands-on practice. "No man becomes a workman by books," he admits, but his book will acquaint practitioners with the "terms of the art" and "the knowledge and use of the instruments" so that dissection becomes eminently possible.[97]

When we understand translation as a knowledge-making act that prompts—or even substitutes for—physical bodywork, Pembroke's ars moriendi translations become an important foundation for her postmortem poetics. Composed between 1592 and 1600, *The Triumph of Death*, *Antonius*, and *A Discourse of Life and Death* all explore different representations of death and dying, and all three, like "Angell spirit," incorporate Pembroke's own experiences of loss and mourning. Jonathan Goldberg describes Pembroke's translation of Petrarch's *Triumph of Death* as "a poem of mourning for Philip Sidney," and her translation of Robert Garnier's *Antonius* begins at the end, the battle of Actium decided, Antony preparing to end his life, and Cleopatra enclosed in the monument that will become her tomb.[98] Pembroke scholars often connect *Antonius* and *A Discourse of Life and Death* to Pembroke's own mother's death, famously recorded in Holinshed's *Chronicles* as an exemplary instance of dying well, but Mornay's text also foregrounds an inquiry we will see often in Pembroke's psalms: where does death reside in and on the body?[99] Mornay outlines his ars moriendi philosophy most explicitly about two-thirds of the way through the text when he argues that "life and death in man is all one" (248). In Pembroke's translation of this argument, "we think we die not, but when we yield up our last gasp. But if we mark well, we die every day, every hour, every moment" (248). If "living is but a continual dying," as the text goes on to argue, then readers should embrace death as a "haven" rather than approaching it "with horror and trembling" (248; 229).[100]

As Goldberg notes in his argument on Pembroke's translations, these texts have long been praised on the grounds of accuracy, for their adherence to the original texts. Admonishing critics for "reading each other, and not Mary Sidney's poems," Goldberg pointedly asserts that "for Mary Sidney to rewrite it faithfully word for word would nevertheless mean for her to have rewritten it."[101] With this important reminder in hand, and building on my reading of "Angell spirit" as embalming work, I want to consider how Pembroke's experiences with grief and mourning infuse her psalm translations, specifically the penitential psalms, numbered 6, 32, 38, 51, 102, 130, and 143.[102] These were, arguably, the most well-known of the biblical psalms in early modern England, and certainly the most fashionable to translate, with Thomas Wyatt, George Gascoigne, John Bale, Anne Locke, King James I, Thomas Carew, and John Milton all producing English versions of at least one of the collection. Reformation writers used these psalms as vehicles for "theological, political, artistic, and poetic engagement," but they were also a rich site for studies in grief—the most elegy-like of all the psalms as the psalmist reflects on loss as a gateway to penitence.[103] Pembroke translated four penitential psalms (we have already seen her version of the spiritual cleansing rituals in Psalm 51) and, in this cluster of poems, she displays a penetrating interest in grief's empirical features and physical effects. That is, Pembroke's penitential psalms consistently raise the question: what does grief feel like in and on the mourner's body?

Psalm 102 sets up an analogy between the poet's mourning body, "wast[ing] away" (Ps 102.20) as a result of her grief, and the ruined city of Zion, laying "like a carcass . . . stretched forth in ruin's bier" (Ps 102.50–51). For Anne Lake Prescott, this psalm is representative of Pembroke's interest in "the language of ruination"—her fascination with "Time's speedy consumption or gradual erosion."[104] One characteristic mark of Pembroke's "ruination vocabulary," Prescott argues, is her tendency to include language about time's effects on material structures.[105] In Psalm 102, Pembroke applies such language not only to the ruined city of Zion, but also to the poet's body.

> My days as smoke are past:
> my bones as flaming fuel waste:
> mown down in me (alas)
> with Scythe of sharpest pain,
> my heart is withered like the wounded grass;
> my stomach doth all food disdain. (Ps 102.7–12)

Pembroke marks time's effects on the poet's body with the striking image of her "withered" heart, an adjective that conveys not only the quality of dryness but also the process of drying out over time. Later in the poem, her entire being takes on the "withered" texture of her heart: "I as hay am dried" (Ps 102.39).[106] Pembroke initially links the term to vegetation with the analogy to "wounded grass" in line 11, but "withered" also evokes the dryness associated with the shrunken, shriveled flesh of a preserved body, an association made explicit in the opening of the next stanza as the poet describes her flesh "cleav[ing]" to her bones.

> So lean my woes me leave,
> that to my flesh my bones do cleave:
> and so I bray and howl,
> as use to howl and bray
> the lonely pelican and desert owl,
> like whom I languish long the day. (Ps 102.13–18)

In this stanza, time's effect on the body is also grief's effect on the body. At the stanza shift between lines 12 and 13—"so lean my woes me leave"—Pembroke manipulates the line's syntax to metrically pair "lean" (the bodily effect) with "leave" (a time marker), using "my woes" to link adverb and verb, triangulating grief, time, and the poet's body.[107] Over time ("as day to day succeeds" [Ps 102.25]), the poet's woes leave her lean and "wrack[ed]" like the ruined city. Grief feels hot ("flaming"), dry ("wither[ing]"), and sharp (like a "Scythe of sharpest pain"), and makes the poet feel as if she is dying. "I waste away" (Ps 102.20), she claims, "irked alone with dolor's deadly fits" (Ps 102.23).

As we read about time and grief's "wither[ing]" effects, the poem itself seems to "languish" as it keeps turning back on itself, an effect that first flickers in the repetition at the center of line 5: "in this my most most miserable day." Throughout the psalms, Pembroke "shapes an oration that relies heavily on parallel structure and repetition," and we will continue to see Pembroke wield these poetic techniques in the service of her thanatological work.[108] In Psalm 141, she uses chiasmus and anadiplosis to enact the inevitability of death—"meanwhile my bones the grave, the grave expects my bones" (Ps 141.19)—and her compressed syntax ("my bones the grave") evokes the inquiry taken up so directly in Mornay's *Discourse of Life and Death* (where does death reside in and on the body?). Repetition is a formal device of doubling, coupling, and, crucially, preserving. Turning back to Psalm 102, the striking chiasmus at lines 15–16 ("bray and howl" and "howl and bray"), raw sounds from a body in distress, prompt us to notice the more subtle repetitions in the poem's mirrored syntax ("my flesh my bones") and consonance and assonance ("flaming fuel," "mown down," "howl"/"owl," and "in watch I waste away"). All of these doublings have the effect of slowing down the poem's progression, preserving the poet in the state of ruination brought on by her "woes." The linear time of the poem breaks and stretches as both poet and poem "languish long the day" and "languish so the day" (Ps 102.18–19).

Notably, Pembroke locates the poet's grief in Psalm 102—she is "bereft of spouse and son" (Ps 102.22). With this added detail, not in the biblical psalm or any of her source texts, Pembroke again leverages her own experience for epistemological ends, as an anchor point for the poems' inquiries surrounding death. The death of her spouse and son, referenced only briefly, catalyze her sustained experiment in how to convey the physical effects of grief poetically. We might read this poem, then, as a continuation of Pembroke's elegiac project in "Angell spirit." There, her poem functioned to preserve Sidney's dead body and investigate the relationship between physical and poetic postmortem practices; here, the site of investigation shifts to the body of the mourner rather than the mourned. Leveraging the highly physical language of this penitential psalm, Pembroke gives a texture and a weight to the embodied experience of grief.

The later penitential psalms take that weight and situate it physically in space—specifically, as *depth*, emulated in the stanzaic structure of Psalms 130 and 143. At the start of Psalm 130, the poet seems entombed, "drowned" in a "depth of grief."

> From depth of grief
>   where drowned I lie,
> Lord, for relief
>   to thee I cry:
> my earnest, vehement, crying, praying,
> grant quick, attentive hearing, weighing. (Ps 130.1–6)

The poet's desperate pleas at the bottom of the stanza—made more frantic by Pembroke's withholding of the conjunctions—are crushed by the weight of the four lines above.[109] This poem has an unusual formal structure that is unique to Pembroke's version of the psalm. Writing about George Herbert's shape poems (particularly useful models for thinking about poetry with weight and depth), Kimberly Johnson argues that poems like "Easter Wings" and "The Altar" "demand[ ] that we confront form *qua* form, that we register the presence of the poem as a material artifact."[110] In many of Pembroke's psalms, we are confronted with the poem's material heft—the dense sestina blocks of Psalm 49, for example, and the acrostic structure of Psalm 117. Here, in Psalm 130, each stanza repeats the gravestone-like structure of the opening, with the compressed first four lines pressing on the bottom two lines, so the poem's shape imbues grief with a weight that enhances Pembroke's representation of the "depth of grief." In

Psalm 143, Pembroke takes these experiments with poetic matter even further, constructing the poet's grief as a very literal descent as she is chased into an early grave.

> O, rather look with ruth upon my woes,
> whom ruthless foes
> with long pursuit have chased,
> and chased at length have caught,
> and caught, in tomb have placed
> with dead men out of thought.
> aye me! what now is left me?
> alas! all knowledge reft me,
> all courage faintly fled,
> I have nor heart nor head. (Ps 143.11–20)

In the first part of the stanza, Pembroke establishes a relationship between the poet's woes and her entombment as the lines move downward and inward on the page, culminating in the poet's cries—"aye me!" and "alas!"—as if from in a tomb. The rhythm of lines 12–16 and the repeated conjunctions "with" and "and" at the start of each line create a cascading effect, so that the poetic lines, like the poet, seem "chased" (Ps 143.13) into the stanza's depths.

Pembroke's penitential psalms function as thanatological practice centered on the somatic experience of mourning. Pembroke works to translate the physical effects of grief, even as she experiments with how her own experience fits into the broader theological and thanatological inquiries of the penitential psalms. More important than the passing references to her own experience in Psalm 102 (her loss of "spouse and son" [Ps 102.22]) is that she uses it to theorize how experience writ large offers a way into a particular kind of experimental knowledge work—the "as if" of death (Ps. 119D.1). As historian Piero Camporesi has observed, the "tactile association and intimacy with the products of death—corpse, bones, the diseased and dying" results in "a different relationship with the human body."[111] In Pembroke's case, experience with death, dying, and loss results in a different relationship to poetry, a unique position from which to explore poetry's capacity for representing the intimate matters of death. We must recall, though, that Pembroke's unique position as postmortem poet has another register because her writing is always already framed as a collaborative practice. She is only one part of a whole, and the material and physical poetics I am reading in her psalm translations are a "coupled worke," a collaboration between her "penns impression" and Sidney's reanimated "blest hand" (110.3–4).

Anatomists of the Soul

The most famous account of Pembroke and Sidney's collaboration comes from John Donne in his laudatory poem "Upon the Translation of the Psalms by Sir Philip Sidney, and the Countess of Pembroke, His Sister."[112] In the poem, Donne figures the psalm translations as both a divine project and a complex collaboration between Pembroke, God, and the dead. God cleaves his "Spirit" in two and divides it between the siblings, but the siblings' shared "blood" and the elemental unity of God's spirit transforms two back into one: "so thou hast cleft that Spirit, to perform / that work again, and shed it here, upon / two, by their bloods, and by Thy Spirit one."[113] Further extending the collaborative network, Donne claims that Pembroke and Sidney's shared labor mirrors the "double power" at work in the original composition of the psalms—a collaboration between King David and "heaven's high holy Muse."[114] David's "cloven tongue" becomes Pembroke and Sidney's cloven tongues and the psalm translations—"the highest matter in the noblest form"—are a result of divine alchemical mixing.[115] Donne's representation of the siblings' work is not unlike Pembroke's formulation in the opening stanza of "Angell spirit," in which she depicts a kind of alchemical transformation as two substances merge into one. "Rais'de by [Sidney's] blest hand," the psalm translations have been "imprest" with his "secrett power" (110.3–4). Pembroke's "mortall stuffe" then combines with Sidney's "divine" substance (6), and the poetic psalms are the "coupled worke" ("joynt worke" in the Daniel Variant) that results from this alchemical process (2).

Donne figures the product of the siblings' collaboration as an aural one (first "sung" by David, Pembroke and Sidney are now the "organ" while God provides "the harmony").[116] For Pembroke, though, the collaborative product is distinctly *textual*, a result of the "physical movement of the poet's hand and the materiality of the text it produces."[117] Demonstrating what Ryan Croft calls the "*topos* of poetic physicality," Pembroke establishes a highly material and tactile connection between living and dead, between mortal and divine "stuffe" when she frames the psalms as an act of shared labor—the coupling of her "penns impression" and Sidney's "blest hand."[118] The physicality with which Pembroke saturates her poetry is not the same physical dimensionality we saw in Margaret Cavendish's poetics, where poetic matter functioned as combinatory building blocks, flexible and in motion. Writing and circulating her poems only in manuscript, Pembroke attends much more closely to the physical materials required to make poetry. We are in the realm of paper and ink rather than dimensional fancies. The anatomical forms of Pembroke's poetics are simultaneously textual and somatic in their attention to how Pembroke's mourning body informs and mingles with the labor of her poetic work—two become one in her "sadd Characters" and "wounding lynes of smart" (112.81–82).

As we have seen, for Pembroke the Sidney Psalter is a "peec't" textual product. Even as she offers the psalm collection as a whole ("this finish't now" [110.23]), she conceives of it as a whole in parts ("the rest but peec't, as left by thee undone" [110.24]). In a particularly famous account of the biblical psalms' function, John Calvin described them as an "anatomy of all the parts of the soul," a devotional corollary to the anatomist's epistemological project.[119] In "Angell spirit" Pembroke pleads for Sidney's spirit to "receive the meanest part" of her "full soule" (112.84), figuring the text as both one part of her soul and her soul in parts, with the parts of Sidney's body preserved in the very same text that is described as the "meanest part" of her soul. Pembroke's complex representation of her and Sidney's collaboration is one that layers, mingles, and pieces texts, souls, and bodies. If we read Calvin's description as a methodological provocation—a prompt for devotional poetic practice to align with anatomical work—then we must see it as one that Pembroke's psalms directly pursue.

For many early modern anatomists, the practice of dissection served as a conduit between the living and the dead and between the mortal body and God's divine project. "Whosoever will attain unto the knowledge of the soul," Helkiah Crooke asserts at the opening of *Mikrokosmographia*, "it is necessary that he know the frame and composition of the body."[120] For Crooke, the study of anatomy brings the practitioner closer to both self and divine knowledge. "The image of the divine nature," he continues, "is most lively imprinted in [man's] soul and in his body."[121] If the soul is "cast into this prison of the body," we must know the body to know the soul and know the soul to know God.[122] Both Crooke and Calvin's claims for knowledge of the body and soul attempt to "integrate sacred mystery with corporeal mechanism," a project Katherine Rowe calls anatomists' "divine complicity."[123] In the heightened physicality of Pembroke's postmortem poetics, we are seeing a parallel project—a deep curiosity about how poets might contribute to the shared work of science and devotion. By framing her psalms as an intimate collaboration between the living and the dead, between her "mortal stuff" and Sidney's "divine" power, Pembroke prompts us to shift our understanding of the collaborative work beyond what John Donne outlines in his famous tribute.[124] Pembroke is not only collaborating with the dead, but also staking a claim to expertise within the liminal space between flesh and spirit that greatly enhances her ability to serve as an "anatomist of the soul."

Pembroke's own act of divine complicity emerges quite explicitly in Psalm 139, oft-cited in critical work on Pembroke's scientific investments.[125] Drawing on her source texts, Pembroke's version of this psalm employs anatomical language to explore the relationship between divine and poetic projects, specifically the intersection between divine and poetic modes of knowing. As anatomists were arguing that knowledge of the body leads to knowledge of God, Pembroke's Psalm 139 asks: how can the poet's work contribute to this process of knowledge making? How can poetry be a tool for studying and knowing "the frame and composition of the body" as a way into divine knowledge?[126] Pembroke begins her investigation into these questions with a meditation on God's omniscience. In fact, through much of the poem, knowledge belongs solely to God, which would seem to undercut any claim for poetry's capacity as epistemological tool. "Stanza after stanza" of Psalm 139, as Beth Wynne Fisken notes, "reveals yet another layer of God's penetrating understanding."[127] Pembroke iterates on the verb "to know" throughout the poem, employing polyptoton to describe her fruitless efforts to "shun" God's "notice" (Ps 139.22) and "lie / unknown" (Ps 139.27–28); to assert that God "know'st in highest rate" (Ps 139.83); and to acknowledge that her words, "yet unuttered," God "dost know" (Ps 139.14).

God's omniscience—from the Latin *omni* (all) and *sciēns* (from *scīre*, "to know")—finds its fullest expression when Pembroke uses anatomical language to describe God's forming of the poet's body in the womb.

> Thou, how my back was beam-wise laid
> and raftering of my ribs dost know:
> know'st every point
> of bone and joint,
> how to this whole these parts did grow,
> in brave embroidery fair arrayed,
> though wrought in shop both dark and low. (Ps 139.50–56)

Pembroke's knotty syntax and carefully crafted line breaks draw heightened attention to the doubled knowing in lines 51–52 ("dost know / know'st") that defines God's relationship to the anatomical parts anchoring the stanza—back, ribs, bones, and joints. Also doubled in this stanza is the interrogative adverb form of "how," first emphasized through the assonance of "thou how" at the stanza's opening and repeated in line 54 to represent God as anatomist, as one who knows "the knowledge of the frame . . . and the use of each part."[128] In both instances, "how" modifies the act of knowing and calls our attention not just to God's knowledge but also to his "know-how," the means through which God knows.

Immediately after the poet describes God's anatomical knowledge of her body, however, there is a crucial turning point in the poem's second half. Pembroke tempers her initial representation of God's omniscience when the poet's labor becomes another source of knowledge formation.

Nay, fashionless, ere form I took,
thy all and more beholding eye
my shapeless shape
could not escape:
all these, with times appointed by,
ere one had being, in the book
of thy foresight, enrolled did lie.

My god, how I these studies prize,
that doe thy hidden workings show!
whose sum is such,
no sum so much:
nay summed as sand they sumless grow.
I lie to sleep, from sleep I rise,
yet still in thought with thee I go. (Ps 139.57–70)

The language in the first of these two stanzas (lines 57–63) focuses on God's work to form and order the human, his knowledge about how to shape the poet's body, predetermined and already outlined in "the book / of [his] foresight." But the syntax of line 57 allows for an alternative reading. Pembroke ends the line with the verb "took," which has both active and passive modes: to take actively is to *seize or grasp*, to take passively is *to be taken with*. The poet's "shapeless shape" cannot escape God's "beholding" eye, but is it specifically God who *forms* here? Pembroke's poetic form is hyper-conspicuous in this stanza, even as she invokes processes of "fashioning," "forming," and "shaping" in the poem's language.[129] The stanza is full of enjambed lines (58–59; 59–60; 62–63) and medial caesuras, which call heightened attention to Pembroke's own formal work as a mode of making alongside God's forming work. Her formal technique is even more intricate in the next stanza (lines 64–70), where she again employs polyptoton, one of her favorite figures of repetition, to model God's "wrought" work in her poetry. Praising God's "hidden workings," Pembroke plays on the root word "sum" to describe their infinite number: "whose sum is such, / no sum so much: / nay, summed as sand they sumless grow" (Ps 139.66–68). She compares God's work to "brave embroidery fair arrayed," and then enacts that "wrought" work poetically, weaving the word "sum" through the stanza and shaping it into different forms to echo God's labor in the act of poetic formation.[130]

The poet stakes her own claim to divine knowledge by showing how her poetic practice aligns with God's work of form. In the poet's devotional praise—"My god, how I these studies prize, / that doe thy hidden workings show!" (64–65)—she marks shared modes of knowing in her slippery syntax. Whose are "these studies"? God's study of the poet (his "beholding eye") or the poet's course of study (the knowledge-making work of Pembroke's poetics)? If anatomists claimed the project of "show[ing]" God's "hidden workings" as the project of anatomical science, here Pembroke suggests that it is *also* the project of poetry. With the poet's agency more firmly established in the poem's second half, the knowledge work represented in Psalm 139 becomes a collaborative act cultivated around shared processes of fashioning, forming, and shaping. Pembroke deftly asserts her expertise as she illuminates how form anchors both anatomical study and poetic craft, and then situates both epistemological projects in the liminal space between life and death, between flesh and spirit. The poetic labor of the "anatomist of the soul," Pembroke's Psalm 139 argues, resides in the work of form.

Pembroke's *scientia mortis*

Throughout this chapter I have been attending closely to the intricate forms of Pembroke's poems, but it is worth remembering that Pembroke's formal experiments far exceed the bounds of close reading in a single book chapter. She used 126 different verse forms in her psalm translations, meticulously cataloged by Margaret P. Hannay, Noel J. Kinnamon, and Michael G. Brennan.[131] In their editorial introduction to Pembroke's work, they survey her uses of chiasmus, different forms of repetition, parenthetical statements, enjambment, and alliteration, along with the distinctive composition practices on display in her prose translation works.[132] Hannibal Hamlin also reminds us that the "metrical psalm" was a challenging and rigorous poetic project, rivaling "the Petrarchan lyric as the dominant poetic mode of the English Renaissance."[133] Tracing Pembroke's thanatological work through her poetic psalms allows us to better understand just what Pembroke was using her complex formal experiments to *do*.

One of the central tenets of *Anatomical Forms* is that early modern form "operates as an ontological category."[134] For Pembroke, as we have seen, poetic form is, first, capable of serving as an instrument for ontological inquiry: is death a fundamental phenomenon of life? What happens when we die? Form is also itself a concept and process subject to ontological investigation: what does it mean to form something or to subject something to form? Wielding poetic form for ontological inquiry, Pembroke remains invested in the process of inquiry itself. Her poetry, that is, simultaneously asks, *can poetry be a postmortem practice?* and *how does poetry become a site for postmortem practice?* How does the poet's somatic work ("hee did warpe, I weav'd this webb to end" ["Even now that Care," 102.27]) combine with the materials out of which poetry is made ("theise dearest offrings of my hart / dissolv'd to Inke" ["Angell spirit," 112.78–79]) in order to create a physical site for preserving, embalming, and memorializing? The gerund form of these verbs is important, because Pembroke's poetics is always a process-driven (or practice-oriented) poetics.[135] This is true for how her poetry exists in the archive as a "peec't," in-progress, and collaborative collection of extant writings, and it is true for how she studies and enacts postmortem practices in her poetry (110.24).[136]

What we see in Pembroke's postmortem poetics is "the tension between form as a fixed shape and form as the act of giving shape" that Mann and Sarkar find to be "widespread in early modern literature and science."[137] Because of its process-oriented epistemological and ontological work, poetry offered a prime site for not only processing one's experiences with death and loss, but also for turning that experiential processing toward epistemological ends. As a site of active *forming*, poetry resists representing death as a fixed form and instead allows for systematic inquiries on the actions and feelings surrounding death. And, perhaps most importantly for Pembroke, poetry refuses to preserve the dead in a fixed, static state, especially within the manuscript coterie culture within which her poetry circulated.[138] Poetry allows for a preservation of the dead that remains active, dynamic, always in process and always closely linked to the embodied practices of the one living, writing, and mourning. Here, then, we return to one of my opening questions: what does women's mourning poetry contribute to the history of science? For Pembroke, infusing poetic work with postmortem care leverages the physicality of poetic form in order to do what the work of embalming and burial cannot. Poetry is a space for the investigation of death, dying, and loss that relies on the embodied labor of the mourner and offers a highly physical and material site in which to preserve—and revisit, re-touch, recirculate, even reanimate—the body of the dead.

Or a physical and material site in which to imagine *oneself* dying. I want to look at one final psalm—Psalm 88—that returns us to the "as if" speculation of Pembroke's Psalm 119D ("dead *as if* I were") with which I began this chapter. Psalm 88 reads as an end-of-life record, and it is also an intricate, thoughtful formal experiment. This poem is about the fear of the physical experiences that accompany death.[139] It begins with a plea to God, a "cry" for "help" and "health" (Ps 88.1) as the poet's "life draws nigh the grave" (12) and her soul "so flows, so overflows" (11) with "ills and woes" (10). Pembroke goes on to describe her place "among the dead" (Ps 88.18), using repetition throughout the first five stanzas to emphasize the "darkness deep" (26) in the "grave of graves" (25) and the feeling of being "quite, quite cut off" from God (24). The poet's fear of death is particularly palpable in the poem's second half, which begins with the poet trying to persuade God to save her from death.

> Shall buried mouths thy mercies tell?
> dust and decay
> thy truth display?
> and shall thy works of mark
> shine in the dreadful dark?
> thy Justice where oblivions dwell? (Ps 88.49–54)

In these rhetorical questions, phrases like "buried mouths," "dust and decay" (50), and "dreadful dark" (52) betray the poet's fear of the actual, physical experiences of death. Pembroke foregrounds this fear in other psalms too, such as Psalm 55 where the poet speaks in "trembling pants" in the face of "deaths terrors" (Ps 55.11). She describes acts of self-violence ("my body I no less torment and tear" [Ps 55.6]) as she prays for deliverance and struggles to express the precise embodied experience of her distress: "I fear, nay shake, nay quivering quake with fear" (Ps 55.12).

The palpable fear of death Pembroke conveys in these psalms seems far from the ars moriendi philosophy in Mornay's *Discourse of Life and Death*. Recall that in Mornay's text, death is a haven—the "entrance of the port where we shall ride in safety from all winds" (247). But what if we consider these different perspectives as parts of Pembroke's thanatological study? If we think about ars moriendi literature as theory—treating "the problem of death as a philosophical and theological issue," as Pembroke's editors note of Mornay's *Discourse*—we might think of the psalms as an exercise in practice, a kind of *scientia mortis* in which Pembroke impresses a *form* onto the work of ars moriendi.[140] The scientia mortis mode of Pembroke's psalm translations uses the philosophical foundations of ars moriendi, but imposes structural complexity on its inquiries and infuses the theory with the embodied handwork of poetic practice, the poet's work as "anatomist of the soul." Pembroke's thanatological project blurs the space between ars moriendi (the art of dying) and scientia mortis (the science of death), modeling an interdisciplinarity that must inform our approach to the practices surrounding death, dying, and loss in early modern England.

Through Pembroke's persistent exploration of the experiences of grief and fear across the psalms, these affective states become topics for scientific study, systematically (and repeatedly) investigated throughout her poetry. In Psalm 88 Pembroke attends to the tangible feelings that accompany death, shifting from the inquiry central to her penitential psalms (what does grief feel like in and on the mourner's body?) to a more experimental and speculative "as if": what does death feel like in and on the body of the dying?

> Aye me, alas, I faint, I die,
> so still, so still,
> thou dost me fill,
> and hast from youngest years,
> with terrifying fears,
> that I in trance amazed do lie. (Ps 88.61–66)

The repetition of "so still, so still," isolated on line 62, rhetorically evokes the feeling of paralysis. Both poet and poem take on a corpse-like stillness evoked again in Pembroke's description of the poet lying "in trance amazed" in line 66. The entire poem is full of such linguistic doublings (recall "the grave of graves" [Ps 88.25] and the poet feeling "quite, quite cut off" from God [Ps 88.24]), a formal technique that keeps the poem churning over the same language. The poem, like the poet, feels stuck "among the dead at rest" (Ps 88.18). To enhance that effect, Pembroke uses her rhyme scheme (*abbcca*) to create capsule-like stanzas that close in on themselves as the reader moves through the poem. She employs the poem's form not only to contribute to but also to *actively construct* the palpable feeling of "darkness deep," the claustrophobic effect of the "buried mouths" and suspended state of stillness ("so still, so still") documented in this end-of-life record.

The forms of repetition that saturate Psalm 88—the same forms that power Pembroke's postmortem poetics throughout the psalms—transform the poem into an experimental and experiential site for studying the relationship between fear and death. Unlike the philosophy of Mornay's *Discourse*, this experiment finds no hope or comfort in death. There is only a "darkness" marking the poet's fear of dying and the affective weight of loss and grief. Constructed as a firsthand account of the felt experience of death ("now my life draws nigh the grave" [Ps 88.12]), Pembroke offers a scientia mortis that is deeply connected to the affective and psychological experiences surrounding grief. We glimpse this connection in stanza 6, where dying is figured as loss: "Who did know me, whom I did know, / removed by thee / are gone from me / are gone?" (Ps 88.31–34). With the repetition of "are gone," the poet checks the certainty of her loss, and this question immediately gives way to a description of her grief (her "wasted eye . . . melt[s] away" in "streams of pain" [Ps 88.37–39]). The poet's extended and painful mourning throughout this psalm exists simultaneously with the poem's claustrophobic sense of entombment, a coupling that Pembroke renders most palpably in the poem's final stanza.

> Who erst to me were near and dear
> far now, oh, far
> disjoined are,
> and when I would them see,
> who my acquaintance be,
> as darkness they to me appear. (Ps 88.72–78)

Pembroke isolates the phrase "disjoined are" on its own line, and uses the stanza's opening lines to establish the scene of separation and parting ("far now, oh, far") also evoked by her unusual use of "disjoined" as an adjective. Death, dying, and loss, like the poem itself, all end in "darkness" (78). This poem poses thanatological questions that rub against Pembroke's ars moriendi work: What does death feel like? Does the physical experience of dying feel like grief and mourning? It also poses aesthetic inquiries: How can the embodied experience of death be translated into poetry? How can the physicality of poetic form enhance our understanding of death? Most simply, the poem asks: what *is* death?

Pembroke's scientific and poetic investments co-emerge in her psalm translations. The resulting anatomical forms challenge us to read (and experience) poetry as a very literal and physical postmortem space. Pembroke writes from the position of an early modern woman intimately familiar with death, and then repeatedly leverages her experience for epistemological ends. Through her poetics, she establishes an experiential mode of postmortem knowing, and her poetry prompts us to shift from long-standing feminist idealizations of experience to a more critically robust understanding of experience as an evidentiary category that requires us to interpret, complicate, and study its constructedness. Pembroke does not take her own experiences with death, dying, and loss as subjective (individual, unique, and originating in oneself), but rather as *situated* within the broader social, material, and devotional apparatuses surrounding death. We have, therefore, decidedly moved away from locating Pembroke's science in a speculative laboratory at Wilton House, or in her social connections to various members of the Royal College of Physicians. In that version of the story, with science looking like we expect it to look, Pembroke was a chymist and an alchemist. And perhaps she was. But the concrete and overwhelming evidence within her writing shows her to have been an anatomist of the soul and a student of the dead and dying body.

If, as Pembroke's poetry suggests, poetic form opens space for affect and alternative modes of embodied practice within the experimental work of science, we need to reassess what we think we know about the history of science archive. In this chapter, I add Pembroke's writings to that archive and insist that her thanatological project be read alongside Philbert Guibert's and Peter Forestus's treatises on embalming and the pervasive triumph over death heralded by the period's anatomical treatises. Coupling scientific inquiry with poetic technique, Pembroke's writing begins to reform "the fragmentary and elusive sources that reflect actual practice" surrounding postmortem care in the early modern period.[141] It also expands our ability to comprehend poetic form as a significant scientific tool, one of my central objectives in this book. As we have seen, it is the *form* of Pembroke's poems through which she explores the complexity of death as a textual construction. Poetic formation is death in suspension, death as an active state of becoming, still connected to the flesh even as it leaves the body behind for the spiritual realm. For Pembroke, this scientific work occurs in "Inke"—in the repeated, propagative inquiries of her "penns impressions" (112.79).

CHAPTER 3

Blank Matter

Copperplate Engraving, Paper Skins, and Mary Wroth's Anatomical Poetics

Engraving is an act of forming through incising that relies on surface dynamics—the relationship between cut and uncut surface. It is a form that activates both positive and negative space. What is there has been cut into what is not there. In this *intaglio* printmaking technique (*intaglio* from the Italian word for "incision") usually practiced using a copperplate, the plate is carved with fine lines that then get filled with ink. The ink sits just below the surface of the plate and, when the plate is run through the press, the pressure of the printing press lifts the ink out of the incisions and onto the paper. An important part of the final rendering is the uncarved sections of the plate, which, when printed, manifest as blank space on the paper. This blank space around and between the cuts is just as potent, just as substantive, just as material as the engraved lines themselves. In the sixteenth and seventeenth centuries, anatomists increasingly turned to copperplate engraving—works "composed entirely of lines"—as their preferred mode of illustration because engravings, rather than woodcuts, allowed for more precise rendering of the body's subtle details.[1] In 1545, only two years after Andreas Vesalius's famous *De Humani Corporis Fabrica* was published with over two hundred woodcut anatomical illustrations, Thomas Geminus and his printer, John Hereford, republished all of Vesalius's full-page illustrations as copperplate engravings, advertising this important difference in the book's title: *A Complete Delineation of the Entire Anatomy Engraved on Copper*. In 1556, Juan Valverde de Hamusco introduced his own series of Vesalian-inspired engraved anatomical plates, and declared that the art of engraving allowed him to demonstrate "the progression of the filaments of the flesh" as they "run through each muscle."[2] For Geminus, Valverde, and the many other anatomists who employed engraving, this mode of illustration "imbibe[d] the two-dimensional drawing with a sense of three-dimensional form and depth" through its chiaroscuro effect—the interplay between light and shadow produced by the relationship between cut, inked lines and blank space.[3]

This chapter, like the act of engraving, turns around what is there and what is not there in the poetry of Lady Mary Wroth (1587–1652). First, a bit about what is there. Wroth's poetry is extent in several forms—one manuscript of 117 poems and songs held at the Folger Shakespeare Library (MS V.a.104); a printed sonnet sequence titled *Pamphilia to Amphilanthus* published with the first part of her prose romance, *The First Part of the Countess of Montgomery's Urania* (hereafter, *Urania*) in 1621; and a collection of poems incorporated into the *Urania* itself, both the printed first part and the unpublished manuscript of the second part (Newberry Library VAULT Case MS folio Y 1565.W95).[4] Wroth's texts make for an especially generative material archive, in part because all of her writing exists in a liminal, in-process state between manuscript and print, finished and unfinished, poetry and prose.[5] True to early modern romance form, Wroth's *Urania* is long, windy, and full of characters, including Pamphilia, the romance's most exemplary female poet who then becomes the poetic speaker of the *Pamphilia to Amphilanthus* sonnet sequence. All of Wroth's writings are heavily influenced by the English Petrarchan tradition, a product of the Sidney and Herbert literary lines from which her writings stem. Wroth was niece to Sir Philip Sidney and Mary Sidney Herbert, Countess of Pembroke; eldest daughter of Sir Robert Sidney; and mistress to William Herbert, Earl of Pembroke.[6] Not unlike her aunt's poetry, which we are now familiar with from Chapter 2, Wroth's poetry is quite dense, saturated with wordplay, tightly compressed syntax, and intricate formal experiments—features that have led scholars to decry the "many challenges and uncertainties" the poetry poses for its readers.[7] Mary Moore notes the "labyrinthine texture" of Wroth's sonnets; Paul Salzman, in an essay cheekily titled "Not Understanding Mary Wroth's Poetry," claims that her poetry performs a "calculated obscurity"; and Roger Kuin asserts that Wroth's language is "so torturous and so involuted that even [her] contemporary readers must have found these texts hard work."[8] In other words, there is a lot *there* when we engage with Wroth's writing.

There is also a glaring *not there*: the blazon, a poetic technique of fragmenting and cataloguing the physical attributes of a subject, usually a woman. The blazon was a pervasive early modern English poetic mode, particularly in poetry like Wroth's, steeped in the Petrarchan tradition. In Heather Dubrow's nuanced argument that Wroth's writings align "with other, male speakers of the Petrarchan discourse and counterdiscourse" while simultaneously establishing "a distinctively female idiolect," she traces Wroth's literary influences to her uncle and father, but also to Shakespeare, a supremely Petrarchan poet.[9] As we will see, Wroth uses the blazon mode throughout *Urania*, but when literary scholars have looked for a blazon in her poetry, specifically a blazon of the male beloved by the female poet, they find a conspicuous omission—a "blank," in Laurie Maguire's sense of the word.[10] Instead of a satisfying, proto-feminist response to those violent "gesture[s] of display," in which male poets dissect women's bodies and then "extend[ ] the privilege or pleasure of 'seeing'" to their readers, Wroth's poetry focuses almost exclusively on its poetic speaker (Pamphilia) rather than her beloved (Amphilanthus).[11] The oft-noted absent blazon is frequently met with a corresponding desire to explain and fill the blank space. In Ilona Bell's introduction to her critical edition of Wroth's poetry, for instance, she emphasizes that Wroth's title "announces that Pamphilia's poems are addressed *to* Amphilanthus," which is why we do not find poems *about* Amphilanthus.[12]

Throughout her poetry, Wroth focuses on the effects of love on Pamphilia's mind and body with an intensity that results in a brooding, twisting, deeply reflective poetic collection. Poem after poem finds a tortured Pamphilia ("My paine, still smother'd in my grieved brest, / seekes for some ease, yett cannott passage finde" [MS 18.1–2]), embracing despair and darkness ("Night, welcome art thou to my mind destrest / darke, heavy, sad, yett nott more sad then I" [MS 43.1–2]) as she works through the absence and neglect of her beloved ("Gon is my joy while heere I burne / in paines of absence, and of care" [MS 58.1–2]).[13] One of Wroth's central questions in both her romance and her poetry concerns how to represent Pamphilia on the poetic page, and I read this inquiry as one rooted simultaneously in the poetic blazon tradition and the developing material-textual techniques of Renaissance anatomical study. Picking up on some threads related to preservation and paper from Pembroke's poems and the possibilities for textual dimensionality activated in Margaret Cavendish's poetic fancies, my analysis in this chapter brings these theorizations to bear on the representation of women's bodies in texts. In Wroth's writing, we will find an alternative mode for crafting knowledge about women's bodies that renovates the violent tendencies of the blazon tradition. Wroth's poetry, I will demonstrate, is about the blazon as a knowledge-producing inscriptive form. She writes around and about the blazon in order to investigate its inscriptive project—the way it inscribes women's bodies into narratives of penetration, fragmentation, and, above all, absence. Implementing the knowledge gained from that investigation, Wroth develops an innovative anatomical poetics that shapes the female form through the materials of her poetic practice.

Wroth is not a writer of poems, I argue in this chapter, but an *engraver* of poems. Throughout her texts, she pays extraordinary attention to the act of writing poetry, as well as to the inscriptive surfaces and "material supports" that hold her poems.[14] Like anatomical engravers, Wroth finds rich potential in the unmarked portions of her manuscript pages. The early modern anatomy books that use engraving do so for specific reasons tied to both the representational precision of the resulting illustrations (a perspective we heard above from Valverde) and the creative potential of the engraving process itself. As we take a closer look at the step-by-step process of engraving in this chapter, we will come to understand that there was a direct correlation between the "technique of representation" and the represented "features—specifically textural ones—of a body part."[15] Employing this new technique as part of the broader bookmaking project of Renaissance anatomy, anatomists conveyed a very particular approach to the science of the body, an approach driven by heightened attention to surface and the desire to infuse representational content with the tactile methods of making. This coupling will be especially useful in understanding Wroth's highly material poetics, her exploration of paper as a multimodal technology. Situating Wroth within this book's web of women's scientific work, like Cavendish and Pembroke, she intervenes in the ongoing theorization of the relations between bodies and texts. Unlike the work of these other thinkers, though—Pembroke and her postmortem practice, Cavendish and her science of bodies in parts—Wroth's writing does not exhibit specialized anatomical knowledge. Instead, Wroth's scientific engagement aligns with that of contemporary anatomical engravers, steeped in conventions and traditions of artistic representations of the body that gave them a unique perspective on the developing science of the body.

The Blazon as Blank

In the first hundred pages of the printed *Urania*, we are introduced to Melasinda, Queen of Hungary, through the words of her pining lover, Ollorandus. Ollorandus claims that Melasinda appeared to him in a dream vision, "like a Picture well drawne," pleading with him to rescue her.[16]

> Eyes, like the perfect'st mixtures of heavenly powers, not to be resisted but submitted to. Lipps fully commanding the plenty of duty, when they

> seem'd to demaund obedience: Her neck the curiousest pillar of white Marble, breast of Snow, or smooth waves of Milke, swelling, or falling, as the sweet gale of her most sweet breath did rise, or slacke. All other parts so exquisite as none, save onely she, can be so excelling. (I.78)

Ollorandus vows to "seeke" this fragmented vision of loveliness (I.78), travels from Bohemia to Hungary, presents himself to Melasinda as an "adventrous Knight" (I.79), and joins her service against a rebellion. Even though Ollorandus's brave deeds eventually win his beloved's favor, the rebellion ends with Melasinda marrying another man (the rebel leader, Rodolindus) for the sake of peace in her kingdom. Ollorandus and Melasinda do spend a short time as lovers, meeting in secret after her marriage, but he is forced to leave when her husband begins to suspect. When Amphilanthus first encounters Ollorandus in *Urania*, he is traveling, pining, and writing poetry about Melasinda and his frustrated desire.

Melasinda and Ollorandus's story is a minor subplot in *Urania*, but it packs a strong Petrarchan punch. This subplot contains many elements that signal Wroth's attentiveness to the Petrarchan tradition and its central feature, the inaccessible female beloved. Melasinda is a woman out of reach in several different ways. Appearing to Ollorandus first in a dream vision, she is both distant from him (in a far country) and, quite literally, untouchable—a vision without substance, inaccessible because imagined.[17] Wroth stages an encounter between a male lover and a female beloved who is *intact*, drawing on the term's original etymology from the Latin *in* (not) and *tactus* (touch).[18] In Thomas Wyatt's "Whoso List to Hunt," one of his most direct adaptations from Petrarch's *Canzoniere*, Wyatt's poetic speaker hunts a doe that, like Melasinda, is both out of reach and fleeting vision. The doe "fleeth afore" as the "fainting" poet follows, seeking "in a net . . . to hold the wind." Those who do reach the doe, the poet warns, will still "spend [their] time in vain," for "graven with diamonds in letters plain / there is written, her fair neck round about: / *noli me tangere*, for Caesar's I am."[19] Plucked from its biblical origins, *noli me tangere* gets rescripted within the English lyric tradition to mark "an untouchability that is, somehow, the property of 'woman.'"[20] Wyatt adapts "Whoso List to Hunt" from Petrarch's Sonnet 190, "Una Candida Cerva" ("A White Doe"), in which the poet encounters a doe with a diamond and topaz inscription "around her lovely neck" that reads "nessun mi tocchi" ("let no one touch me").[21] Both Petrarch and Wyatt's poems turn on a phrase—*noli me tangere, nessun mi tocchi*—that denies touch, keeps the woman intact, and incites violent inscriptive alternatives. Faced with an inaccessible female beloved, the male poet responds with the blazon, fragmenting the woman's body in poetry as compensation for a frustrated desire that can find no tangible relief.

As Wroth's readers, we come to know Melasinda through a series of representations that begin with Ollorandus's blazon, and then continue with her portrait ("he led me into a Gallery where he shewed mee the picture of that distressed Princesse" [I.79]). Wroth cleverly frames this subplot so that we encounter Ollorandus *after* he is forced to leave Melasinda (who, at that point, is inaccessible because she is married to Rodolindus—"for Rodolindus's I am") and, consequently, our knowledge of Melasinda is mediated several times over. We hear of the dream vision and the portrait not in real time but in retrospect, a nod to the retrospective nature of Petrarch's representation of Laura, and Ollorandus's account is mediated again through Wroth's narrative voice. These highly mediated representations circulate through *Urania*'s pages—all before we actually encounter the character Melasinda—and invite us to think critically about how our knowledge of this woman is crafted. Melasinda does not remain inaccessible to Ollorandus (eventually, her husband dies and she and Ollorandus marry), but this subplot, saturated with Petrarchan elements, signals Wroth's attention to the literary techniques through which women's bodies become inscribed into the Petrarchan narrative. The *noli me tangere* imperative scripts a mode of knowing that relies on women's absence and manifests on the page as a fragmented, objectifying representation of their bodies driven by phallocentric scopophilia. The exaggerated layers of mediation in the romance's introduction to Melasinda's character prompt us to think not just about the content of the blazon but also the inscriptive acts that form it.

Urania is full of blazonic language, which Wroth uses most often in her initial descriptions of the women characters.[22] The romance's women have skin "as white and soft as Swannes downe on the breast" (I.423), eyes "gray, and shining like the morning" (I.342), foreheads that radiate with "azure veined purenesse" (I.639), lips "delicate, and redd" (I.342), and hair "bright and delicate" as "straw in the Sunne" (I.638). There are moments when the partitioning of women's bodies verges on satire, such as when Veralinda hikes up her skirts to run from a bear and Leonius is "conquer[ed]" by her "dainty leggs" (I.426).[23] And there are moments when Wroth draws on the language of dissection, explicitly acknowledging the discursive links between anatomy and love poetry. At one point in Amphilanthus and Ollorandus's travels together, they overhear a "dainty maide" singing in the woods, who calls on her beloved to "peruse" and "seeke" inside her body to "find" proof of her love.

Pray desect me, sinewes, vaines,
Hold, and loves life in those gaines;
lying bare
to despaire,
When you thus anotamise
All my body, my heart prise;
being true
just to you.
Close the Truncke, embalme the Chest,
Where your power still shall rest,
joy entombe,
Loves just doome. (I.289)

The maid's song evokes the "riot of aesthetic and scientific" impulses central to Jonathan Sawday's argument on the Renaissance "culture of dissection."[24] In fact, the connection between the blazon and dissection has long anchored scholarly accounts of the relationship between literature and science in the early modern period. This connection is based on a substantial archive of both literary and scientific texts, of which the French poetic genre of the *blason anatomique*, or "anatomical blazon," stands out as a particularly potent example. The pages of the definitive edition of *Blasons Anatomiques des Parties du Corps Fémenin* (*Anatomical Blazons of the Female Body*, 1543) feature sonnets and short poems about a very specific female body part ("Blason du Tetin" [nipple] and "Blason de la Cuisse" [thigh], for example) paired with a woodcut image of the isolated body part. The poets writing in this genre, Nancy J. Vickers reminds us, called themselves "*anatomistes*" and participated in the same text-based project as those "more concretely grounded practitioners," circulating "printed texts that displayed the private, particularized body to the public gaze."[25] Wroth's poem here fits neatly into this entrenched understanding of anatomical poetics, even as it upends some well-worn conventions. This is a female poet lover writing about a male beloved (certainly of interest as we consider Pamphilia), and the poem itself is not a blazon per se, but it does call attention to the fragmenting act of the blazon in its scientific form. By dissecting and anatomizing the woman's body, the male beloved can gain knowledge of her constancy.

One character who is notably immune to this fragmenting rhetorical treatment (although, at one point, her body is literally dissected in a dream vision) is Pamphilia, first described only as "Sister to the noble Prince Parselius" (I.37) and "the most silent and discreetly retir'd of any Princesse" (I.61).[26] Rather than blazon Pamphilia's body, Wroth uses scenes in the romance to establish Pamphilia's skepticism toward the desires underlying the blazon tradition. In one early scene that anticipates the structure of Melasinda's introduction, we learn about an encounter with Pamphilia through the voice of a male character, Steriamus, Prince of Albania. As he tells it, he revealed his love to Pamphilia using a mirror rather than words. In characteristically Petrarchan fashion, he approaches Pamphilia as a "dead man," so consumed by love that there is "nothing left in [him] but empty hope, and flourishing despaire" (I.69). Pamphilia asks if there is a "cure" for his state, and Steriamus "look[s] about" and finds a "faire Glasse" hanging on a nearby wall. "I tooke it up turning it to her, mine eyes onely speaking for me," he relates. "She (with seeing her face, saw my cause of torment) said as little as I: onely taking the Glasse turn'd the other side, which was dull like my gaines" (I.69). Steriamus's desire, we might interpret, is not Pamphilia herself, but the mirrored reflection of Pamphilia—the likeness without the substance. Pamphilia, who, up to this point, has entertained him "modestly and gravely" (I.68), for the first time becomes "cruell" and rebukes Steriamus for the "great[ ] offence" of his "glasse discovery" (I.70).

This scene does not read as a blazon (where is the fragmentation? where is the dissective act?), but it does play with the inscriptive logic underlying the blazon form. Steriamus does not use language to describe his desire; he instead transfers that rhetorical display to a signifying object—the mirror. To borrow Vickers's language, Pamphilia's face reflected in the mirror becomes "the shining and resistant . . . surface" that must be penetrated in order to gain access.[27] As the text's central and most celebrated poet—and, ostensibly, a cipher for Wroth herself—Pamphilia is cued to recognize and call out conventions that inscribe women's bodies into particular kinds of narratives and onto particular kinds of surfaces.[28]

Both the blazon form and Wroth's skepticism of it weave through the *Urania*, supporting Heather Dubrow's argument for Wroth's "juxtaposition of the most conventional Petrarchism with its most rebellious counterdiscourses," and compelling us to revisit the critical tendency to dwell on the absent blazon of Amphilanthus in Wroth's poetry.[29] Such a tendency distracts from a much more complicated point of departure for studying Wroth's relationship to this literary form. How, I want to ask, does the blazon function as an absent presence or, drawing on Laurie Maguire's work, a kind of "blank" in Wroth's poetry? In Maguire's study of the typographical and literary "blank," she takes printed texts as her focus to demonstrate how gaps and absences marked or constructed through the material of a text—white spaces, empty parentheses, *etceteras*—function as integral, creative elements of early modern pages. "Blanks are never truly blank,"

Maguire argues, "because they always call attention to what is not there."[30] The "blank" invites readers to fill a gap and calls attention to the acts and materials of representation rather than the thing being represented, recalling Mary Sidney Herbert's instrumentalization of the white space between stanzas as the space of postmortem care (the place, remember, where Philip Sidney's "fresh bleeding" wounds are cauterized).[31] For Maguire, blanks are an important material trace of the meticulous ways early moderns attended to the "rhetoric of the page." Developing from "a typographical shorthand to a ludic creativity," the form of the blank becomes especially useful for reading poetry, always already about the "interchange of energies" between the ink marks and blank spaces of the page.[32]

The poetic mode of the blazon is a mode of blanking, a projection of an identity out of the gaps and spaces between fragmented parts. By evacuating the conventional blazon mode from her poetry, Wroth amplifies the blanking function of the blazon—outdoing the form with its own technique. Maguire's analysis of the "blank," though, gets us closer to just how Wroth then renovates the blazon form through her highly material poetics. As part of the "rhetoric of the page," the absent blazon of Wroth's manuscript—the collection's most prominent "blank"—takes form materially, and such a formulation (and formation) entails a new set of questions about her poetics, questions about surface, materiality, white space, and absence. For Wendy Wall, Wroth's *Pamphilia to Amphilanthus* sequence thematizes absence. Consequently, "as readers, we are directed not only to discover and give voice to the silences within the work," Wall urges, "but also to understand how silence, absence, and vacancy themselves define both text and speaking subject."[33] While Wall ultimately argues that this thematic absence "expels materiality" from Wroth's text, leaving only the trace of an abstracted female poetic speaker, I want to linger in the realm of book history that anchors Maguire's concept of the "blank," because Pamphilia is not simply represented as an exemplary female poet in *Urania*.[34] Wroth represents her as a poet particularly interested in the materials of her craft, persistently attentive to the process of poetic inscription. And, in the first of Pamphilia's many scenes of writing in the romance, this interest manifests around a kind of blank.

Shortly after Pamphilia's first appearance in the romance, she is alone in her bedchamber, "heavie" and "turning . . . in her bed" as she tries to sleep, looking for ways to "expresse her woe" (I.62–63). Pamphilia's "impatient desires" lead her to her "little Cabinet" filled with poems she has written, but none of her past poems convey her current feelings, so she pens a new one: "she took pen and paper, and being excellent in writing, writ these verses following" (I.62).[35] The eighteen-line lyric Pamphilia writes is about the pain of unrequited love, which she describes as "knowledge" that "cloyes [her] brest with woes" (I.63). Drawing on Petrarchan conventions, Pamphilia blames her eyes as the "instruments of woe" that "did begin the act," and describes both her eyes and heart as "weeping," her body dissolved to liquids as she pines for her beloved (I.62). Then, only several lines after the poem itself, Wroth tells us that Pamphilia "tooke . . . the new-writ lines, and as soone almost as shee had given them life, shee likewise gave them buriall" (I.63). Immediately after encountering this poem in the romance, we read of its destruction (its "buriall") within the fictional world of the text. Much later, Pamphilia will show the poems in her "little Cabinet" to Amphilanthus, but the narrator is quick to remind us that she cannot show all her poems, only those "shee had saved from the fire" (I.320). Wroth conspicuously recalls Pamphilia's earlier act of destruction and the limits it places on what Amphilanthus can read.

Pamphilia's poem exists within the pages of *Urania* (as physical book), but is expunged from the narrative world of the romance. Simultaneously marking paper's preservative capacity and its material fragility, Wroth plays in the space between the permanent and the ephemeral, and *paper* is both the site of play and the medium under scrutiny. This is an instance, within the pages of a book, where poetry becomes hyper-visible as "writing matter" and we as readers, to borrow Juliet Fleming's formulation, "confront the material properties of the graphic trace."[36] Our introduction to Pamphilia's poetic work grafts Wroth's representation of this female poet to the materiality of her writing medium. In these two early encounters with Pamphilia in the pages of *Urania*—first, her response to Steriamus's "glasse discovery" and, second, her absent present lyric—Wroth establishes her interest in both the processes and materials of inscription. Turning back, then, to the absent blazon of Amphilanthus within an archive of writing so conspicuously Petrarchan, this gap challenges her readers "to respond imaginatively to what is absent."[37] For my purposes, this absence offers an opportunity to imagine otherwise in our study of early modern anatomical poetics, a chance to destabilize the entrenched coupling of blazon and dissection. Wroth's poetry prompts us to turn our attention instead to anatomy's inscriptive experiments, the techniques through which anatomists and illustrators rendered fragmented bodies in the pages of their texts.

The Art of Copperplate Engraving

Copperplate engraving, or chalcography, was originally an artistic practice, used by artists like Antonio Pollaiuolo ("Battle of Nude Men," 1470) and Domenico

Fiorentino ("Two Flayed Men and Their Skeletons," ca. 1540–1545) who were experimenting with new ways to represent dimension and texture. For these Florentine artists, historian of science Domenico Laurenza argues, the body was "clothed in skin, alive and in motion," and their focus was on the "variations in the surface of the body."[38] Engraving allowed for more precise renderings of texture than "the coarser woodcut technique," and there was significant overlap between anatomical and artistic study in the period, evidenced, for example, in Leonardo da Vinci's extensive dissection work and intricate anatomical sketches.[39] Andreas Vesalius's influential illustrations were originally woodcuts, but by the mid-sixteenth century, as Laurenza documents, "the demands of science called for a new vocabulary and new techniques capable of expressing the fine structure of anatomical forms through printed images."[40] As copperplate engraving increasingly replaced woodcuts as the desired medium for anatomical illustration, the science of the body became infused with a "sensuous attachment" to lines and flesh.[41] "To really see the inside of the body," as James Elkins argues of this historical shift, was "to risk falling in love . . . with the seductive textures of the smooth and sensitive membranes—more delicate than ordinary skin, more sensitive and vulnerable, and above all more redolent of the most intense pain."[42]

From early modern accounts of engraving, such as William Faithorne's *The Art of Graveing and Etching* (1662), we learn that this technique relied on multiple, layered acts of inscription.[43] To engrave a copper or brass plate for printing, Faithorne outlines, first the engraver traced the illustration onto tracing paper, then covered the metal plate with a very thin layer of wax (it must be "the whitest Virgin-wax," according to Faithorne), and overlaid the tracing paper onto the wax layer.[44] Using a light, fine-tipped stylus, the engraver then transferred the illustration from the tracing paper onto the wax layer, also lightly scratching faint lines into the metal plate below the wax. To carve the final rendering onto the plate, the engraver would then set the plate on a sand-filled pad and use a burin (or "graver") to make deep etches into the surface of the plate. With this important new anatomical illustration technique, textual representations of the body became not only more accurate but also more textural, more tactile. The engraved print, in other words, rendered texture in ways that more directly aligned with (and more fully conveyed) the work of the anatomist's hands. The systematic, highly artistic process Faithorne outlines shares a methodological lexicon with the act of dissection—carving, cutting, incising—a semantic coincidence reinforced by the French term for copperplate engraving, *taille douce*, which John Evelyn glosses as "sweet or tender cut" in his *Sculptura: or the History and Art of Chalcography and Engraving in Copper* (1662).[45] The process of engraving transferred the actions of the dissection room into the printing room, acting on inscriptive surfaces—tracing paper, wax, metal plates, and paper—and bringing the dissective lexicon to bear on texts rather than bodies.

Describing engraving as a "visual language" attentive to "tone, texture, and volume," art historians emphasize the dimensionality of copperplate illustrations, particularly suited to "subtly render[ ] the chiaroscuro variations of anatomical forms and details."[46] As an inscriptive mode, copperplate engraving animates the surface of paper as substrate—as both a material that "underlies or forms the basis of another" and "the substance which a particular agent or reagent acts on."[47] Like other intaglio techniques, engraving makes hyper-visible the interactions on the page surface between inked lines and blank space, even as it relies on carving, cutting, and etching into a surface in order to form the representation. The concurrent work of carving *into* and inking *onto* are particularly apparent in the final steps involved to prepare early modern copperplates for printing. Once the engraving was complete, the printer used a dauber to spread and push ink across the surface of the plate, forcing ink into the finely engraved lines. By "sliding, pressing, and dappling in sundry ways the [ink] ball, all over the engraved surface of the plate," Faithorne explains, "you make the black [ink] enter and pierce into all the strokes."[48] The printer then used thick cardstock to pull excess ink off the plate from top to bottom, removing the thick layer of ink while exerting additional pressure on the plate's surface to force more ink into the engraved lines. Finally, the printer cleaned and polished the plate with muslin cloth and his own hands, using a process known as "hand-wiping" to buff the plate with "the fleshly side under the little finger."[49] Faithorne emphasizes that the printer must leave "neither black nor spot" on the plate so that the unengraved sections of the plate will be "as white as the margin of the paper when the sheet is printed."[50] The inked copperplate, ready for printing, will seem as smooth, polished, and clean as the original, unengraved plate, but this intaglio form holds ink in the fine lines of its surface (ink "pierced" into its surface, to use Faithorne's language), only revealed through the pressure of the printing press, which lifts the ink out of the incisions and onto the paper.

I began this chapter with the assertion that Wroth was not a writer but an engraver of poems. With a better understanding of the intricacies of the copperplate engraving process, it is now time to clarify that point because, certainly, it is hard to align the technique I have just described with what we think we know about Wroth's compositional practices: she penned both her poetry and prose romance in manuscript, and then revised and rearranged her poems before sending some of that material off to the printing shop. But let us return to Pamphilia, the "prototype for female authorship" in *Urania*.[51] We have already been introduced to Pamphilia as a poet keenly interested in writing matter and, in one scene, this interest takes a distinctly anatomical turn when Pamphilia carves a sonnet into "the barke of [a] fayre and straight Ash[ ]" (I.92). In this sonnet, Pamphilia calls on the tree to "beare part" of her pain, demanding, "keepe in thy skin this testament of me" (I.92). Pamphilia is not writing on a surface here, but rather carving (Wroth uses the term "ingraving") *into* a skin. She describes her act of engraving on the tree as an act of wounding (recalling Faithorne's language of "piercing") that corresponds to love's penetrative wounding of her body. She uses a "knife" instead of a pen to describe how love has "ingraven" and "cutt" her body "with miserie" and "griefe," as she in turn "ingrave[s]" and "cutt[s]" a "testament" of her pain into the tree's bark.[52]

> Thy sap doth weepingly bewray thy paine,
> My heart-blood drops with stormes it doth sustaine,
> Love sencelesse, neither good nor mercy knowes
>
> Pitiles I doe wound thee, while that I
> Unpitied and unthought on, wounded crie:
> Then out-live me, and testifie my woes. (I.93)

In these final lines, Pamphilia's engraving draws the tree's interior to its surface, filling her carved lines not with ink but with sap as the tree "weeps" in tandem with the poet's tears and "heart-blood drops." The tree's skin and Pamphilia's inscribing act work together to "imitate the Torments of [her] smart" and "testifie [her] woes" (I.92–93), blurring the boundaries between Pamphilia's own bodily surface and the tree's "skin" and transforming the tree into "a living codex," to borrow Vin Nardizzi and Miriam Jacobson's description.[53] While the tree's bark is initially the writing medium of interest in this poem, the analogous, imitative interactions between the poet's body and the tree's "skin" transform the tree into a substitute for the poet's body, while simultaneously turning the poet's skin into an inscriptive medium.

Pamphilia's act is one of wounding on "skin" that is a "testament" of her body, an act of penetrative self-wounding that Wroth describes in language that evokes sexual violation.[54] This is not a poem about the poet's pain; it is a poem about how pain comes to be inscribed. When Pamphilia "tak[es] a knife" rather than a pen as her instrument to "finish[ ] a Sonnet, which at other times shee had begunne to ingrave in the barke" (I.92), the act of writing is the act of cutting, incising, and penetrating. Wroth's theorization of the translation of pain here comes close to Elaine Scarry's formulation that "physical pain does not simply resist language but actively destroys it."[55] In Wroth's translation of Pamphilia's pain onto tree bark, and the near-simultaneous grafting of that bark with the poet's own skin, the violence and trauma that Pamphilia both works to document and inflicts in this poem manifests in the *act* of writing rather than the content of writing. Wroth urges us to consider how inscriptive medium and writing practice work together with the "finished" sonnet to inform the reader's knowledge of Pamphilia's pain. She uses the tree's "skin" as a medium on which to represent and a technology with which to interrogate the representational process, merging inscription and inscriptive act as she explores the very idea of surface.

In Evelyn's treatise on chalcography, he tracks the parallel histories of the "art of engraving" (which he refers to as "sculpture") and the art of writing. "Whoever was the inventor of letters," he insists, "was also doubtless the father of sculpture," an equation that recalls the "sculptural" flap anatomy figures of Chapter 1 and Cavendish's insistence on the animated dimensionality of texts.[56] As evidence for his claim, Evelyn reminds his readers that all writing surfaces before paper and parchment required engraving. Cataloging a variety of writing surfaces (including marble, wooden tablets, palm leaves, "rinds of trees," silk, and "sheets of lead, linen, wax, and ivory"), Evelyn argues that the same act would have been required to "write" on all of these materials—"it was still by insculpting, scarifying, and making a kind of incision into it."[57] "Writing with ink, in paper and parchment," he reflects, is "altogether a novelty in comparison to the more ancient forms and materials."[58] The intersecting histories of engraving and writing raise questions about the materiality of texts that guide my analysis of Wroth's poetry through the rest of this chapter. What can a writing medium (or "skin," for Wroth) be or do if it is not simply a passive surface?[59] How does writing matter hold within itself traces (or "scars," for Evelyn) of inscriptive acts? If Wroth conceived of the act of writing as a kind of engraving, how does that shift her understanding of *paper* as the medium on which and into which she penned her poems?

Inscriptive Skins

With this extended investigation of anatomical illustrations, inked copperplates, and tree carvings, we might seem very far from the question of absence (specifically, the absent blazon) with which I began this chapter. We return quite organically to absence, though, when we look at the pages of Wroth's extant poetry manuscript. Wroth was a poet who thought with and through the materials of her poetic practice, as Pamphilia so often does in *Urania*, and one of the most noteworthy features of the manuscript is Wroth's luxurious use of page space. A holograph manuscript in Wroth's neat, italic script with numerous revisions throughout, Folger MS V.a.104 is a quarto volume (5.5″ × 7.5″) of sixty-six leaves containing one hundred ten poems and songs.[60] Wroth most often penned only one poem per page, leaving a significant amount of white space around the poem and between the quatrains and tercets of her sonnets, written in the Petrarchan form. Because this practice resulted in a visually tidy, clean manuscript copy, scholars have speculated that the Folger manuscript was a presentation copy (portions of which may have been meant for William Herbert) or a late draft of the *Pamphilia to Amphilanthus* sonnet sequence that Wroth revised for publication with the first part of *Urania*.[61]

The seemingly tidy pages of Folger V.a.104, however, mask the messy materiality of its making.[62] As scholars have increasingly demonstrated, the Folger manuscript is an extremely fluid text, a collection of "adaptable groupings and sequences of poems" that resist "fixed formation" and obscure any transparent relationship to the printed *Pamphilia* sequence.[63] Rather than presentation copy or early draft for printing, this manuscript asks to be read, studied, touched on its own terms as, to borrow Cavendish's phrase, a paper body. Marta Straznicky's study of the extant manuscripts of Wroth's play *Love's Victory* has been especially illuminating on Wroth's "patchwork" composition and revision processes. Straznicky finds evidence in the Huntington manuscript of *Love's Victory* (HM 600) that Wroth "composed . . . in 'loose' sheets" and was very likely "copying" and "reassembling" separate documents to piece together the play.[64] Applying this "patchwork" technique to Wroth's composition of her poetry manuscript, there are several significant features that have captured scholarly attention, particularly Wroth's arrangement of poems throughout the volume and her decisions about where and how to break different poetic sequences. To this list, I add the manuscript's blank spaces or, more precisely, Wroth's use and curation of the *mise-en-page*, a material feature with much to tell us about how Wroth frames Pamphilia in and through this text. In the manuscript's blanks, we find the intersection of textual and somatic forms that shapes Wroth's anatomical poetics.

As a material formal feature, Wroth's extensive use of blank space aligns with her exploration of the expressive capacity of "whiteness" throughout the manuscript. One might look, for example, to the third sonnet of the *corona* ("black desernd by white, / Error by truthe, and darknes knowne by light" [MS 84.6–7]), or the heightened attention to lightness and darkness in the poem that begins, "which should I better like of, day, or night" (MS 20). This sustained exploration of whiteness is especially concentrated in one of the manuscript's most oft-cited sonnets, "Like to the Indians, scorched with the sunne" (MS 75). In this poem, which focuses on skin color and the contrast between blackness and whiteness, Wroth introduces an important theorization of surfaces that we can then trace through the rest of the manuscript (Figure 3.1). "Like to the Indians" is a reflection on how internal desire becomes legible (that is, available for reading) through external bodily markers. The Indians "scorched with the sunne" are marked "in skin" with "blacknes" (MS 75.13; 5), while the poet, "us'd by love" (3), is instead "pale, and white" (6), unmarked and, therefore, unknown because of her "whiteness" (6). Lamenting that her sacrifice is "hid as worthles rite," Pamphilia initially seems to envy the black Indians through the poem's first three stanzas ("better are they who thus to blacknes runn" [MS 75.5]). In the closing stanza, however, we find the "fundamental urge for whiteness" that Kim F. Hall traces through the period's English sonnet cycles when Pamphilia voices her desire to wear her "blacknes" on the inside rather than the outside: "lett mee weare the marke . . . in hart as they in skin" (MS 75.12–13).[65]

There are two interconnected impulses at work in this sonnet that help us understand the link between Wroth's representation of Pamphilia and her inscriptive work on the poetic page. First, both blackness and whiteness signify, but in importantly different ways. Blackness, Wroth asserts, is a literal mark—an external, highly visible sign—of the Indians' adoration, caused by the object of their desire. Drawing on the prevailing climatological theory of skin color difference,[66] Wroth describes the Indians as "scorched" by the presence of their beloved ("the sunn, which they doe as theyr God adore" [MS 75.2]), and they are blackened (marked) in the act of desiring it. In contrast, the poet's object of desire is absent and, therefore, she cannot be marked by it (she is "pale, and white" [MS 75.7]). But her whiteness signifies all the same, a mark of how she is "us'd by love" (MS 75.3), so Pamphilia, "like to the Indians," is marked by her desire. Whiteness and blackness both become signifiers, which allows the poet to maintain, as Hall argues, "her racialized judgment of the Indians," while still employing the language of racecraft as a poetic conceit.[67] Second, Pamphilia portrays her heart as a surface to be marked just as the Indians' skin is a surface to be marked. The closing stanza's declaration equates heart and skin as equally inscriptive surfaces: "Then lett mee weare the marke of Cupids might, / in hart as they

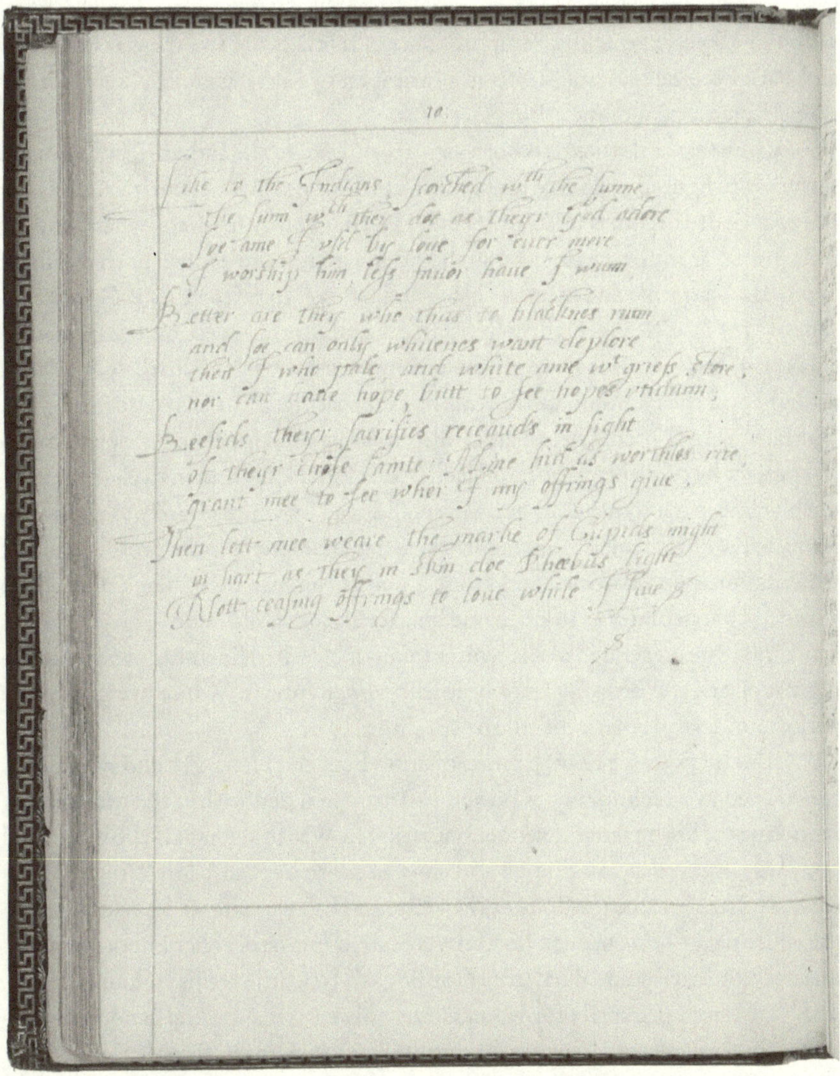

Figure 3.1. "Like to the Indians, scorched with the sunne" (MS 75). Lady Mary Wroth, *Pamphilia to Amphilanthus* (c. 1614–1621), sig. 39v. Call #: MS V.a.104. Image #: 23465. Folger Shakespeare Library.

in skin doe Phœbus light" (MS 75.12–13). Notably, both surfaces are not marked *on*, but marked *in*, recalling Pamphilia's act of tree carving and the way both skin and bark become material substrates inscribed by love's wounds ("keepe in thy skin this testament of me" [I.92]).

Whiteness and blackness both signify in "Like to the Indians," and somatic surfaces are figured as layered and multiple. If this poem is a theorization of how internal desire becomes legible through external bodily markers, what happens in the shift from body to page as the poet attempts to render both kinds of markings—external and internal, black and white—in text? Miles P. Grier's concept of "inkface" is especially helpful in relation to this question because, with this term, Grier unites in complex and nuanced ways "the histories of racial thought and the technologies of reading and writing."[68] In Grier's formulation, "inkface" conscripted African and Indigenous peoples into the European imperial project by tying black skin to black ink, symbolically and materially bounding blackness as a signifier to be written and read by white experts. Building on art historian Michel Pastoureau's work, Grier characterizes the proliferation of print as one of transformation from a polychromatic to a "black and white world,"[69] particularly as ink became blacker in tone. Whiteness, Grier argues, was similarly exaggerated as the whiter tones of bleached linen-rag paper increasingly replaced the browner hues of medieval parchment. White became "mute background, while black did all the speaking."[70]

Grier helps us see how the interrelation between blank ink and white page functioned as a technology of power, instrumentalized within the intersecting economies of both race and gender operating in Wroth's poem. If the blackened body was "figured as ink-dipped and thus, *made* to be read,"[71] in Wroth's "Like to the Indians" we see the body of the white English woman—and, by extension, the white page—*also* made to be read. As Hall illuminates in her important work on the "fair texts" and "dark ladies" of English lyric, the terms "white," "light," and "fair," used primarily for women, not only carried "specific gendered judgments" but also marked "a desire for a stable European linguistic order."[72] Wroth's writing displays an active interest in the political capital of whiteness, as well as in England's contemporary projects of nation and empire building.[73] Simultaneously, she is hyperaware of the vulnerability of the "fair" woman of Petrarchan poetic tradition and she consistently calls attention to whiteness as a mark, a tension that speaks directly to, as Bernadette Andrea puts it, "sixteenth- and seventeenth-century English women's contradictory position as both agents of imperialism *and* subjects of patriarchy."[74] In one later sonnet, Wroth's condensed syntax equates "purest white" with "staine": "noe staine is ther butt pure, as purest white" (MS 84.2). In another, Pamphilia's life "spills" from her "hartles" body as a sign of her "pure, and spottles love" (MS 68.8; 13). Wroth's poems consistently fuse the race-making technologies of reading and writing with the inscriptive conditions of Petrarchan conventions within which "fair" women's bodies are written into narratives of inaccessibility, penetrability, and purity. For Wroth, the fair woman's body ("pale, and white")—as lover, as beloved, as writer of her own body, as reader of the Indians' bodies, as body to be read—is a body subject to inscription and highly attuned to the penetrative possibilities of surface, to the erotic logics of wounding as violation that power the Petrarchan love poetry mode. As Pamphilia takes on whiteness as the "marke" both of and on her interior in this sonnet, Wroth cues her readers to attend more closely to the white space of the page around and within the poem—to the interaction, that is, between black ink and white space that forms her representation of Pamphilia.

Let us recall an important detail from William Faithorne's description of the copperplate engraving process. When wiping clean the copperplate before printing, the artisan must leave "neither black nor spot" on the plate so that the unengraved parts of the plate will be "as white as the margin of the paper, when the sheet is printed."[75] Framed in such a way, the technique of copperplate engraving relies on the interaction—really, the stark distinction—between "black" and "white," which here, as in Grier's argument, manifest in ink and paper respectively. Both black "spot[s]" and white page are required in order to render the engraved anatomical form, and it is this key surface dynamic—the body in parts formed through the interaction between ink and paper—that allows us to understand Wroth's epistemological study of surface in her manuscript poetry. Particularly in the more aesthetically rich anatomical treatises of the period, illustrators used the interactions between engraved lines and white spaces to theorize surface and to foreground the significance of surface (and surface penetration) to anatomical study. As Elkins points out, there is a history of medical illustration in which anatomical parts are "seen with an artist's eye, which is to say that they are seen too well, with useless precision."[76] This is nowhere more evident than in *Anatomia Humani Corporis* (1685), a large and especially striking anatomical treatise by anatomist Govard Bidloo and artist Gerard de Lairesse. Bidloo and Lairesse published *Anatomia Humani Corporis* (hereafter, *Anatomia*) toward the end of the seventeenth century, as anatomists increasingly turned away from texts to instead rely on wax models and preserved specimens for their study.[77] *Anatomia* is a persuasive invective against this trend. The large folio (roughly 14″ × 23″), replete with dramatic, full-page plates, makes a tour de force argument for the advantages of paper and ink in representing knowledge of the human body—the superior capacity of paper, that is, as a medium for anatomical preservation and study.[78]

The fastidious details of Bidloo and Lairesse's text, rather than a "useless precision," make an argument about the *forming* work of anatomical study that is especially useful for scholars of Renaissance poetics. Most crucially for my argument here, the *Anatomia*'s engravings repeatedly betray Bidloo and Lairesse's fascination with skin and surfaces.[79] The text's first dissection plate, for instance, is a dissection of the skin itself, with fifteen different illustrations of the "parts" of skin (Figure 3.2). The thumb is the only immediately recognizable part, but Bidloo and Lairesse position it in the bottom third of the page, prioritizing the dissected skin components within the page's visual hierarchy. The plate depicts several skin hairs as viewed through a microscope and, more predominantly, a collection of skin fragments, some of which lay flat on their own and others that are stretched and held with small pins. Bidloo's corresponding index explains that we are looking at multiple layers of the skin (hypodermis, dermis, and epidermis) and different sections of skin (arm, back, bottom of the foot) produced using various techniques of preparation and observation. The table's first three illustrations (numbered 1.1, 1.2, and 1.3), for instance, depict the skin's outermost layer ("*depicta est cuticulae, ab exteriore*"), which William Cowper, in his English edition of Bidloo's atlas, translates as the "common covering to the whole surface of the skin."[80]

Bidloo and Lairesse's skin plate, essentially an anatomization of the sense of touch, reflects what historian of science Dániel Margócsy has called Bidloo's "complex paper epistemology."[81] Margócsy defines this concept as the anatomist's investment in "the visualizing power of . . . paper," its ability to "reflect[ ] the variability of nature" and "accommodate different methods of visualization."[82] Building on Margócsy's analysis, we must also account for the important function of copperplate engraving in Bidloo and Lairesse's anatomical plates. Bidloo and Lairesse orient us to their anatomical study—this is the book's first dissection plate, remember—by drawing heightened attention to the surface on which, and practices through which, their anatomical knowledge is rendered. The skin plate invites us to consider the page as a conceptual skin and material substrate that retains form, function, and meaning on its own terms even when marked or covered with ink. Like the "holes, cuts, scrapes, tears, or stretch marks" on medieval parchment wherein Sarah Kay reads "traces of the living animal," the surface of Bidloo's page is made hyper-visible by its content. This is a page surface imprinted with representations of the body's surface.[83] Later plates in the *Anatomia* reinforce such a reading, particularly when Bidloo and Lairesse use paper as

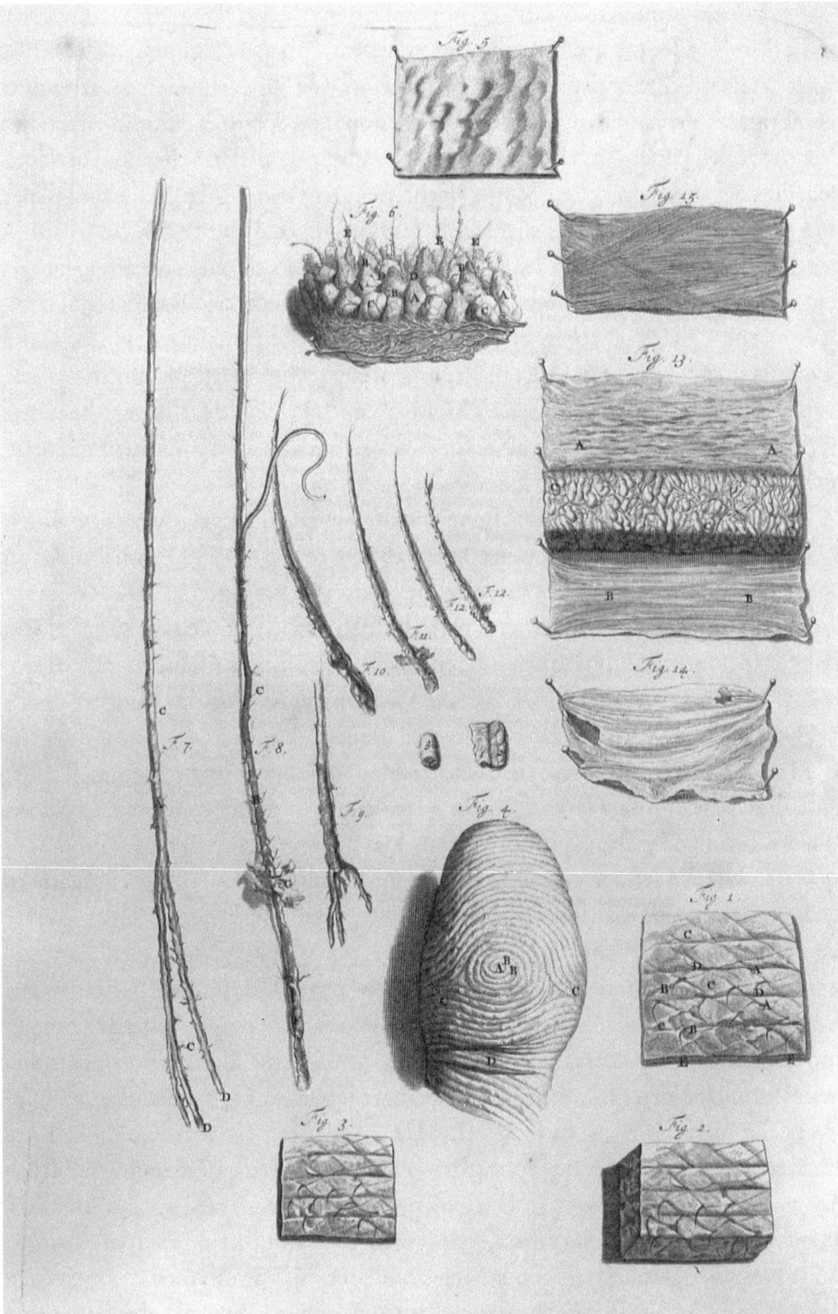

Figure 3.2. Dissection of the skin. Govard Bidloo and Gerard de Lairesse, *Anatomia humani corporis* (Amsterdam, 1685), T. 4, Wellcome Collection, EPB/F/2051.

an anatomical prop. Amid a dissection tableau on the placenta, for instance, we find a thin bladder membrane stretched over a roll of paper (Figure 3.3). The rolled paper creates a sense of depth on the page surface as it slides through the stretched membrane to evoke the membrane's dimensionality. As the membrane stretches over the rolled paper—so thin it is almost membranous itself—the sense of depth becomes obscured on the right side of the representation. Membrane and paper fuse into a two-dimensional surface that is both textual and fleshly. In this slippage, Bidloo and Lairesse construct the anatomical knowledge on offer in their text through the relations among content, form, *and* method. The material practices of the dissection room and, most crucially, the transformation of those practices into text become part of the anatomical work on display in this book. As a result of Bidloo and Lairesse's highly material thinking, the page becomes not only a surface on which to write (or engrave) but also a dynamic material technology that works with and on bodies.

Bidloo and Lairesse's use of paper aligns with the myriad ways in which Wroth foregrounds paper's materiality in her texts. As Elizabeth Hodgson notes, "tangibility" is a "key element of [poetry's] presence in *Urania*," with sonnets "constantly being burned, torn, hidden, erased, copied, or sung."[84] Recall, for instance, Pamphilia's first poem, simultaneously permanent and ephemeral within the book's pages—burn after (writing and) reading. Another of the romance's poems, penned by a character named Perissus, becomes a material agent as it "suffer[s] itselfe patiently to receive the discovering of so much of it" (I.2). One pair of lovers in the romance are caught by a jealous husband because of a wayward paper ("most unfortunately lost") on which are written "lines [of] love" (I.187). And for a particularly potent example, we can return again to Melasinda and Ollorandus's subplot. At one point, Melasinda receives a letter from Ollorandus, exclaiming "O deare Paper!" as she kisses and reads (and rereads) the "words written by the best of men" (I.272). Resigning herself to the fact that the letter must be destroyed for the lovers' "safetie," she burns the paper and then carefully places the ashes "in a daintie Cabinet" along with a poem she writes while the letter is burning ("witnessing the sorrow for the burning, and the vowes she made to them burned" [I.272]). As readers, we never read the lines written by Ollorandus, a key contrast to Pamphilia's first poem (withheld from Amphilanthus but included in *Urania*'s pages), but Wroth accentuates Melasinda's excessive act of preservation, reinforced by a kind of meta-inscription as Wroth tells her own readers about Melasinda's actions. Wroth includes the poem that Melasinda writes while the letter is burning, in which she pleads, "blame me not dearest lines, / that with loves flames your blacknesse twines" (I.273). This scene draws heightened material attention to a poem we cannot read, even as it invokes the discourse of surfaces I traced in "Like to the Indians" (MS 75) in Melasinda's reference to the "blacknesse" of the poem's lines. It is often not the content of the poems so much as the papers—those material technologies of writing and reading—that hold significance throughout the *Urania*, bringing us back to the material conditions of Wroth's own poetic practice.

Wroth's Paper Epistemology

The spaces around and between Wroth's poems in the Folger manuscript are not just white, they are also *blank*, defined both as un-inked paper and, recalling Laurie Maguire's definition, a textual feature that calls attention to what is not there—in Wroth's case, the blazon as the most conspicuously absent piece of her poetry collection. Notably, even though Wroth does not include a conventional blazon of either Amphilanthus or Pamphilia in her manuscript poetry, the *language* of the blazon is scattered throughout her manuscript's pages. The root word "part," for instance, is central to Wroth's poetic lexicon. If, as David Hillman and Carla Mazzio argue, individual parts of the body and the lexical network of parts gained "more semiotic complexity than they had ever had before" in early modern England, then Wroth's extensive use of *part* deserves our attention as a kind of material element (inked letters) that marks the blazon tradition (and its absence) in her poetry.[85] We find a particularly striking example in one of the manuscript's early songs as Pamphilia directly addresses a departing Amphilanthus, pleading, "lett us nott thus parted bee / love, and absence ne're agree" (MS 28.5–6). Faced with his inevitable leave-taking, she offers her heart as a parting gift: "soe in part, wee shall nott part / though wee absent bee" (MS 28.13–14). These lines play on the double meaning of "part," evoking both the act of parting and the act of partitioning as Pamphilia offers a body part in order to stay present with her absent beloved. Simultaneously, Wroth explores both connotations of "part" in relation to "absence," emphasizing the fact that, although this poem offers a brief glimpse of Pamphilia's beloved (one of the only poems in the manuscript that does), he is present only in the act of becoming absent. In its closing lines, Pamphilia gives voice to the one left behind in John Donne's "A Valediction Forbidding Mourning" when she wonders, "Butt can I live having lost / chiefest part of mee?" (MS 28.19–20).

We might expect this turn inward, a textbook instance of what Jeffrey Masten calls Pamphilia's "interior discourse,"[86] but, more unexpectedly, with

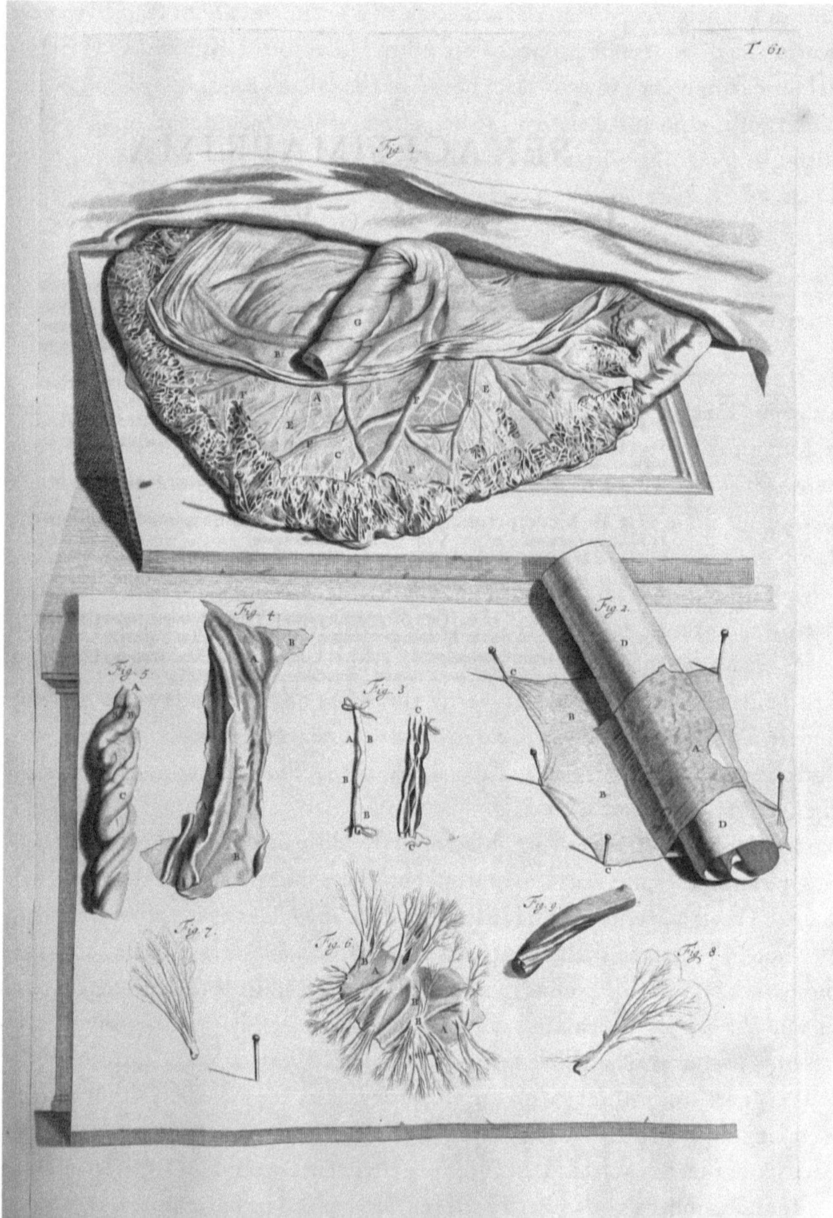

Figure 3.3. Dissection of placenta and membrane detail. Govard Bidloo and Gerard de Lairesse, *Anatomia humani corporis* (Amsterdam, 1685), T. 61, Wellcome Collection, EPB/F/2051.

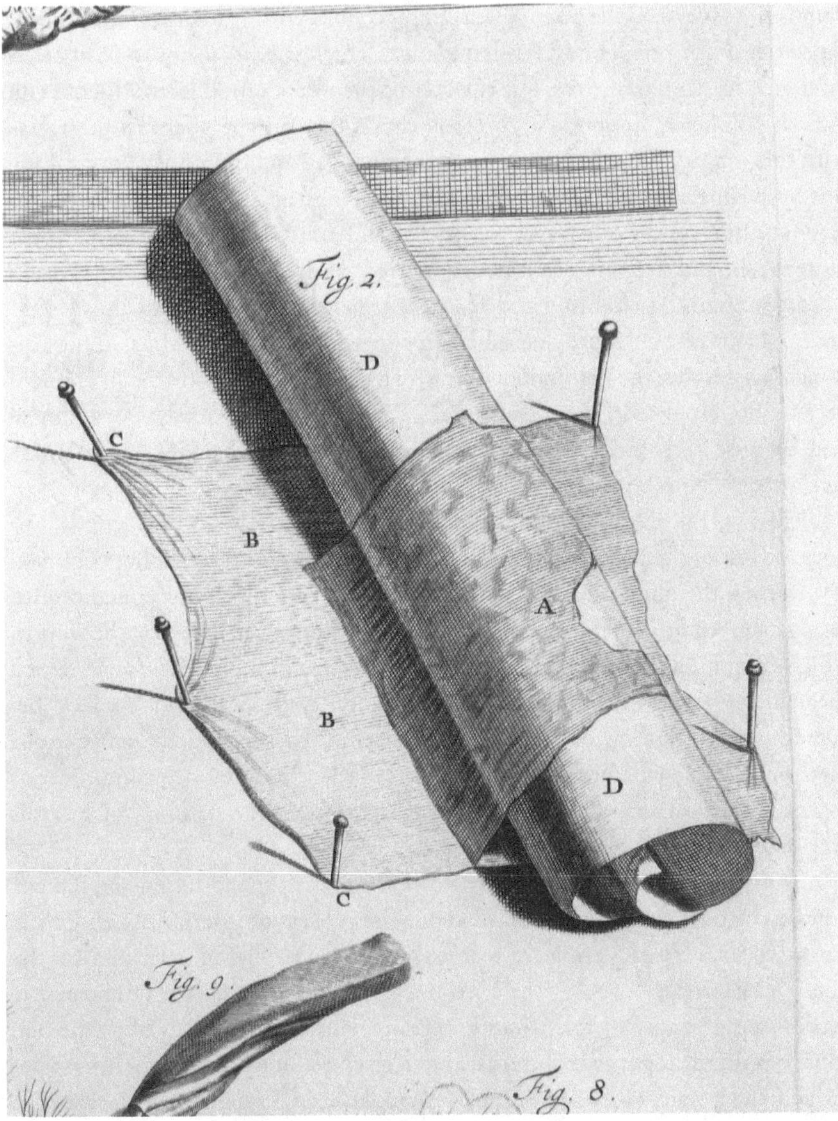
Fig. 2.
D
C
B
A
B
D
C
Fig. 9.
Fig. 8.

Pamphilia's inquiry, the poem's wordplay collides with the actual, anatomical capacities of the poet's body. At first glance, "chiefest part" seems to be another reference to Amphilanthus, but the rest of the poem's final stanza focuses on Pamphilia's heart, her body's "chiefest part." Opening the poem's final stanza with this inquiry, Wroth reveals her interest not in Amphilanthus's leave-taking (his becoming absent), but rather in how the language of "part" and "absence" comes to bear on the poem's representation of Pamphilia's body. In Pamphilia's plea that she and Amphilanthus not be "parted" and, therefore, "absent," Wroth's wordplay aligns partitioning and absence, anticipating some of the most foundational feminist critiques of the blazon convention. Cynthia Marshall, for instance, argues that the Petrarchan narrative figures a female body that is "present because doubly absent," first, because she cruelly rejects the male poet and, second, because his subsequent poetic anatomization marks her lack, or the poet's lack of her.[87] Notably, Pamphilia's closing inquiry—"butt can I live having lost / chiefest part of mee?"—does not involve a rhetorical display of her body, nor does the rest of the poem include any conventional blazonic metaphors or imagery. Rather, the question signals Wroth's interest in a much more nuanced interrogation regarding female embodiment and inscription, invoking the ways in which Pamphilia's character challenges blazon conventions in *Urania*. Reframed through the poem's final question, Wroth's wordplay on "part" and "absence" becomes a confrontation of the kind of embodiment such poetic conventions force onto the female body, an exploration of the blazon mode's simultaneous work to partition and efface. In other words, this poem is not a blazon, but it is very much about the blazon.

The conventional blazon poem circulating in early modern England did not require the poet to think very hard about the process of inscription, about the narrative into which primarily women's bodies were being inscribed by the poet's "scattered rhymes."[88] The blazon was, in many ways, a formulaic poetic mode—a plug-and-play form—and there are countless examples of poetic blazons that simply repurpose and rearrange a given set of tropes. Scholars have described the poems of Philip Sidney's *Astrophil and Stella*, a sonnet sequence steeped in the blazon mode, as "serial forms," iterating one after another on "the active experience of desire."[89] Philip Edwards's *The Mysteries of Love and Eloquence, or, the Arts of Wooing and Complementing* (1685) contains extensive material to help early modern readers compose their own blazons, including a catalog of corporeal features with corresponding lists of descriptive adjectives. The entry for "hair," for example, offers no less than thirty-nine possible descriptors, including *golden*, *radiant*, *curling*, *careless*, *silken*, *disheveled*, *ambrosial*, and (my personal favorite) *crisped*.[90] One could imagine the blazon poems in Thomas Watson's *Hekatompathia* (1582) composed with such a list in hand when we encounter lines like, "her yellow locks exceed the beaten gold" and "her breast transparent is, like crystal rock."[91]

The more complicated blazons of the early modern period, and the ones that have received the most critical attention, distinguish themselves by calling attention to the formula. Shakespeare's Sonnet 130, a familiar example of the anti-blazon, is distinctive in part because it exposes the conventional mechanics of the blazon mode. "My mistress' eyes are nothing like the sun," the poet begins, simultaneously citing and upending a particularly well-worn blazon analogy (comparing eyes to sun, or stars). Wielding the conditional "if" through the poem's opening, Shakespeare prompts us to notice a comparison mode that relies on a series of subjective ("if snow be white, why then her breasts are dun"), even ridiculous ("if hairs be wires, black wires grow on her head"), descriptive logics.[92] Shakespeare takes the tradition of the anti-blazon (going on to insist that there are no "roses" in his beloved's "cheeks" and her breath "reek[s]" rather than "delight[s]") and uses it to pull back the curtain ever so slightly on the inscriptive function of this poetic mode. The poem's closing lines ultimately characterize the blazon as a poetic mode that (problematically) transforms "mistress" into "goddess": "I grant I never saw a goddess go; / my mistress, when she walks, treads on the ground."[93]

If Shakespeare is employing conventional tools (in this case, blazonic language and the form's conspicuous tropes) to expose the inner workings of this poetic form, Wroth is dismantling those tools to study the *how* and *why* of their engineering. Wroth's consistent use of the language of parts draws attention to the acts of partitioning, fragmenting, and effacing ("I, ame that hartles trunk of harts depart," Pamphilia asserts in another sonnet [MS 48.7]).[94] But such language is only one trace of a much more extensive material exploration of the blazon's inscriptive work. To borrow Dániel Margócsy's useful concept, I find Wroth's most innovative critique of and response to the blazon in her "paper epistemology"—the way she manages the mise-en-page and the material conditions of her manuscript.[95] If we use the complex theorization of surface at work in anatomical engravings to read the blank spaces in Wroth's poetry manuscript, her luxurious use of page space becomes a formal technique that harnesses the material affordances of paper in order to amplify and repurpose the inscriptive project of the Petrarchan blazon.

In another of Wroth's poems on parts, a poem that begins, "Eyes, can you tell mee wher my hart remaines?" (MS 96), the lexicon of parts works together with the mise-en-page to articulate the poem's critique. First imagining that her heart is "lodg[ed]" with her beloved, Pamphilia asks if her eyes have seen it in Amphilanthus's "lovely eyes" or "blessed brest." There is a playfulness in the first two quatrains as she speculates on the heart's pleasure if it were to inhabit those spaces. But, by line 9, the poet's inquiry becomes more desperate.

> Iff neither wher? wher lives itt? wher abides
> this careles sprite who from mee closely slides,
> and hartles leavs mee? O, alas I knowe (MS 96.9–11)

The poet's frantic questioning ("iff neither wher? wher lives itt?") comes to an abrupt end with her laconic self-awareness ("O, alas I knowe"), which sits alone at the end of a line, absent any closing punctuation, and followed by blank space isolating the poem's final tercet. This unsettling turn snuffs out the playful tone of the poem's opening, giving way to Pamphilia's wretched conclusion: "Thus while I thought itt sor'de, itt creeps beelow" (MS 96.14).

This manuscript sonnet, which is not included in the printed *Pamphilia* sequence or either part of the *Urania*, is full of body parts. The addressee is a body part (the poet's eyes) that ultimately fails the poet in her quest for her heart. It is not her eyes but her knowing ("o, alas I knowe") that comes to her aide. "Eyes" is the first word at the top of the page, and Pamphilia's heart, "creep[ing] beelow" ends the poem, so the mise-en-page participates in the scattering of body parts across the sonnet (Amphilanthus's "lovely eyes" and "blessed brest" are also here, in the first and second quatrains). The page itself is quite typical of the sonnet pages in Wroth's manuscript (Figure 3.4), with hand-drawn page margins at top, bottom, and left-hand side. This is the only poem on the page and it fills a little over half of the white space within the drawn margins. The poem is unnumbered, headed simply "Sonnet $" at the top of the page, and closed with a semicolon, short slash, and fermesse sign—the slashed "S" Wroth uses throughout the manuscript, most frequently at the end of poems.[96] The first eleven lines of the sonnet are fairly cramped, emphasizing the prominent white space after the volta, "O, alas I knowe," which allows the white space to graphically enact Pamphilia's suspended realization. The fermesse sign, too, is significant here as a mark of rupture that "introduces a temporal pause and a material gap in the manuscript's arrangement."[97] As Megan Heffernan argues, it functions as a "technolog[y] of textual organization that makes possible an evolving, patchwork "arrangement of poems," a network of lyrics (a lyric sequence in movable, animate parts, perhaps) that "consistently unfold[s] in moments of textual making and remaking."[98]

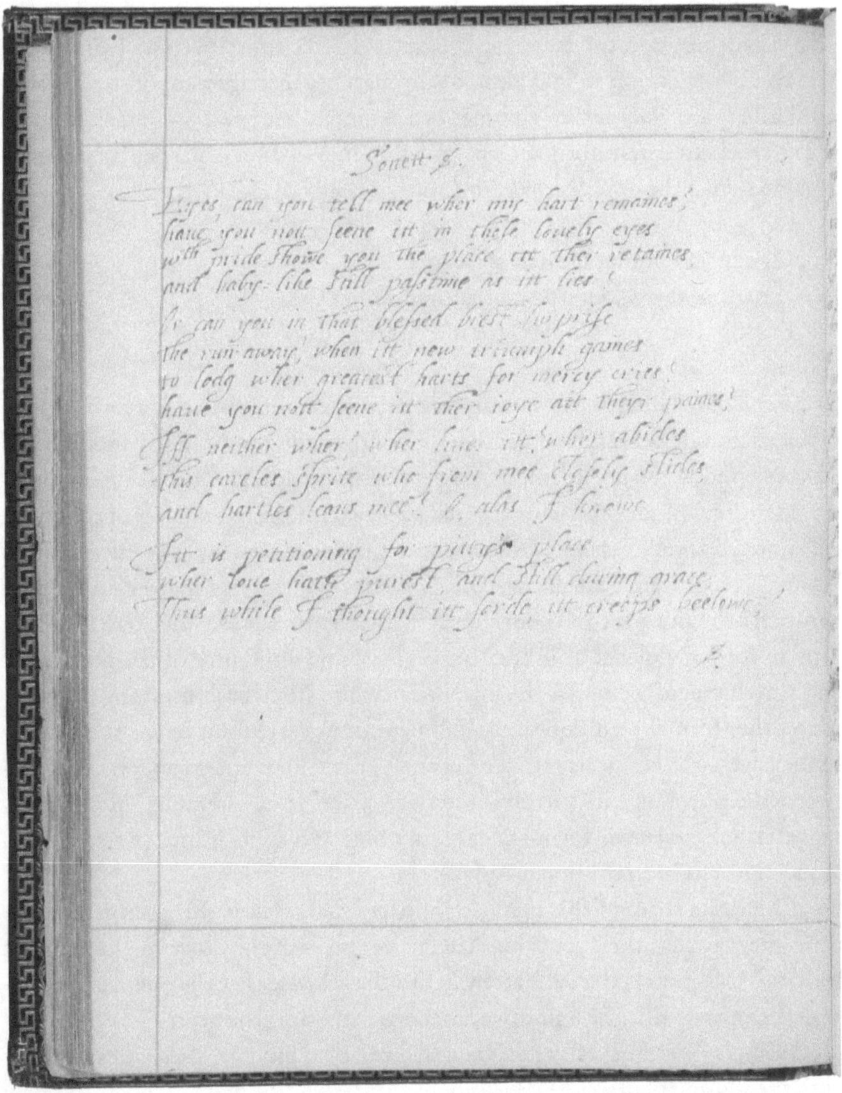

Sonett 8.

Eyes, can you tell mee wher my hart remaines?
have you nott seene itt in these lovely eyes
with pride showe you the place itt ther retaines
and baby-like still pastime as itt lies?
Or can you in that blessed brest surprise
the run-away, when itt new triumph gaines
to lodg wher greatest harts for mercy cries?
have you nott seene itt ther ioye att theyr paines?
Itt neither wher? wher lives itt, wher abides
this careles sprite who from mee sleyly slides
and hartles leaves mee? O, alas I knowe
Itt is petitioning for pittys place
wher love hath purest, and still during grace,
Thus while I thought itt ferde, itt creeps beelowe!

Figure 3.4. "Eyes, can you tell mee wher my hart remaines?" (MS 96). Lady Mary Wroth, *Pamphilia to Amphilanthus* (c. 1614–1621), sig. 47v. Call #: MS V.a.104. Image #: 23481. Folger Shakespeare Library.

These lines vibrate with the presence of what is absent: Pamphilia's heart, Amphilanthus, and the traditional blazon. In her construction of the poem, Wroth calls attention to that absence through her management of the interlinear white space. She uses enjambment throughout her poetry—in fact, it is one of her most consistent formal features—and in MS 96 it is particularly overexaggerated, like the double enjambment here:[99]

> Or can you in that blessed brest surprise
> the run-away? when itt new triumph gaines
> to lodg wher greatest harts for mercy cries? (MS 96.5–7)

These enjambed lines revolve around the medial caesura in line 6, a formal technique that underscores the enjambment's partitioning work. Early modern poets conceived of the poetic line as a tight visual and aural unit, and enjambment is a poetic technique precisely designed to play with the boundaries of that unit. In Samuel Daniel's *Defense of Rhyme*, he defines enjambment as a boundary violation, a technique that allows the "violence" of poetic "matter" to "break through" the line's formal container.[100] The poem's subject "matter" inflicts a kind of formal violence as it breaks the visual and aural unit of the poetic line and simultaneously charges the interlinear white space with meaning. More so than other formal techniques, enjambment draws attention to the materiality of the page—the connective tissue between poetic lines and the interactions between the poem and its surrounding page space. It is, as Daniel's description suggests, a *fragmenting* form—it takes a whole and splits it into parts, blazonesque in its effect on the poem and page.

The white space of this manuscript page "marks" acts of fragmentation, in the same way that the poet is marked "pale, and white" by love in "Like to the Indians" (MS 75). And it calls attention to the absence that the poem itself tries so desperately to fill. The figurative language describing the heart's "joye" in Amphilanthus's "lovely eyes" and "blessed breast" temporarily replaces absence with the illusion of presence. The light playfulness of the first two quatrains bely the wretchedness of the poem's conclusion, as the poet imagines her heart's "pride" and "triumph" at being "lodg[ed]" in Amphilanthus's breast, a place "wher greatest harts" cry "for mercy" (MS 96.3; 6–7). The black marks of ink distract and, at first, elide the white space underneath, within, and around the poem's figurative language. The poem's representative work, we might say, masks its inscriptive work. The strikingly fragmented first tercet, though, signals a breaking point. As the poet's search for her absent heart becomes more desperate, the poetic lines themselves fragment and the white space—the space of absence—asserts itself.

Poems like MS 96 expose Wroth's paper epistemology at the level of the page and demonstrate how that paper epistemology augments her theorization of the textual representation of Pamphilia's body. Such poems (MS 8, MS 28, and MS 48 among them) deconstruct blazonic language and employ formal techniques that animate the interactions between poem and page, transforming the page itself into what Jerome J. McGann calls a "bibliographic feature": "nonlinguistic forms" that "embody ideas and conceptual content."[101] There is also, though, a paper epistemology functioning at the level of the manuscript itself, perhaps even across the poetry and *Urania* manuscripts as Wroth arranges, rearranges, and curates her representation of Pamphilia. Wroth's own poetic craft, to some extent, relied on blanks. There is evidence in the manuscripts of both the first and second parts of *Urania* that Wroth composed the romance by leaving blank spaces for poetry. Flipping through Wroth's *Urania* manuscripts, we encounter blank spaces between sections of prose (Figure 3.5) or instances where a poem is cramped into a small space, suggesting that Wroth originally did not leave enough blank space for the poem. The poems in *Urania*, then, start as blank, negative space held around and within the inked prose—as both not there and there, almost the inverse of Pamphilia's first poem in the romance (first there, then not there). As the poetic persona powering Wroth's poetry collection, Pamphilia's material poetics provide an important framework for reading Wroth's manuscripts, particularly in the moments when this poetic persona bodies forth, moments when Wroth prompts us to notice her own work of formation and her own investment in the materiality of poetic craft.

We find one such moment at the end of "How like a fire doth love increase in mee" (MS 55), a sonnet that ends with Pamphilia's signature (one of two instances of this signature throughout the manuscript). In this particularly violent representation of Pamphilia (Rosalind Smith actually describes this poem as "an anti-blazon of the speaker's own body"),[102] Wroth employs one of her most persistent tropes: the burning heart fueled by the fires of love.

How like a fire doth love increase in mee,
        the longer that itt lasts, the stronger still,
        the greater purer, brighter, and doth fill
        noe eye with wunder more, then hopes still bee

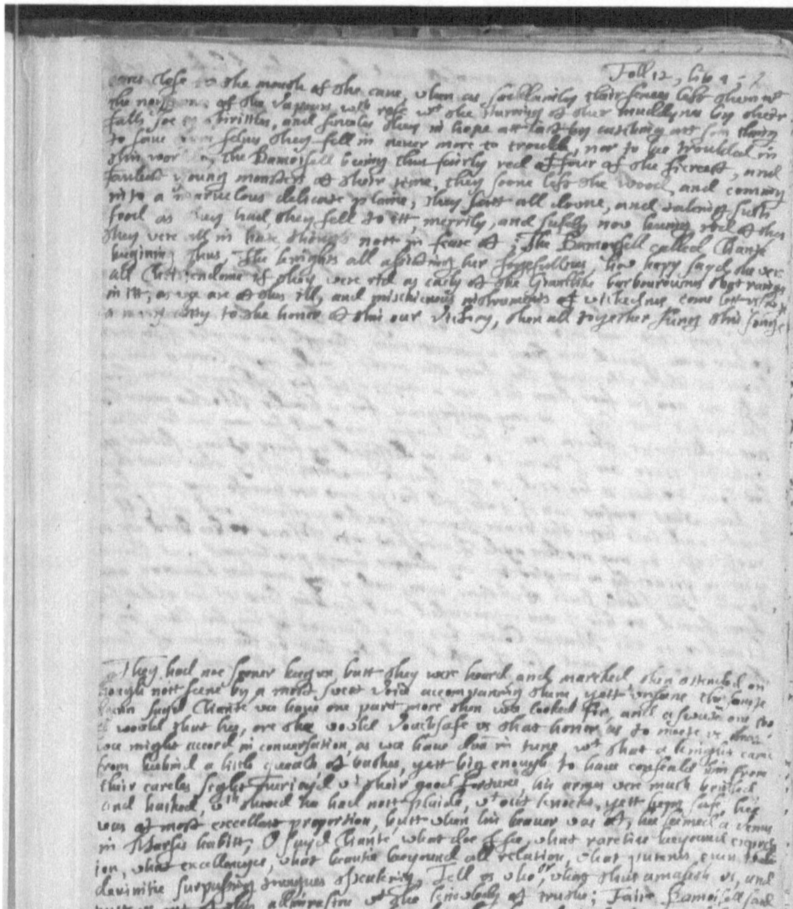

Figure 3.5. Lady Mary Wroth, *The first and secound booke of the secound part of the Countess of Montgomery's Urania* [vol. 1] (c. 1621). Call #: VAULT Case MS folio Y 1565 .W95. Image #: 049. Newberry Library.

bred in my brest, wher fires of love are free
to use that part to theyr best pleasing will,
and now impossible itt is to kill
the heat soe great wher Love his strength doth see.

Mine eyes can scarce sustaine the flames my hart
doth trust in them my longings to impart,
and languishingly strive to show my love;

My breath nott able is to breathe least part
of that increasing fuell of my smart;
yett love I will till I butt ashes prove (MS 55)

In this sonnet, Wroth stages a violent conflict between the demands of Petrarchan lyric conventions and the physical limitations of Pamphilia's body. Love's "fire" swells "greater purer, brighter" in Pamphilia's "breast" and, faced with a "heat" that is "impossible . . . to kill," Pamphilia embraces love's flames and declares that she will "love" until her body turns to "ashes." As she describes the fire's effects on her body, Pamphilia represents her body as a catalog of parts (breast, eyes, breath, heart), reinforcing this anatomization with the repetition of "part" in lines 6 and 12. The enjambment between lines 9 and 10 also calls attention to another form of bodily fragmentation—a disconnect between her heart and her eyes. Pamphilia's heart tasks her eyes with "impart[ing]" her "longings," but her eyes "languishingly strive" to make that passion known. "Mine eyes" and "my hart" are together on the same line, but the hard and soft enjambment in this tercet enacts the disruption between her heart's desire to make known ("impart") and her eyes' inability to "show."

A reading of this sonnet focused just on the poem itself might consider the "dynamic of consumption and renewal" that unfolds within these lines.[103] Smith, along with Madeline Bassnett, connects this dynamic to the phoenix imagery associated with Philip and Mary Sidney, while Heather Dubrow links the poem more closely to its Petrarchan context and reads it as a "repetition of desire" aligned with "the masculine and often masculinist discourses associated with the *Rime Sparse*."[104] Pamphilia's assertion of desire at the end of the poem, however, is a particularly extreme expression of the poet's effort to make her internal desire legible—"love I will till I butt ashes prove." Not only will the fires of love eventually consume her body; they will simultaneously transform her body into an ashy manifestation of her "exemplary constancy."[105] This transformation of body to ashes returns us to Wroth's heightened attention to paper in *Urania* (Pamphilia's first poem and Melasinda's poem "witnessing the sorrow for the burning" [I.272]) and reminds us that the materiality of Wroth's poetics always need to be part of our close-reading work. If we pay attention to the page around sonnet MS 55, we find that Wroth's paper epistemology undercuts Pamphilia's closing assertion. Wroth again cues us to read beyond the bounds of the poem itself with her extensive use of enjambment. In "How like a fire," Wroth uses enjambment, along with the poem's crowded syntax, to formally enact the effect of love's fire on the poet's bodily boundaries. Poem and page together constitute Wroth's exploration of surface. Conceptually, love's fire strains against the bounds of Pamphilia's physical body; formally, the sonnet strains against the bounds of its lines and meter. In the third line, for instance, Wroth omits the conjunction from the list of adjectives—"greater purer, brighter"—allowing the description to run and swell like the fires. Line 3 is also the first of three consecutive enjambed lines, which draw the interlinear white space into the fire's swelling action. Lines 9 and 10 are also syntactically overtaxed (lacking even a natural caesura), which exaggerates the violence of the poetic matter "break[ing] through" the line's formal container.[106] Consequently, the first tercet's last line—"and languishingly strive to show my love"—feels particularly fragmented, a truncated clause far removed from its referent ("mine eyes") at the tercet's opening.

This sonnet, particularly in its final lines, is a poem about *knowing*. Wroth marks this theme with the part-root word "impart," drawing on the term's double connotation, both "to make known, tell, relate" and "to distribute, divide, deal out."[107] On this particular page, let us recall, Wroth's act of *imparting* knowledge about Pamphilia to her reader includes Pamphilia's signature, here surrounded by four of Wroth's distinctive fermesse marks (Figure 3.6). Scholars often read this page as a natural breaking point in the Folger manuscript, one of three "distinct endings" that contributes to the manuscript's "patchwork" arrangement.[108] After this sonnet and Pamphilia's signature, there is a completely blank page before the manuscript continues with another section of poems that again ends with Pamphilia's signature (MS 110). The manuscript's final section includes seven additional poems, one a thirty-nine-stanza pastoral that ends with a flourish mark, the same mark we then find at the end of the manuscript's final poem, "I, who doe feele the highest part of griefe" (MS 117). Beyond the manuscript's three possible endings, Gavin Alexander reads even more frequent "breaks and terminations" throughout the manuscript in Wroth's use of the fermesse, her erratic numbering system, and her inconsistent use of poem headings and catchwords.[109] The manuscript's patchwork construction, we might consider,

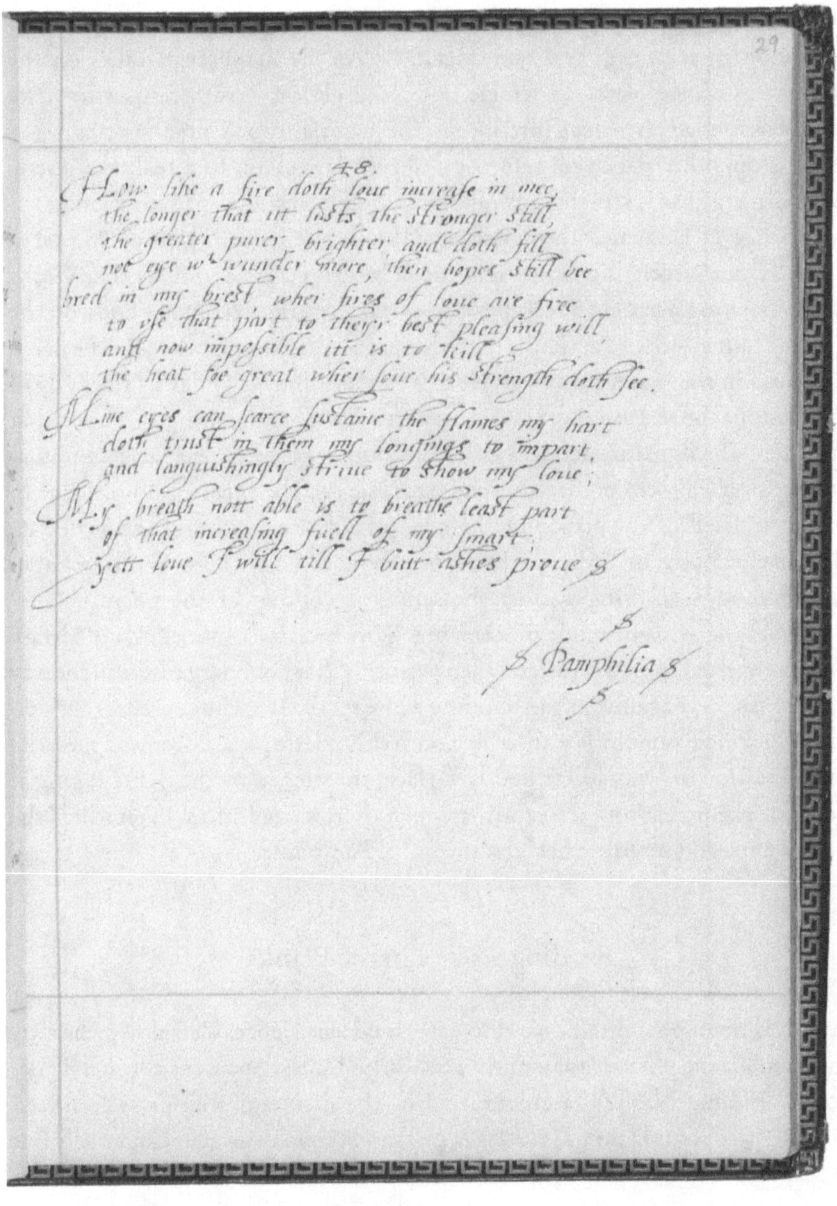

Figure 3.6. "How like a fire doth love increase in mee" (MS 55). Lady Mary Wroth, *Pamphilia to Amphilanthus* (c. 1614–1621), sig. 29r. Call #: MS V.a.104. Image #: 23444. Folger Shakespeare Library.

turns Wroth's poetic text into an assembly of fragmented parts, a paper body "partitioned, arranged, and rearranged."[110] As the manuscript takes on the conditions of the blazoned female body, the blazon Wroth refuses to write representationally is instead inscribed in the materiality of her manuscript pages.

Through her paper epistemology, Wroth lays claim to a realm of poetic embodiment that is semantically unrepresentable, resistant to definition or appropriation by language. The signature at the end of "How like a fire" (MS 55) is, crucially, not Wroth's signature. It is instead a conspicuous reminder that Wroth is constructing a female poetic persona—representing a woman's body on the page. Wroth inscribes Pamphilia onto this page in multiple modes—in the poem itself and in the signature as a glyph that renders unusually fleshly the body of this fictional poet. Pamphilia, Wroth asserts, has a hand that signs, a hand that inscribes. The signature acts as a scribal encounter between textual (Pamphilia) and writing (Wroth) bodies that connects back to the theme of knowledge in the poem itself. If the blazon mode relies on a desire to control and penetrate the female body in order to know the female body, here Wroth's paper epistemology—working against the semantic content of the poem itself—harnesses that violent mode of knowing. Wroth crafts a sonnet about surface pressure exerted from within, and then performs that boundary threat in the way the poem strains against its surrounding white space. The white space, as another mode in which Pamphilia is inscribed onto this poetic page, becomes a material manifestation of Pamphilia's bodily surface, threatened by the desire raging at her body's center. Wroth transforms her manuscript page into skin on which she carves, cuts, and inscribes her "testament" of Pamphilia.

Reading Science in the Blanks

Wroth is not a poet often linked to early modern science. Certainly, when we think about the phrase "anatomical poetics," Edmund Spenser's corporeal allegories, Thomas Nashe's fascination with "the disintegrating or anatomized body,"[111] and John Donne's "erotics of salvation" seem easier associations.[112] But Wroth is, I want to insist, an anatomical poet. The scientific engagement in her writing is located elsewhere, not in the explicit language of dissection or the blazon as its poetic counterpart. It is located in a place so unfamiliar to our reading of anatomical poetics that we have not been able to see it, though it has been there all along in the blanks. In her engagement with the English Petrarchan tradition, she concentrates on the inscriptive logics underlying that tradition, logics that press women's bodies into narratives of penetration, fragmentation, and absence—underlying logics that have long connected this poetic tradition to the Renaissance anatomical archive. What, then, are the implications of reading Wroth as an anatomical poet? How might this new understanding of her epistemological project shift the way we approach and engage with her poetry? And what can she teach us about where and how we look for science in the archives—about the possibility of reading science in the blank spaces of the archives?

The answers to these questions turn us, finally, to Wroth's "crowne of Sonetts," undoubtedly the most famous section of her poetry collection. This group of fourteen sonnets, modeled on the Italian verse form, the *corona*, forms a circular unit linked by concatenation—the closing line of each sonnet serves as the opening line of the next. It is included in both the Folger manuscript and the printed *Pamphilia* collection, but the manuscript version will continue to be my focus. In the manuscript, the corona begins on a new page, set off with the heading, "A crowne of Sonetts dedicated to Love." Unlike the rest of the manuscript poems, Wroth writes the corona sonnets one after another—without page breaks or blank spaces—so the fourteen sonnets together span only eight pages (from sig. 43^r to 46^v), followed by a completely blank page. The overwhelming critical tendency is to read the crown and its central conceit of the "strang labourinth" as a metaphor for physic depth. Mary Moore's influential reading, for instance, tracks how the labyrinth offers a model of female interiority—a "maze of self" defined by "enclosure and complexity."[113] For Clare Kinney, the corona's labyrinth recasts and expands on the poetry's earlier "image of physic imprisonment," when Pamphilia imagines herself locked in a "strang . . . cage" (MS 69.11).[114] Such readings align with Jeffrey Masten's claim that Wroth's poems, "like the anomalous manuscript in which they are inscribed," "encode a withdrawal from circulation" that consequently resists "the inauthentic theatricality of stock Petrarchism."[115]

I have been arguing, however, that Wroth is a poet of surface, a poet deeply invested in acts of inscription. If the corona reads as a concentrated exploration of "inwardness" and "self-knowledge,"[116] a retreat into depth and interiority, it is one that functions like the anatomical *écorché* figure—a figure showing the muscles of the body without skin. The écorché, as Elizabeth D. Harvey notes, is "a body that is both all surface and all inside,"[117] a figure that invites closer investigation of the shifting boundaries and thresholds that constitute the body's surface. Bidloo and Lairesse, those anatomists of surface par excellence, provide an illuminating example in their striking series of four écorché plates in *Anatomia Humani Corporis* (Figures 3.7 and 3.8 portray the first and last plates in the

Anatomia's écorché series). These plates depict the successive dissection of the female back muscles, with each plate in turn removing another layer of muscle, membrane, and tissue to reveal what lies beneath. The technique of copperplate engraving again enhances the intense study of surface and flesh at work in these plates. Lairesse's highly textured crosshatching creates a rawness in the representation, as if each subsequent layer is being flayed as he engraves his copperplate. The dissected layers of skin, tissue, and muscle fold together (which are the surfaces and which the insides?), and each new inner layer becomes a new surface to be dissected for the subsequent plate.

I will return to Bidloo and Lairesse's plates momentarily, but first I want to spend a bit more time with Wroth's corona now that we have this framework of the écorché in hand. In the corona's concatenation, Wroth experiments with a formal technique that functions like the anatomical écorché. Concatenation accentuates, even as it weakens, the thresholds and folds between poems, prompting close attention (on the part of both writer and reader) to where the boundaries of one poem end and the next begin. Choosing this form for what is, arguably, the manuscript's most concentrated investigation of Petrarchan love poetry conventions, Wroth uses her complex paper epistemology to simultaneously confront the inscriptive logics of those conventions and construct an exceptionally unusual female form that takes shape within and around the Petrarchan narrative. In the language and form of the crown's linked sonnets and (even more assertively) in the graphic profile of the corona's pages, Wroth crafts a kind of écorché poetics that exposes the violence and solipsism at the heart of the Petrarchan tradition, the very tradition within which Pamphilia finds herself locked at the start of the sequence.

Framed as a "crowne unto [Cupid's] endless prayse" in the manuscript sonnet leading into the corona, the sequence's early poems (MS 83–88) initially appear to praise Cupid, represented as the "light of true love" (MS 83.7). "When chaste thoughts guide us," Pamphilia asserts in the crown's second sonnet, "love is true vertu," "the shining starr of blessings light," and "the roote of peace" (MS 83.9–13). As she continues her encomium of Cupid's virtues, however, the more sinister conventions of love poetry start to seep into her representation. The ending of the fourth sonnet (MS 85) provides a glimpse of the critique that ripples beneath the surface of these poems. Cupid's arrow is here a "pleasing sting" that will cause "happy smarting" and "smale paine" (MS 85.11–12). "Although, itt pierce your tender hart and burne," Pamphilia assures her reader, "yett burning you will love the smart," a line that recalls the pressurized heat of "How like a fire doth love increase in mee" (MS 55). The violence underlying the Petrarchan narrative starts to come to the fore but, importantly, Wroth's critique is not just of that violence (the "weight of true desire" [MS 86.2]), but also of the creative limitations Petrarchism imposes on the artistic mind. "Love will a painter make you," she cajoles her reader, evidenced by the "millians" who "adorn the throne of love" (MS 88.9; MS 87.13). Drawing on the familiar early modern associations of poetry and craft, she goes on to insist "you / shall able bee to drawe your only deere / more lively, parfett, lasting, and more true / then rarest woorkmen" (MS 88.9–12). It is worth remembering, though, that Wroth's poetry does not "drawe" Amphilanthus in a "lively, parfett, lasting" representation, so the offered promise of these lines rings hollow within Wroth's own sequence. As both "prophet" and "tutor," Cupid starts to bind the poet within a kind of creative cage, barred and locked by the eighth poem of the sequence when the poet warns, "Never to other ends your phant'sies place" and describes the "never changing thoughts" required by the "heavnly fires" of "vertuouse love" (MS 89.13; 4–7). The Petrarchan narrative creatively entraps through the violence it performs as the poet is forced again and again to the well of praise and pain.

The corona becomes less coherent as it progresses, gradually unraveling its initial project of praise. The ninth sonnet (MS 90) is racked with referential instability as the power over the "throne of love"—which, as Daniel Juan Gil argues, becomes an allegory for "the Petrarchan form itself"—changes hands from Cupid to Venus to "lust."[118] The firm, constant Cupid of the sequence's opening sonnets transitions into a "worthles," "lasiviouse" love who "suckd his fleeting pace" from Venus's breast (MS 90.3–8). This "monstrous" child, Pamphilia now asserts, should have been "torne" from the court of love. "To say that the monstrous child must be banished," Gil points out, "is of course to acknowledge that it is in some sense already inside, inside both the court of love and . . . the Petrarchan conventions (the 'stile') Wroth is operating with and operating on."[119] While for Gil this latent monstrosity exemplifies Wroth's "central strategy of self-authorization," constituting *herself* as "the image of the monster which demands that all eyes turn to it,"[120] for my reading this "fruit of a sowre, and unwholsome ground" symbolizes the solipsism that is Petrarchism's ultimate (and only possible) poetic product. This poetic tradition, Wroth argues, is violent and degenerative, "a timeles, and unseasonable birth / planted in ill, in wurse time springing found" (MS 92.5–6). "Hemlock like," the confines of the Petrarchan narrative "feed a sick-witts mirthe," foment an "endles rounde" of "unruld vapors," and "quench[ ] out" "true borne fires" (MS 92.7–8). "Joy wee nott in what wee ought to shun," Wroth finally declares in the corona's eleventh sonnet, which ends with what seems a return to the sequence's initial praise of love ("lett love his glory

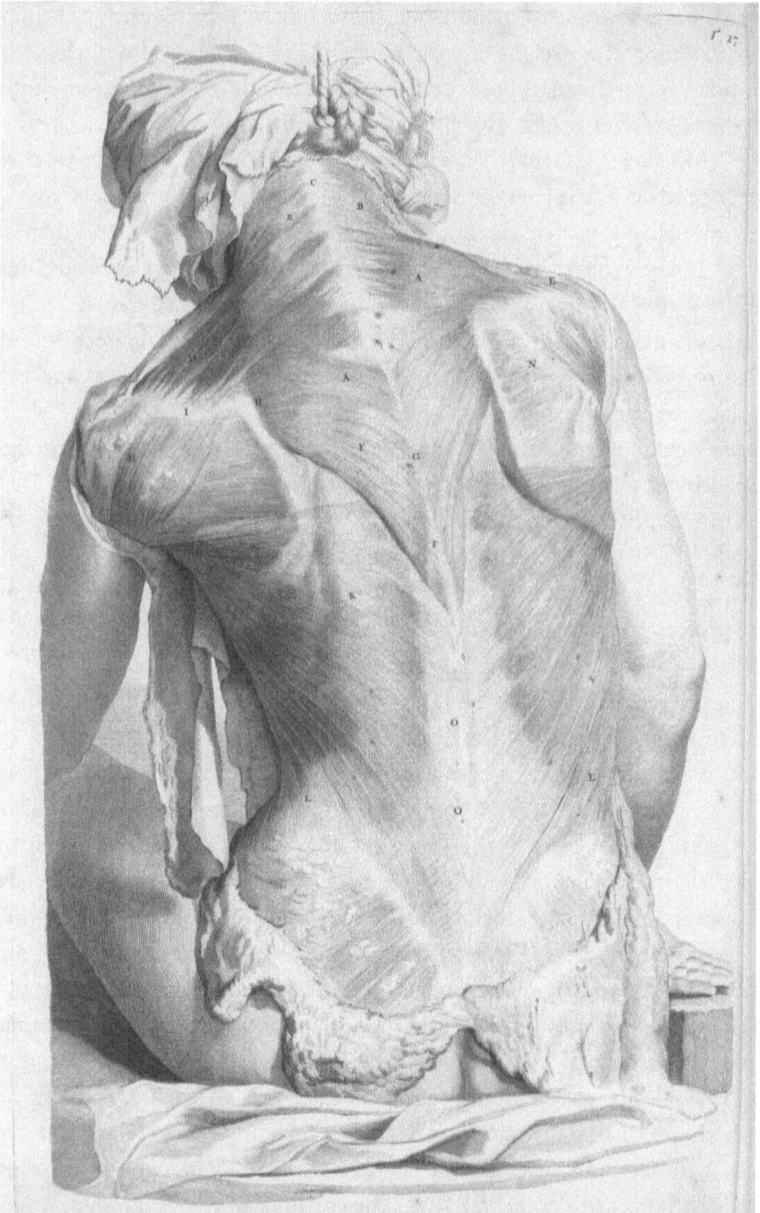

Figure 3.7. First plate in écorché series. Govard Bidloo and Gerard de Lairesse, *Anatomia humani corporis* (Amsterdam, 1685), T. 27, Wellcome Collection, EPB/F/2051.

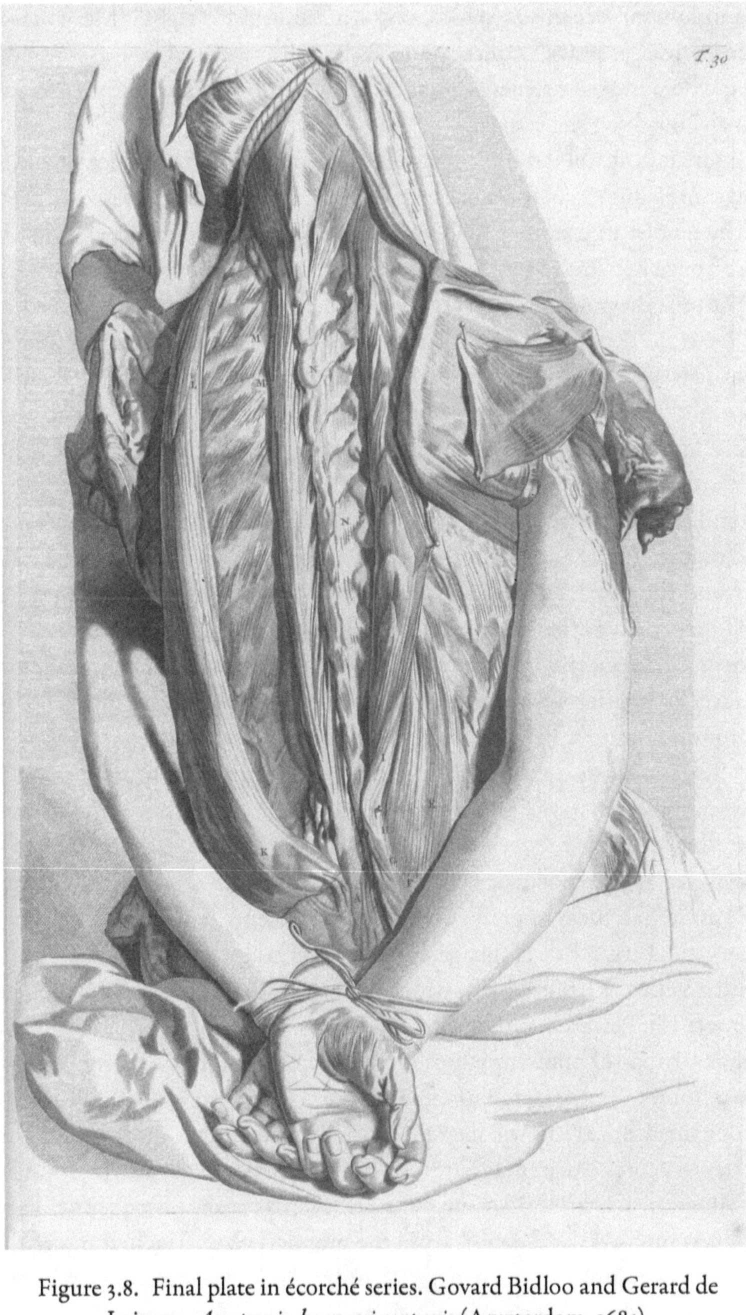

Figure 3.8. Final plate in écorché series. Govard Bidloo and Gerard de Lairesse, *Anatomia humani corporis* (Amsterdam, 1685), T. 30, Wellcome Collection, EPB/F/2051.

have and might / bee given to him who triumphs in his right" [MS 92.13–14]). But this thread of praise is once again, and finally, upended by the corona's last sonnet, where the poet remains trapped and "burn[ing]" in the same "strang labourinth" in which she started.

Before taking a closer look at the final sonnet of the corona, let us return to the beginning in a circular act of close reading prompted by the sequence's form. The crown's opening sonnet is about a body locked into a frame that has been imposed onto her. "In this strang labourinth how shall I turne?" Pamphilia begins. She tests the possible "wayes . . . on all sids," turning first "to the right hand," then "forward," "to the left," and "back" (MS 82.1–6). In each direction, she finds her way barred or constricted by Petrarchan poetic conventions, first by fire ("if to the right hand, ther, in love I burne" [MS 82.3]), then by danger, suspicion, shame, and grief. Pamphilia then reiterates this spatial orienting in the sonnet's third stanza—"thus lett mee take the right, or left hand way; / goe forward, or stand still, or back retire" (MS 82.9–10)—only to finally "leave all" and "take the thread of love" (MS 82.14). Wroth frames the corona, then, as a sequence about poetry and the poetic process as dictated by the English Petrarchan tradition, figured here, first, as the "strang labourinth" enclosing the poet and, second, as the poet's salvation, the "thread of love." In order to work her way through the enclosure, Pamphilia sutures her hand to the page—taking the "thread of love" that promises to move her body safely through the labyrinth.

Pamphilia is not dissected in the pages of the corona; her body is not fragmented or subjected to the acts of exposure and penetration so central to our understanding of early modern anatomical poetics. Instead, a poetic tradition itself is flayed as each poem of the corona peels off another layer from the veneer of the Petrarchan form to gradually reveal the violence at its center—a violence that pervades Petrarchan language, but is also central to its epistemological project, and a violence from within which Wroth is unable to disentangle her female poet. The Petrarchan mode of knowing exposed in Wroth's corona is the anatomical mode of knowing exposed by the écorché figure, a textual body that calls attention to rawness, to vulnerability, to the very act of violent uncovering on which anatomical knowledge relies. Briefly returning to Bidloo and Lairesse's series of écorché plates (Figures 3.7 and 3.8), it is worth noticing that these plates function as a kind of visual concatenation as each subsequent engraving repeats certain forms and details from the previous plate. The first figure shows the outermost muscles of the back, with only skin and underlying tissue pulled away, and then the images progressively expose layer after layer of back muscles to reveal finally the muscles directly attached to the vertebrae. These images are

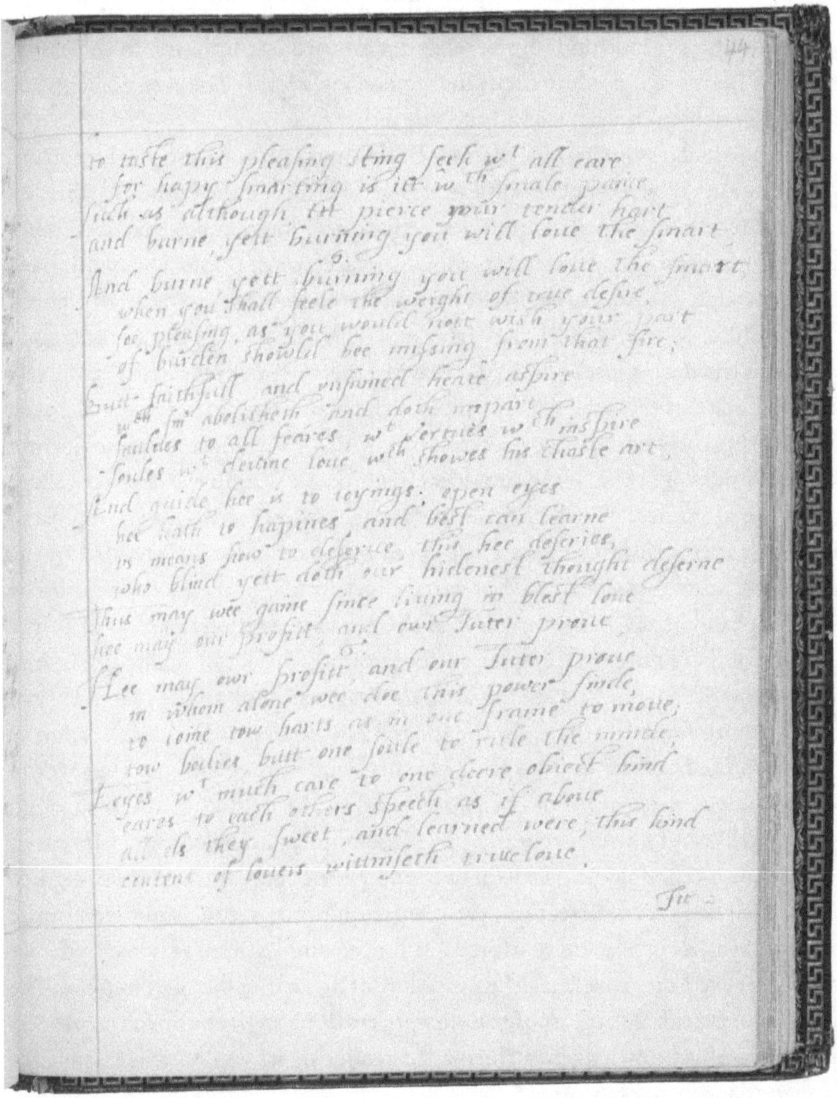
To taste this pleasing sting seek wt all care
for happy smarting is itt wth smale paine,
such as although itt pierce your tender hart
and burne, yett burning you will love the smart.

6.

And burne yett burning you will love the smart
when you shall feele the weight of true desire,
soe pleasing, as you would nott wish your part
of burden showld bee missing from that fire;
Butt faithfull and unfained heate aspire
wch sinn abolisheth, and doth impare
faultes to all feares, wt vertues wch inspire
soules wt devine love wch showes his chaste art,
And guide hee is to ioyings; open eyes
hee hath to hapines, and best can learne
us means how to deserve, this hee descries,
who blind yett doth our hidenest thought deserne
Thus may wee gaine since living in blest love
hee may our profitt, and our Tuter prove.

7.

Hee may our profitt, and our Tuter prove
in whom alone wee doe this power finde,
to ioine tow harts as in one frame to move,
tow bodies butt one soule to rule the minde,
Eyes wt much care to one deere obiect bind
eares to each others speech as if above
all els they sweet, and learned were; this kind
content of lovers wittnesseth true love.

Itt

Figure 3.9. Page from Wroth's corona sequence. Lady Mary Wroth, *Pamphilia to Amphilanthus* (c. 1614–1621), sig. 44r. Call #: MS V.a.104. Image #: 23474. Folger Shakespeare Library.

striking in large part because of the palpable vulnerability of the female cadaver, a vulnerability Bidloo and Lairesse seem determined to emphasize in their staging of the body—the demure tilt of the woman's head, the rope around her neck, the tendril of hair curling around the side of her face.

Of particular note in these plates is Bidloo and Lairesse's use of white space, which they use to represent simultaneously the surface of the body and its internal parts (the fleshly layers of muscles, tissues, and membranes that are, themselves, other kinds of surfaces). In the first plate, for example, the blank page shines through (it seems to catch the light in the engraving's characteristic chiaroscuro effect) as the skin of the right arm, the cloth wrapping the woman's head, and the deep spaces and folds between the woman's muscles (Figure 3.7). The final plate in this series again employs the blank page to represent the skin still covering the woman's undissected arms, which surround and frame the raw, exposed insides of her back and shoulders (Figure 3.8). Within the depths of those raw, exposed insides, the white space of the page bodies forth again to represent the most *inner* anatomical detail, the vertebrae that are finally revealed in the progressive act of dissection.[121] The textual effect created in the successive écorché plates—a kind of textual simulation of the act of dissection itself, akin to the anatomical flap figures of Chapter 1—foregrounds the act of making this anatomical text. The blank spaces, in particular, call attention to the materiality of the engraving act itself (these are the spots, remember, that William Faithorne instructed the engraver to take great care in "hand-wiping" so that they would be "as white as the margin of the paper") and to the textual medium through which Bidloo and Lairesse represent their anatomical knowledge.[122] Moreover, because the methodological lexicon of copperplate engraving overlaps so significantly with the act of dissection, the dissection process, too, is ghosted in the printed pages, a correlation reinforced in Bidloo and Lairesse's choice to depict four successive plates that track the procedure of dissecting skin and muscles. The anatomical plate holds traces of the violent acts done to the body in the process of curating this knowledge for the reader, rendering its anatomical knowledge *through* that record of violence.

In the crowded pages of Wroth's corona, her use of white space mounts a similarly complex argument about the making of the Petrarchan form and the inscriptive project at its center. The corona is a highly intricate, interwoven form, the "thread of love" made literal in the concatenated lines that stitch these sonnets together across the series. Characteristically for Wroth's project, enjambed lines abound throughout the sequence (with more time we might take a look, for instance, at the enjambment at the end of MS 94), but it is the sequence's concatenation that takes center stage formally. As a formal technique keenly interested in boundaries, rendering poetic lines that are intent on "crossing the thresholds that separate . . . lyric spaces,"[123] concatenation meditates on the slippery space between text and textuality. When this formal element also becomes graphically and materially significant, as it does in the pages of Wroth's corona where the concatenated lines seem to fuse at the sonnets' seams, poetic form and paper epistemology emulsify. As I noted earlier, Wroth drastically changes her use of the mise-en-page when she begins the corona. She writes multiple sonnets on each page, filling the white space between the page margins with ink and drawing heightened attention to the sequence's concatenated lines, at times leaving so little room between sonnets that the corona sequence starts to feel like one long, unbroken poem (Figure 3.9).[124] Recall in "How like a fire" (MS 55) that the blank space of the page became a material manifestation of Pamphilia's bodily surface, threatened from within by Petrarchan conventions and vulnerable to the inscriptive acts imposed upon her by those conventions. If, in MS 55, the surrounding white space held the violence of the Petrarchan narrative in check, here in the corona the black ink effaces more and more of the page space, filling the exposed, vulnerable surface of Pamphilia's body with ink, effectively smothering her within the violent inscriptive project of the Petrarch narrative.

Crucially, though, the corona's final poem returns us to Pamphilia's heart, "spotles," "untouch'd" (white and blank) as a mark of her love (MS 95.4; 6). "Thus my harmes I see," Pamphilia calls out at the end of the fourteenth sonnet, even as the poem's closing lines turn us back to the corona's opening: "Soe though in Love I fervently doe burne, / In this strange labourinth how shall I turne?!" (MS 95.12–14). The labyrinthine momentum of the concatenation prompts us back to the first sonnet, to begin again the "endless round" of Petrarchan conventions. But Wroth's resistance—and her anatomical intervention—is in the blank. The ink stops after the fourteenth sonnet's last line, followed by a completely blank page (Figure 3.10). More than any other part of Wroth's manuscript, the corona's pages are about the act of poetic composition itself—how it feels to be a woman poet writing within the Petrarchan tradition and how such a poet might appropriate the violent tendencies of that tradition in the service of aesthetic innovation. Even as the "crowne of sonnetts" mounts its gradual, flaying critique of Petrarchism, Wroth uses the pages of this sequence to *animate* Pamphilia's affective and embodied form as she writes herself out of the Petrarchan narrative and finds not obfuscation but power in absence.

Wroth's innovative anatomical poetics should impress upon us the importance of deconstructing our well-worn narratives about early modern literature

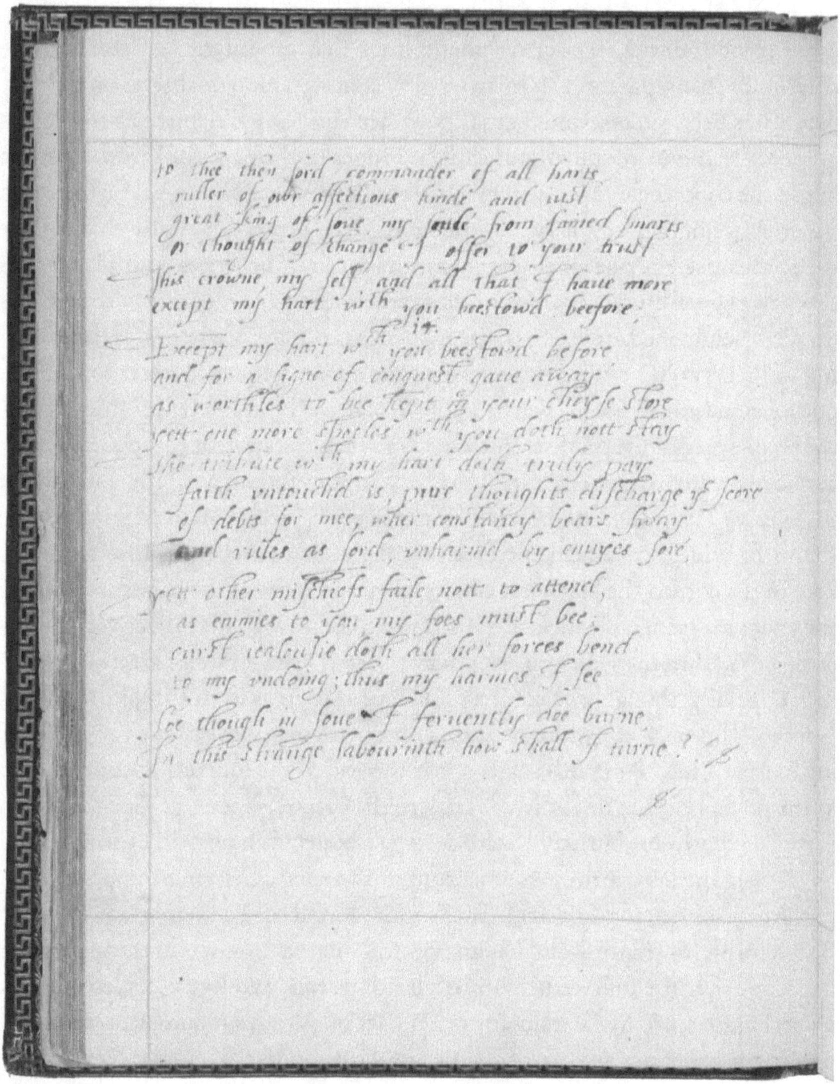

to thee then Lord commander of all harts
ruller of owr affections kinde and iust
great King of Love my soule from fained smarts
or thought of change I offer to your trust
This crowne, my self, and all that I have more
except my hart with you beestow'd beefore.
14.
Except my hart wth you beestow'd before
and for a signe of conquest gave away
as worthles to bee kept in your cheyfe store
yett one more spotles wth you doth nott stay
the tribute wth my hart doth truly pay
faith untouch'd is, pure thoughts discharge ye score
of debts for mee, wher constancy bears sway
and rules as Lord, unharm'd by envyes sore,
yett other mischiefs faile nott to attend,
as enimies to you my foes must bee,
curst iealousie doth all her forces bend
to my undoing; thus my harmes I see
Soe though in Love I fervently doe burne,
In this strange labourinth how shall I turne?

Figure 3.10. Final page of Wroth's corona sequence and subsequent blank manuscript page. Lady Mary Wroth, *Pamphilia to Amphilanthus* (c. 1614–1621), sig. 46v–47r. Call #: MS V.a.104. Image #: 23479 and 23480. Folger Shakespeare Library.

and science. There is much at stake in the almost reflexive critical tendency to connect the blazon and dissection, particularly when we consider its implications for the history of knowledge about women's bodies. The blazon is about knowledge in the service of power, possession, and exploitation.[125] When we preserve the poetic blazon as *the* representative site of literary and anatomical intersections in early modern England, violence to and violation of women's bodies will always be at the fore of our histories of literature and science. Early modern anatomical development marked a significant (if gradual) shift in the history of science about the body. During this crucial shift we find a corresponding and constitutive cultural project intent on the violation of women's bodies, literalized in the bodies of the working class women, sex workers, and otherwise criminalized women dissected in the pages of Andreas Vesalius's *De Humani Corporis Fabrica* (1543), which I will take a closer look at next chapter. Within such a narrative, it is hard—even impossible—for women writers to exist in anything other than a hostile relationship to the history of anatomy and the pervasive Renaissance culture of dissection.

As we have already seen in Margaret Cavendish and Mary Sidney Herbert's poetry, we can find in women's writing a different orientation to emerging scientific knowledges and practices. Wroth upends the entrenched analogy between the poetic blazon and anatomical dissection, and invites us to think instead about poetry and anatomy's shared inscriptive work. Her poetry is about the blazon as inscriptive form—the way it inscribes women's bodies into very particular knowledge systems and the way a poet might creatively investigate that process of inscription to craft a very different kind of anatomical poetics. She offers us another model, a nuanced and materially driven inquiry into the relationship between poetic and anatomical practice, an inquiry that can only be illuminated when we attend to a broader spectrum of Renaissance anatomical study. Once we uncouple our reading of Petrarchan lyric poetry and anatomical science from the dissection table, what new possibilities emerge within the formal, material, and tactile links between poetic and anatomical work?

CHAPTER 4

Writing the Womb

Aemilia Lanyer on the Secrets of Women

The female cadaver on the title page of Andreas Vesalius's *De Humani Corporis Fabrica* (1543) has a secret, one that Vesalius is poised to reveal to the large crowd gathered to watch her dissection. In this famous anatomical illustration (Figure 4.1), the female cadaver is positioned vertically toward the center of the image so that we, the readers, might imagine we are part of Vesalius's audience, standing directly at her feet. The cadaver's naked breasts and exposed legs are prominent, but the most conspicuous feature of the title page is her dissected lower torso. There are no discernible organs; rather, the dissection cut exposes an indistinguishable arrangement of soft, curved folds. Multiple layers of skin, fat, and tissue unfold from the woman's body below her breasts and just above her vagina, creating a flower-like opening at her body's (and the illustration's) center that seems too large for the size of her figure. Her dissected body seems all wound and all womb. Around her, a crushing circle of male observers fill the space near the dissection table and the tiered anatomy theater galleries, jostling one another for a closer look inside the mysterious and powerful space of her body cavity.

The secret this female cadaver holds within her body's folds is one of the oldest of all time—the knowledge of reproduction, increasingly linked in the early modern period to the anatomical structure and physiological functions of the uterus. As historian of science Katharine Park has demonstrated, the phrase "secrets of women" crystallized around the anatomical enclosure of women's bodies in the late medieval and early modern periods because "the male genitals presented themselves for easy inspection and consideration" while women's bodies remained "intrinsically opaque" and thus "required intensive scrutiny and

Figure 4.1. Title page. Andreas Vesalius, *De humani corporis fabrica libri septum* (Basil: Johann Oporinus, 1555), Wellcome Collection, EPB/D/6562/1.1.

study."[1] In Book V of Vesalius's *Fabrica*, we come to learn that the female cadaver on the book's title page was authorized for dissection by the Venetian magistrate precisely because of her secrets. This Paduan woman, Vesalius reports, "pled the belly," claiming she was pregnant to delay her execution after criminal sentencing.[2] When midwives were brought in to examine her, they ruled that she was not in fact pregnant, so the woman was hanged and her body turned over to Vesalius for autopsy to verify the midwives' ruling. Vesalius and his printer, Johann Oporinus, knew what they were selling with this frontispiece illustration. Female cadavers, particularly pregnant female cadavers, were extremely difficult to procure for dissection, so the *Fabrica*'s title page employs the rarest specimen of all to advertise the book's content to its audience of wealthy, educated physicians and collectors—all the better that the woman's pregnant status remains ambiguous amid the messy folds of her dissected torso.[3] Vesalius's corresponding story, tucked at the end of his discussion of the uterus, encourages further speculation about this woman's secrets ("she was never willing to say how long she had gone without menstrual purgations").[4] In his account, Vesalius also cleverly positions himself as the authority verifying the knowledge of the midwives, evidence of the period's ongoing "negotiation of competing modes of reading and interpreting the secrets of the female body."[5]

Undoubtedly the most influential Renaissance anatomical treatise, Vesalius's *Fabrica* was begrudgingly hailed as an "immortal book" even by some of its contemporary critics.[6] Historian of medicine Vivian Nutton has observed that "there are few books so famous and yet so unfamiliar as the *Fabrica*."[7] One reason for this, Nutton argues, is the complex relationship between the book's form and its content.[8] Vesalius's argument is "nothing less than a demand for an anatomical understanding of the human body based on observation and dissection of human bodies" mounted in the pages of a book.[9] Consequently, this is a text constantly forced to reconcile with its own "book-ness" as it experiments with how to bring the physical, tactile, dimensional work of dissection to its pages. It is not insignificant, therefore, that Vesalius begins this task with a dissected uterus. The *Fabrica*'s famous title page not only ushers the reader into a large, lavish anatomy book that aims to "place the dissected body, as it were, before the eyes of those studying the works of Nature," but also indoctrinates its readers into the *Fabrica*'s reverse incarnational project.[10] To *incarnate* is to "embody in flesh" or, in the Christian connotation, "to be made flesh," as in the transformation of Jesus Christ from "word to flesh" through the womb of the Virgin Mary.[11] In the pages of Vesalius's text, we are invited to participate in and experience a process that reverses the Christian narrative of incarnation—a transformation not from word to flesh but from flesh (the cadaver's body and the embodied work of dissection) to word (the pages of the *Fabrica*). And in this book dedicated to Holy Roman Emperor Charles V, the foremost site of that reverse incarnation is the womb.

The dissected uterus on Vesalius's title page is a hyper-concentrated example of the matrix of text, flesh, and women's reproductive bodies at this chapter's center—a matrix I take in hand to read Amelia Lanyer's *Salve Deus Rex Judæorum* (1611), a poetic text with an incarnational project and many wombs at its center. In *Salve Deus*, I find this book's most complicated and striking instance of anatomical form, one that aligns with the theological and devotional investments of Pembroke's psalm translations, and reverberates with the theorizations of textual dimensionality so central to Cavendish and Wroth's poetry. At its core, Lanyer's text is a poetic retelling of the final hours and crucifixion of Christ (known colloquially as the Passion of Christ). As Barbara Bowen describes, though, this poem "is actually not one work but many, a series of texts folded in on a disappearing central narrative of the crucifixion."[12] Bowen's description of the poem being "folded in" on itself refers both to the digressive, nonlinear structure that nests different parts of Lanyer's Passion poem inside other parts and to the substantial textual apparatus that works *around* the central Passion poem. Scholars often comment on the book's striking amount of component parts—its lengthy section of dedicatory poems (almost half of the entire text); its extended meditations (within the central Passion poem) on the Virgin Mary, Eve, and a number of "famous women elder times have knowne" (l. 1465); its appended country-house poem on the estate at Cookeham; and its epilogue "To the doubtfull Reader" in which Lanyer declares that she was divinely "appointed to performe this worke" (137).[13] Taking Bowen's description of the "folded" text one step further, Jacqueline Pearson argues that Lanyer's narrative "functions as a series of enfoldings themselves modelled on pregnancy,"[14] depicting the world as a "web of maternities" spun out from the central Passion poem and Lanyer's famous vindication of Eve, to the dedicatory epistles and their celebration of maternal lineage, to the country-house poem where Lanyer describes a fertile, pastoral scene "sprung" from the body of Cookeham's mistress, the Countess of Cumberland (134).

In this chapter, I take up the implications of such a formal claim within the context of early modern anatomical science.[15] What does it mean to argue that the very form of *Salve Deus* relies on the female anatomical form, particularly at a time when the female reproductive body was such a complex and charged site of scientific and devotional speculation? As we will see, the word "womb" occurs only twice in Lanyer's text, and even its early modern synonyms ("uterus" and "matrix") are distinctly absent.[16] Yet, if we consider how the womb functions as a formal structure rather than semantic element, excavating what Pearson describes as the "wombs within wombs" organization of *Salve Deus*,[17] we find a complex poetic experiment that corresponds to the inquiries surrounding text and body, word and flesh that animate Vesalius's *Fabrica* and its descendants. Drawing on the early modern understanding of the womb as a folded, nested organ with a powerful secret at its center, Lanyer wields the fleshly and textual potential of "the privitays of women" in order to employ the womb as anatomical form, as the structural frame for her poetic devotional project.[18] Lanyer was, I argue in this chapter, a gynecological and obstetrical thinker, fully participating in the epistemological investigation surrounding the womb, its form, and its function in early modern England. By the end of the chapter, we will see that Lanyer's womb-oriented poetics are also an important reclamation (and innovative repurposing) of the pervasive literary trope of male poets birthing texts, a trope in which Vesalius's title page conspicuously participates.

The Virgin Mary's Enigmatic Womb

Printed by Valentine Simms in 1611 and extant in only nine copies, Lanyer's quarto volume begins with eleven prefatory dedications, all addressed to white noblewomen, including an extended poetic dream sequence to Mary Sidney Herbert and a prose dedication to Lady Margaret, Dowager Countess of Cumberland, Lanyer's primary dedicatee throughout the text.[19] The book continues with Lanyer's adaptation of the Passion narrative, an 1,840-line poem composed in *ottava rima*. Finally, the volume concludes with Lanyer's country-house poem ("The Description of Cooke-ham"), an unbroken 210-line poem of pentameter couplets dedicated to the Countess of Cumberland, and followed by Lanyer's short epilogue ("To the doubtfull Reader"). The book's title page includes a table of contents focused on the Passion poem, and then signals that the volume also includes "divers other things not unfit to be read" (Figure 4.2). As Danielle Clarke keenly observes, this title page "gestures towards the miscellany," advertising *Salve Deus* as "a text constituted of extractible parts which nonetheless accrue meaning and significance via patterning, arrangement and relationship, linear and otherwise."[20] The organization of Lanyer's volume is particularly significant, and as I read individual pieces in the following pages, we will return again and again to how these pieces are arranged and positioned—nested and folded—into the whole.

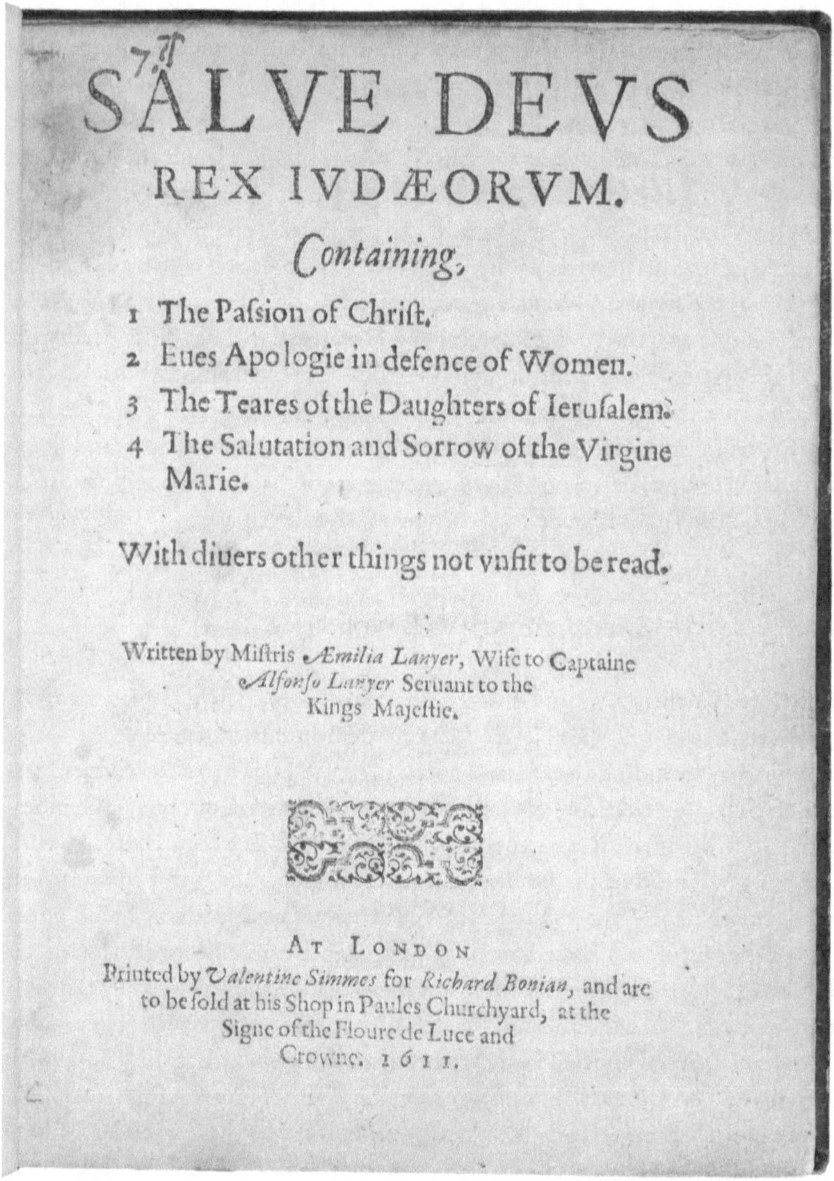

SALVE DEVS
REX IVDÆORVM.

Containing,

1 The Paſsion of Chriſt.
2 Eues Apologie in defence of Women.
3 The Teares of the Daughters of Ieruſalem.
4 The Salutation and Sorrow of the Virgine Marie.

With diuers other things not vnfit to be read.

Written by Miſtris *Æmilia Lanyer*, Wife to Captaine *Alfonſo Lanyer* Seruant to the Kings Majeſtie.

AT LONDON
Printed by *Valentine Simmes* for *Richard Bonian*, and are to be ſold at his Shop in Paules Churchyard, at the Signe of the Floure de Luce and Crowne. 1611.

Figure 4.2. Title page. Aemilia Lanyer, *Salve deus rex judæorum* (London: Valentine Simmes, 1611), Call #: STC 15227.5 copy 1. Image #: 42487. Folger Shakespeare Library.

We begin our investigation with the section of Lanyer's text that is, arguably, the most interested in the womb, both semantically and formally—the sixteen stanzas on the Virgin Mary that Lanyer inserts just after the halfway point of the Passion poem. The Marian stanzas (ll. 1009–1136) begin during Christ's procession up to Golgatha, the site of his crucifixion. Mary is one of a large crowd of mourners accompanying the procession, and Lanyer introduces her by describing Mary's "extreme" grief at the sight of her son's "bleeding body" (1011–12). For Lanyer, Mary's grief is a continuation and extension of the collective female "labour[ing]" of the "daughters of Jerusalem," who "by teares, by sighs, by cries . . . labour" to stop the "spitefull men" tormenting Christ's body (996–98). As many scholars have noted, Lanyer's depiction of this "woefull mother," "drownd" in "depth of sorrow" (1010–11), recalls medieval representations of the "hysterical virgin," which often linked Mary's suffering at the crucifixion to a kind of temporary madness (a "self undone," in Lanyer's words [1013]).[21] The crucial detail here is that Lanyer's Mary faints at the sight of her son's torment: "to see his bleeding Body oft she swouned" (1012). "Comfortelesse" in her "griefes extreame," Lanyer's Mary is an embodied, feeling mother who crawls "upon her knees" in the "open street" (1020) to "wash away" her son's "pretious blood" (1017).

Lanyer's representation of Mary directly counters the "controlled and silent" Mary of the late sixteenth and seventeenth centuries, both within the Protestant context of Jacobean England and in the aftermath of the Catholic Council of Trent (1545–63).[22] Roughly a century before Lanyer published *Salve Deus*, the Catholic Church printed an official ruling on the *lo spasimo* controversy, a debate surrounding the manifestation of Mary's grief at the crucifixion: did she, or did she not, faint at the sight of her son's crucified body?[23] Thomas de Vio, Cardinal Cajetan, ruled on the canonical status of Mary's grief that "not only is the swoon contrary to the text of the gospel, for John (19:25–26) says that the Blessed Virgin 'stood near the cross' but 'in the strict sense used by physicians' the swoon is a morbid state resulting from the contractions of the sinews, and 'therefore it is not proper to attribute this to the Blessed Virgin.'"[24] In Cardinal Cajetan's ruling, we glimpse the delicate relationship between theological and medical discourse often traversed in medieval and early modern treatments of Mary. Here, Cardinal Cajetan draws on anatomical language to define Mary's body *against* such discourse, but, much more often, anatomy lessons were wielded to speculate on the function of Mary's reproductive parts and the remarkable paradox of a virgin mother. Gary Waller uses the term "gynotheology" to describe the ongoing debate, from as early as the fourth and fifth centuries, surrounding the "physical details of the Virgin's sexual and reproductive anatomy in the context of her perpetual virginity."[25] Mary's "womb, birth canal, vagina, hymen, and capacity to generate moisture," Waller catalogues, "were all legitimate subjects of pious speculation."[26] Such speculation continued into the sixteenth and seventeenth centuries, even as the Reformation gave way to increased emphasis on Mary's immaculateness and the "profound spiritual resonance" of Christ's birth.[27] In John Pettus's 1674 *History of Adam and Eve*, for example, he grapples with anatomical specifics as he compares Mary's painless labor ("without fraction of any membrane, which causeth that virginate effusion of blood, and consequently pain") to the "affliction and pains" all other women experience as a result of Eve's sin, another important biblical narrative centered on the womb, and one to which we will return.[28]

For now, though, we linger with Lanyer's depiction of Mary, which not only includes the detail that Mary "oft . . . swouned" during the crucifixion (1012), but also detours into an extended reflection on the Annunciation, structurally linking Mary's embodied maternal grief to her reproductive anatomy. After two stanzas on Mary's grief, Lanyer disrupts the Passion narrative to tell the story of how and when Mary became God's "faire chosen vessell" (1030). This is one of several lengthy detours Lanyer makes throughout the Passion poem, resulting in what Clarke describes as "a highly digressive narrative method."[29] I will take a closer look at this narrative technique later in the chapter, but for now it is important to note that Lanyer's shift from the *now* of Mary's grief to the *then* of the immaculate conception is a moment of striking temporal folding that brings different parts of the familiar gospel stories into direct contact for her readers. For thirteen stanzas, Lanyer turns her attention to Mary's womb, beginning with the angel Gabriel's message:

> Loe, this high message to thy troubled spirit,
> He doth deliver in the plainest sence;
> Sayes, Thou shouldst beare a Sonne that shal inherit
> His Father Davids throne, free from offence,
> Call's him that Holy thing, by whose pure merit
> We must be sav'd, tels what he is, of whence; (ll. 1049–54)

For Lanyer, there is much packed into the "of whence" at the end of these lines. "Of whence," that is, the origin point, "from or out of which," the source and circumstance of Christ's "blessed body" (661), and a persistent theme in *Salve Deus*. These Annunciation stanzas call back to Lanyer's prefatory material where she reminds her readers that Christ was "begotten of a woman, borne of a woman, nourished of a woman" ("To the Virtuous Reader," 43–44), showcasing, in Victoria Brownlee's reading, "how pregnancy, labor, and breastfeeding are intrinsic to the Bible's ultimate act of salvation."[30] Throughout the thirteen stanzas on the Annunciation and immaculate conception, Lanyer repeatedly references Mary's status as mother and child bearer, describing her as the "Deere mother of our Lord" (1031), "maiden Mother" (1083), and "Servant, Mother, Wife, and Nurse" (1087). Gabriel announces that Mary "shouldst beare a Sonne" and, in Lanyer's direct address to Mary, she refers to Christ as "this blessed Infant borne of thee, / thy Son" (1071–72) and the "sweet of-spring of thy body" (1129–30).

For all of Lanyer's emphasis on Mary as mother, though, the gynotheological focus of these stanzas is not the moment of childbirth but rather the moment of conception—Mary's gravid uterus rather than her maternal body. Gabriel's "of whence" returns us to the origin point of Christ's Passion, the *how* of it all and, most crucially, the moment at which word became flesh ("*verbum caro factum est*" [John 1:14]). Throughout her interaction with Gabriel, Lanyer's Mary is deeply curious about the anatomical circumstances his words portend. Mary "humbly . . . demand[s]" how she is to have a child ("how that should be?"), struggling to translate Gabriel's words into fleshly action (1074). Noting that Mary's "pure thoughts could hardly apprehend / this salutation," Lanyer describes Mary's reaction as one of wonder and confusion.

> Nor couldst thou judge, whereto those words did tend;
> His pure aspect did moove thy modest cheere
> To muse, yet joy that God vouchsaf'd to send
> His glorious Angel; who did thee assure
> To beare a child, although a Virgin pure. (ll. 1059–64)

Lanyer's syntax in line 1061 ("muse, yet joy") as Mary wonders "whereto those words did tend" foregrounds Mary's musing (with connotations of both "ponder over" and "wonder at") even as it simultaneously corrects, overwrites that musing with joy.

At the end of this stanza, Gabriel reassures Mary that she will remain "a Virgin pure" even as she "beare[s] a child," but Mary's skepticism at the anatomical feat continues when she reminds Gabriel in the next stanza that she is "farre from the desire of any man" (1077).[31] Responding to Mary's skepticism, Gabriel provides further details about just how she will be both virgin and mother, the paradox at the center of theological speculation about Mary's anatomy. Gabriel assures her again that she "a blessed Virgin shouldst remaine," and then offers this description in response to her "mus[ing]": "Yea that the holy Ghost should come on thee / a maiden Mother, subject to no paine, / for highest powre should overshadow thee" (1082–84). Even though this is not, perhaps, the most detailed of explanations, Lanyer's Mary seems satisfied by these words, breaking into "teares of joy" and finally responding with a "humble answer" instead of more questions (1085; 1091). Our poet, however, still has doubts. In the stanza after Mary gives her "humble answer" to Gabriel, Lanyer cannot help but continue to "muse": "what wonder in the world more strange could seeme, / than that a Virgin could conceive and beare / within her wombe a Sonne" (1097–99). As Lanyer, echoing Mary, expresses skepticism, curiosity, and wonder, the anatomical language of the womb bodies forth in her interjection.

It is not so much the language of wonder and skepticism surrounding Mary's womb that interests me here as it is the combination of this thematic thread with the womb-like structure of the Marian stanzas. The stanzas on the Annunciation and immaculate conception dilate, in Patricia Parker's sense of the word, the narrative of Christ's Passion just at the moment when he is about to be crucified.[32] Pontius Pilate has made his fateful decision and Christ is carrying his own cross up to the crucifixion site when his "woefull mother" makes her entrance into Lanyer's poem. And, as we will shortly see, Lanyer picks up the crucifixion narrative right where she left off after the Marian stanzas. Structurally, then, the story of Christ's crucifixion encircles Mary's grief, which encircles Lanyer's reflection on Mary's virginal womb, which births the fleshly body of Christ that can then be crucified. In this concentric circular structure, with Mary's womb at its center, we find the "wombs within wombs" organization that Jacqueline Pearson attributes to the *Salve Deus* volume as a whole.[33]

For Pearson, this striking formal observation is a brief aside, peripheral to her central argument that *Salve Deus* celebrates and authorizes "women's roles as both writers and readers."[34] The trajectory of Pearson's argument is an all-too-familiar one in critical work on Lanyer's writing, which most often situates the woman-centric nature of *Salve Deus* as a proto-feminist project, evidenced in the savvy ways Lanyer gathers and addresses women readers and, most powerfully, in her emphatic defense of Eve's original sin.[35] I am also interested in the woman-centric nature of Lanyer's text, but want to insist on two important warrants for our readings of *Salve Deus*. First, let us consider how the woman-centric nature of the text interacts with its formal structure. Thinking with Katharine Eisaman Maus's speculation that "perhaps . . . the womb is another of those small enclosures in which so many seventeenth-century poets discover their poetic identity and freedom; like Donne's little room, Carew's hidden garden, Lovelace's prison cell,"[36] what would it mean for the womb to be Lanyer's "well-wrought urn"? Second, let us take seriously the epistemological as well as the political implications of Lanyer's woman-centric project. If she wields the womb as textual form, then there is knowledge-making work embedded in the very structure of *Salve Deus*.

Maus's speculation on the womb's formal potential is included in her analysis of male poets' use of child-birthing metaphors to express creative activity. For poets like Philip Sidney, John Donne, and John Milton, the womb can function like a "little room" or a "hidden garden" because theirs is a disembodied understanding of this anatomical part, conceptualized in their writing as spatial enclosure rather than somatic substance. For Lanyer, however, as we are starting to see in her extended reflection on Mary's immaculate conception, the womb is an embodied part, one piece of a structural whole that contributes to the messy, curious, joyful, painful, vulnerable physical realities of women's reproductive activity. Lanyer's own experience with pregnancy and miscarriage is certainly relevant here (Simon Foreman records details of her illegitimate pregnancy by Henry Carey, Lord Hunsdon, and suggests that she lost several children during and after birth),[37] but rather than take these autobiographical reports as grounds for considering Lanyer's maternal themes in *Salve Deus*, let us learn from Mary Sidney Herbert's smart theorization of experience (Chapter 2). Pembroke's experience with death, dying, and loss opened space for her extended study of experiential modes of knowing—the "as if" of her thanatological science. If early modern male poets' ability to understand the womb as a metaphorical vehicle is necessarily limited, Lanyer has the precise understanding of the anatomical womb that comes from having been pregnant and giving birth, coupled with the knowledge gained from the communal space of the early modern birthing chamber.[38] This is not a claim about specialized or distinctly female knowledge, but rather a claim about the critical capacity that comes from having done something—experience as doing rather than being—and, following Pembroke's example, the subsequent potential to think critically about how to leverage that experience for epistemological and aesthetic ends.

In *Salve Deus*, I am suggesting, the womb is a fecund poetic form. Lanyer uses it not as a metaphor or poetic conceit, but rather as a formal paradigm, always already inscribed (in Christian devotional contexts) with the complicated incarnational potential that manifests in Christ's immaculate conception. The Marian stanzas are the best introduction to the womb's formal function in Lanyer's text because their structural placement—the concentric circles of Mary's womb enfolded into the Passion narrative—does incarnational work. In the third stanza on Mary, when Lanyer turns from describing Mary's grief to addressing her as "faire chosen vessell" (1030), the poem's narrative of Christ's Passion swells to accommodate the stanzas on Mary's maternal (conceiving, pregnant, dilating) and reproductive body. Lanyer also uses the transitions into and out of those stanzas to blur the boundaries between Mary and Christ's bodies, formally enacting the intimacy between mother and child that is so central to the divine incarnational project. In only her second stanza on Mary, flesh collides with word in Lanyer's depiction of a grieving, embodied Mary who holds knowledge of Christ's ancestry as written in the Bible.

> Her teares did wash away his pretious blood,
> That sinners might not tread in under feet
> To worship him, and that it did her good
> Upon her knees, although in open street,
> Knowing he was the Jessie floure and bud,
> That must be gath'red when it smell'd most sweet:
> Her Sonne, her Husband, Father, Saviour, King,
> Whose death killd Death, and tooke away his sting.
> (ll. 1017–24)

The referents and actions throughout this stanza are a bit indistinct. At first, Mary seems to be "wash[ing] away" Christ's blood, but her action (or is it the action of the "sinners"?) shifts to "gath'r[ing]" in the stanza's second half. Within these blurry lines—confused and clouded, almost as if the poet herself is seeing the scene through Mary's tears—Mary's grief mingles with the reference to Christ's human ancestry, "the Jessie floure and bud," anticipating the poem's subsequent turn to the Annunciation.[39] Crawling on her knees to protect and collect her son's blood, Mary's tears mingle with Christ's blood to remind readers that her flesh is Christ's flesh, her loss of son is a loss of self (a "self undone"). In the stanza's final couplet, Lanyer reinforces the intimate connection between Mary and Christ with a condensed catalog of relations. If Christ is "her Sonne, her Husband, Father, Saviour, King;" then Mary is simultaneously mother, wife, daughter, sinner, and subject.[40]

In the final Marian stanza, Lanyer uses her transition from Annunciation back to crucifixion to again blur the boundaries between Mary's and Christ's bodies.

> How can'st thou choose (faire Virgin) then but mourne,
> When this sweet of-spring of thy body dies,

When thy faire eies beholds his bodie torne,
The peoples fury, heares the womens cries;
His holy name prophan'd, He made a scorne,
Abusde with all their hatefull slaunderous lies:
Bleeding and fainting in such wondrous sort,
As scarce his feeble limbes can him support. (ll. 1129–36)

In this final direct address to Mary, Lanyer reemphasizes the complicated status of her womb—Mary as "faire Virgin" mourning the "the sweet of-spring of [her] body." Christ's "bleeding and fainting" body in this stanza recalls Mary's "swoun" and her care for Christ's blood in the "open street." The mother "mourne[s]" her son's death, even as his "torne" body evokes the opening of the mother's body in labor, soundtracked by the collective laboring of "the womens cries."[41] This stanza is both a fleshly and temporal compression. Mary and Christ's flesh, blood, and tears—ghosted by the absent liquid signifier, milk—mingle formally through Lanyer's intricately woven referents, and their two bodies transform into one wounded, grieving body.[42] The poetic dilation does the work of fleshly dilation as these stanzas enact a womb-oriented transformation, transporting the reader back to the moment of Christ's birth and regrafting the body of the son onto and into the body of the mother.

The Anatomical Matrix (or, the Anatomist doth protest too much)

The gynotheology surrounding Mary's body was one manifestation of a widespread cultural interest in the womb as *the* site of women's fleshly secrets, an interest that stretched from the medieval period through early modernity.[43] As Elina Gertsman documents, for example, the popular medieval Shrine Madonna figures—sculptures of the Virgin Mary that conceal within their bodies carved or painted iconographies—are functionally quite similar to the anatomical Venus figures that became popular in eighteenth-century anatomical study.[44] The mechanics of both objects (equipped with hidden, internal parts that the handler must reveal) inscribed women's bodies within a narrative of secrecy. As Katharine Park argues, the female body became "the paradigmatic object of dissection" for Renaissance anatomy because of its "hidden and secret interior."[45] The "secrets of women" that Park takes as the title of her book on the social and devotional origins of dissection refers to two separate but interrelated "secrets":

first, the "the secrets of women as an epistemological concept, focused on different ways of gaining knowledge about the body" (for instance, the woman-centric practices surrounding postmortem care we saw in Chapter 2); and, second, "information relating to generation and the female genitals."[46] In this second connotation, women are not *possessors* of knowledge but rather *objects* of knowledge, specimens of great interest to the developing science of the body.

Of great interest, but also uneasy interest. In early English anatomy books, anatomists often voice their reticence to publish descriptions of women's reproductive organs. John Banister's treatment of the female "organs of generation" in his 1578 treatise is illustrative here, both in what it does and does not say. Banister's text, we might recall from Chapter 2, is primarily a translation of anatomy texts printed in other languages ("Sucked from the Sappe of the Most Approved Anathomistes"). In his account of the uterus, Banister wields his privilege as translator to selectively exclude certain information. The section on reproduction, already quite short in this otherwise lengthy text, begins with a marginal note that "the generative partes of the Female are omitted."[47] The male reproductive parts, Banister explains, are "commodiously" described, but that is all he "thought . . . good to translate into English." At the end of the section on reproduction, Banister reaffirms his decision, explaining:

> Because I am from the beginning persuaded, that, by lifting up the veil of Nature's secrets, in women's shapes, I shall commit more indecency against the office of decorum, then yield needful instruction to the profit of the common sort, I do here ordain the [Book] vi. rest of these my labours. An ende of the Histories of the Generative partes.[48]

In Banister's coded language ("Nature's secrets, in women's shapes"), along with the book's more explicit marginal notes (another reads, "why the partes of women are not here spoken of"), we hear the anxious rhetoric that accompanies the treatment of women's reproductive anatomy in early modern texts. There is, in the magisterial folios of learned anatomists, a great desire to know and be able to accurately represent the "secrets of women" juxtaposed with a palpable apprehension that one is committing a serious breach of propriety.[49] But Banister's text reveals the especially interesting by-product that emerges from this contradiction. In their treatment of women's reproductive anatomy, anatomists become hyper-conscious of the fleshly work their texts are performing. That is, in anatomists' textual representations of the womb and their overelaborate attempts to properly scaffold those representations, the womb gets entangled in the language of the book. And it is here, at this intersection of women's fleshly secrets and the textual spaces that purport to hold them, that the womb's potential as an anatomical form comes to the fore.

We have already glimpsed how important this intersection was for Vesalius's famous title page, the first illustration (but decidedly not the last) to advertise anatomical study *through* the womb.[50] As historians of science have noted, the *Fabrica*'s title page is ultimately false advertising because the discussion of female reproductive organs is one of the weakest sections of Vesalius's book.[51] In the 1543 edition of the *Fabrica*, Vesalius relies heavily on animal dissection for his description of female reproduction and, consequently, his knowledge of women's anatomy echoes much of Galen's second-century accounts. Vesalius's desire for knowledge of women's secrets, however, is evident throughout his pages on the "organs of generation," and his interest in female cadavers begins and ends with their utility as a source for uterine anatomy. In the section on the uterus, Vesalius admits that he has only dissected five different female cadavers, and three of those bodies had wombs of particular interest.[52] The woman on his title page, as I discussed in this chapter's introduction, "falsely stated that she was pregnant" in order to avoid hanging.[53] Vesalius describes a second cadaver as "a prostitute of outstanding appearance and in the prime of life," a subject of special interest because of both her fertile age and the state of her reproductive organs after years of sex work.[54] A third woman, this one exhumed by Vesalius's students ("snatched from her grave monument and brought in for public dissection"), purportedly "died from strangulation of the womb or a stroke."[55] The students, concerned that the woman's body would be identified (she was, Vesalius reports, "the elegant mistress of a monk at the church of St. Antony"), skinned her dead body, severed her urethra, "encircle[d] the external genitalia with a knife, split the symphysis and excise[d] the vagina and uterus in one piece."[56]

These reports of how Vesalius and his students treated the bodies of female cadavers support the "ferocious intensity" Jonathan Sawday attributes to the space of the early modern anatomy theater, wherein women "were the victims not only of the emerging science of biology, but of a European artistic tradition which located within the female body the source of a disturbing and dislocating power."[57] In Sawday's reading of the anatomy theater depicted on Vesalius's title page (Figure 4.1), the womb powers the "strong vertical axis" at the center of the image, which Sawday traces from the cartouche at the foot of the page, through the vertical position of the corpse and the skeleton ("its wand gesturing back towards the female body from which, in symbolic terms, it once emerged"), to the suspended armorial crest.[58] "What is depicted," Sawday argues, "is no less than a demonstration of the structural coherence of the universe itself, whose central component—the principle of life concealed within the womb—Vesalius is about to open to our gaze."[59]

For Sawday, the title page's vertical axis is doing purely symbolic work, and we certainly must attend to the violating, scopophilic project he and other scholars have read in this visual representation. As Katharine Park argues, "this use of the female body—which was by no means limited to the *Fabrica* or the sixteenth century—not only deprives her of subjectivity but also . . . sexualizes the act of vision itself."[60] We also, though, must account for the bookwork of Vesalius's title page. How does the opened womb function as a material-textual feature to support the epistemological project of Vesalius's folio? Two of the four parts of the title page's "strong vertical axis" are the cartouches at the top and bottom of the page, which present, respectively, the book's title and author and the book's publication year and printing privileges. Immediately below the bottom cartouche, still along the central vertical axis but just outside the border of the frontispiece illustration itself, the title page also includes the location and name of the printer, Johann Oporinus, who was, arguably, as important to the success of the *Fabrica*'s argument as Vesalius himself. This book has long been of particular interest to both history of science and history of the book scholars because the anatomical knowledge on offer in the text's pages is inextricable from the structural and formal features of the material book. The title page invites us to think fleshly content and textual form together, and the book as a whole offers rich insight into Renaissance experiments with the relations between text and reader, book and body, word and flesh.

This is a book that invited *use*, in Bradin Cormack and Carla Mazzio's sense of the term, in which early modern books were both instruments ("intricate tools that needed to be used in order to become useful") and instrumental ("directing their readers and users . . . toward some more or less practical end").[61] Vesalius wrote a detailed letter to Oporinus before the first printing of the *Fabrica*, and in it he outlined precise instructions for font size, illustration placement, and woodcut inking. "You will readily see in the printer's copy where the type style must be changed," Vesalius notes, directing Oporinus to pay attention to the difference between roman and italic parts of the text—the former font style used for "account[s] of the organs" and the latter for explanations of "the characters engraved in the illustrations."[62] Working closely with Oporinus, Vesalius leverages the affordances of print—"the actual labor of making a codex"—to transform his book into a site where readers must then bring different modes (even *technologies*) of reading to bear on the printed text.[63] Vesalius engages his readers by presenting options for their interactions with the book, even as he systematically makes his argument for hands-on dissection—the book's reverse incarnational project—through those interactions. Turn to the "last figure of Bk I" to see an illustration of the sacrum, instructs a marginal note next to Vesalius's description of the uterus's position.[64] When readers turn to that figure, we are then presented with more options: we could attend closely to the visual details of the image, we could follow the prompt of the diagram's letters to track each anatomical part in the corresponding index, or we could read Vesalius's caption to the figure for additional context on the cadaver's positioning. Such textual features construct knowledge rather than simply presenting it and, for Vesalius, the key to that construction of meaning is the organization of anatomical material, the arrangement of the book's different parts and the many ways those parts come together through embodied acts of reading.

The dissected uterus on the *Fabrica*'s title page is an assertion of the formative work that occurs in the book's pages (the forming, or bringing into being, of anatomical knowledge) and a constant reminder of Vesalius's creative, productive act in offering the text as fleshly body to his readers. The illustration's visual argument is enhanced by the language of "rebirth" we find throughout Vesalius's lengthy dedication to Emperor Charles V, the preface immediately following the famous title page. Referring to the *Fabrica* as "the fruit of my toil," Vesalius makes the ambitious assertion that his project is none other than "the resurrection of the science of anatomy," a resurrective act that relies on the constant negotiation between flesh and text.[65] As he catalogs the many public dissections he has performed in Paris, Louvain, Bologna, and Padua, Vesalius situates the *Fabrica* as mutually constitutive work. "I have now freshly organized the knowledge about the parts of the human body into seven books," he advertises, "in the same order which I am accustomed to use . . . in the assemblage of learned men at Bologna."[66] The *Fabrica* is a book infused with the work of the dissection room and is meant to be used not just alongside but as part of (even as a means of) hands-on anatomical study. He goes on to claim that the *Fabrica*'s seven books, organized according to Renaissance understandings of the body's systems, "include pictures of all the parts inserted in the text in such a way that they place the dissected body, as it were, before the eyes of those studying the works of Nature."[67] Vesalius has taken flesh and produced text and, with the *Fabrica* in hand, his readers can move seamlessly from text back to flesh. Vesalius's project, in other words, lays claim to a version of the incarnational narrative centered on Mary's womb. The book his readers hold is a site for transformational work, channeled through the most mysterious and powerful of all organs—the womb.

If we read the *Fabrica* as the child Vesalius manages "successfully to both father and deliver,"[68] then the title page not only exploits and appropriates women's (pro)creative act, but also takes the womb as an access point into this "learned booke" (a phrase we will later see Lanyer use in a particularly key moment of *Salve Deus*). Readers viewing Vesalius's frontispiece and then flipping the page to enter the *Fabrica*'s interior literally *descend* into the woman's body, which is precisely the language Helkiah Crooke uses in his introduction to the "natural parts belonging to generation," Book IV of his *Mikrokosmographia*. "And so we descend unto our history," Crooke dramatically begins, noting that he will first describe "the parts of generation belonging to men and then proceed to those of women."[69] Crooke, who directly cites John Banister as anatomical forefather, does include a lengthy section on the female reproductive system, but he frames it with a similar gesture of unease to the one we saw in Banister's text.[70] Next to a margin note reading "how cautelous we have been" ("cautelous," meaning cautious or heedful), Crooke confesses an internal dilemma: "being arrived at this place in the tract of my Anatomical Peregrination, I entered into deliberation with my selfe, whether I were best silently to passe it by, or to insist upon it."[71] True to form (recall his argument for anatomical study as the key to "divine nature" in Chapter 2), Crooke turns to theology for justification. "God that Created them," he reasons, referring to reproductive parts, "did he not intend their preservation, or can they be preserved and not known? or known and not discovered?"[72]

Crooke does not wield his prefatory material in the same way Vesalius does (although he does employ the same child-birthing tropes to introduce the text as the "first fruits" of his "untainted fame"), but he does link his treatment of the uterus and reproductive organs directly to the structure of his book. "We have so plotted our business," Crooke assures his reader before discussing the organs of generation, "that he that listeth may separate this Booke from the rest and reserve it privately unto himselfe."[73] Crooke suggests that the *Mikrokosmographia* is organized and arranged (perhaps even *gathered*, depending on the way it was printed and bound for any given edition) in such a way that the reader can extract the pages containing the womb ("the ancient habitation or mansion house of mankind") without compromising the structure of the whole.[74] Crooke's pages on the organs of generation, we might consider, function like Lanyer's stanzas on Mary as an integral, but structurally distinct, insertion into (or dilation of) the whole, out of which they could be pulled. Like Vesalius before him, Crooke emphasizes the fact that hands-on dissection has been his primary method for studying women's bodies. In Chapter 2, we heard Crooke's lament about the lack of dissection access for English anatomists, but in his section on reproduction he flaunts knowledge gained from hands-on practice. Dismissing classical Greek and Roman writers on the structure of the uterus and women's contribution to generation, Crooke mounts "two mighty arguments" against them: "one is taken from the ἀυτοψία in dissection, another from reason."[75] Both Vesalius and Crooke are invested in how the technology of the book aligns with the *techné* of dissection, how the book's form is shaped through the anatomist's work in the space between flesh and text—and the womb is consistently the place (and part) where the fleshliness of their bookwork bodies forth.

What we are seeing in these anatomy texts is a gynotheology—in this case, not specifically focused on Mary, but treated with the same kind of reverence, wonder, and power—that lives in and functions through the structural form of the book, specifically the organization of the book's parts and the different ways readers are invited to handle and interact with them. The secrets of women fuse with the secrets of the anatomical book into which the reader must descend in order to gain knowledge. Particularly in anatomical texts focused entirely on the process of reproduction, such as William Harvey's *Anatomical Exercitations* (1653), this conflation of women's secrets and book's secrets powers the acts of both writing and reading. Harvey uses his preface to usher his readers into an investigative adventure: "Now we must search farther" to discover "what he was in his Mother's Womb, before he was this Embryo, or Faetus; whether three bubbles? or some rude and indigested lump? or a conception, or coagulation of mixed seed? or whether any thing else?"[76] Harvey's speculations hail the reader to participate in his womb-oriented investigation, and introduce the book itself as a labyrinthine "organ of generation" that the reader will explore. The womb is central to the reverse incarnational work of these anatomists (their transformation of flesh into text) and, as a result, it also becomes a site where the reader is invited to engage (or, in Banister's case, is barred from engaging) most fully with the fleshliness of their textual projects.

Framing Lanyer's "Little Booke"

Lanyer is particularly alive to the fleshliness of the book as an object crafted (and birthed) by the writer and an object to be touched, held, and consumed by readers. *Salve Deus* is, very deliberately, a book—a carefully arranged codex for the early modern print marketplace. Lanyer refers to it over and over again as such—"this booke" [9]; "this little booke" [15; 17; 38]; "this blessed booke" [41]; "this small volume, or little booke" [48]—constantly framing it as an object for her readers to "cherish, nourish, and increase" (50). Vesalius-like in her engineering of the text as a technology, Lanyer does a great deal of epistemological work in her prefaces, which scholars have described as her text's "most remarkable feature."[77] As Erin McCarthy notes, the "average poetry book was made up of approximately 8.3 sheets of paper," and "Lanyer's dedications take up six of the twelve sheets used for *Salve Deus*."[78] Scholars often read these prefaces as an extended bid for patronage, tracing Lanyer's connection to the dedicatees through the sparse biographical information we have on her social networks.[79] Reading with the womb, however, we also find in Lanyer's prefaces the most sustained articulation of her innovative formal experiment—how, that is, she employs the *form* of her book to accomplish her incarnational project. The prefaces orient her readers to the anatomical form of the womb around which her "little booke" is constructed and they craft a female-centric community within which the fleshly experiments of her book can unfold.

Complete extant copies of *Salve Deus* include eleven dedications—nine written to specific women (Queen Anne; Princess Elizabeth; Lady Arabella Stuart; Susan Bertie, Countess of Kent; Mary Sidney Herbert, Countess of Pembroke; Lucy Harrington Russell, Countess of Bedford; Margaret Russell Clifford, Countess of Cumberland; Katherine Howard, Countess of Suffolk; and Anne Clifford, Countess of Dorset and Montgomery); one dedicated "to all vertuous Ladies in generall"; and one addressing "the Vertuous Reader."[80] Throughout this series of primarily poetic dedications, Lanyer invites her community of women readers to "reade," "feed upon," "kisse," and "imbrace" the "rude unpollisht lines" of her text.[81] In the calls to read that echo throughout the prefaces, Lanyer draws on familiar devotional language to frame reading as a distinctly somatic practice, a technique that activates the bodies of her female readers while bolstering what Femke Molekamp calls Lanyer's "poetics of incarnation," her "bold claim . . . that Christ is not merely represented but actually incarnated in her text."[82] "I present unto you even our Lord Jesus himselfe," Lanyer asserts in her preface to Lady Margaret (34). In a later preface to Lady Susan, Lanyer invites her to take Christ as a "faire bridegroom" into her "soules pure bed" (42). The dedication "To all vertuous Ladies" (12) directs readers to take specific, embodied actions to prepare their bodies to meet Christ: "put on your wedding garments every one" (8), "adorne your temples" (22–23), and "annoynt your haire with Aarons pretious oyle" (36). Simultaneously part of the reading experience and part of *preparing* for the reading experience, the women must ready their bodies in order to "be transfigur'd with [their] loving Lord" ("To all vertuous Ladies," 51).[83]

These prefaces call on readers to do particular kinds of tasks—to activate their minds and bodies for the hands-on work of reading and the devotional knowledge on offer in this text. As Femke Molekamp argues, "the Bible lay at the heart of early modern female reading culture, and women can be seen to have participated in multiple modes of reading it."[84] Whitney Trettien's research on the Little Gidding religious community offers especially striking evidence that women were not simply reading the bible, but also participating in "radical book-work" as a regular part of their devotional practice.[85] The "creative labors of the women of Little Gidding"—Mary Ferrar, Mary Collet, Anna Collet, and their extended family of female relatives—involved "cut[ting] apart printed Bibles and engravings and glu[ing] the pieces into elaborate concordances, or 'harmonies.'"[86] The Passion narrative was of particular interest to this community of readers and makers, and Trettien documents the careful precision with which the family cut apart and rearranged the four gospels—the books of Matthew, Mark, Luke, and John—in order to craft a "single chronology," enabling them "to read aloud the story of Christ's life in its entirety . . . from birth to resurrection."[87]

In Lanyer's book of many parts (recall Barbara Bowen's description of the poem as "a series of texts folded in on a disappearing central narrative"),[88] her readers are invited into a similarly active, intimate, and polyphonic experience with the biblical stories. Rather than stitching the pieces of Christ's Passion narrative into an unbroken chronology, however, Lanyer maintains—even exaggerates—the nonlinear biblical form. We have already started to see, in the Marian stanzas and their position in the text, how Lanyer spatially and temporally folds different parts of the narrative together, bringing points to touch across the pages of her book. In the prefaces, these acts of folding importantly involve her readers opening their bodies in some way so that reading, to borrow Karma Lochrie's language, "proceeds from a place of rupture."[89] Lanyer's dedication "To the Lady Lucie" represents Lucy's breast as a locked "closet," which "faire Virtue" stands ready to unlock with the "key of Knowledge" (1–3). Anticipating her later defense of Eve, in which Adam "takes" knowledge from Eve's hand, Lanyer figures knowledge here as a tangible thing, a key to be held and used, able to open Lucy's closed "Cabbine" ("To the Lady Lucie," 4). In this tableau the connection between the female body and knowledge involves an explicit opening. The preface takes an erotic, even abject, turn when Christ as "dying lover" then enters the "blessed bowre" of Lucy's breast and she is invited to read his "most pretious wounds . . . while he (dying Lord) doth bleed" (16; 24; 13). Across her prefaces, Lanyer imbues her text with a fleshliness predicated on the meeting and combining of opened bodies (here, Lucy's unlocked chest and Christ's wounds). This is also, again, a moment of temporal and spatial folding, a kind of queer temporal dragging of Christ's "dying" body into the text's prefatory space, couched in the language of what Richard Rambuss would call "sacred eroticism."[90] Moreover, Lanyer promises that this mingling of opened bodies will be a source and site for devotional knowing. It is the "key of knowledge" that opens Lucy's breast and enables her to "reade / salvation" in Christ's body, "stucke with pale deaths arrows" (11–13).

Lanyer's prefaces, like Vesalius's title page, function as a threshold, an access point—even a kind of skin, primed for rupture—that shapes her readers' movements into the text. To this end, they are actually quite iterative, riffing one after another around a cluster of metaphors and imagery that link women's bodies (particularly their reproductive bodies) to devotional knowledge that seems located simultaneously in Christ's wounded body and in Lanyer's "little booke." If we think about Lanyer's text as a theorization of the incarnational possibilities inherent in the form of the book—anchored in the intimacy between the Virgin Mary and Christ filtered through Mary's womb—the prefaces scaffold that theorization in their heightened attention to how female readers' bodies (flesh) are activated by and interact with Lanyer's book (word). This scaffolding work is especially concentrated in Lanyer's dedication to Lady Katherine, which addresses both Katherine and her "noble daughters" (49), reinforcing what Naomi J. Miller calls the "power of maternity" that runs through the book's prefatory material.[91] As Miller notes, Lanyer often pairs dedications to mothers and daughters. Her preface "To the Lady Anne, Countess of Dorcet," for instance, directly follows a preface to her mother, Lady Margaret, Countess of Cumberland. Lanyer draws on familiar biblical and reproductive rhetoric (Margaret has "sowne" the "seeds of virtue" in Anne [58]) as she praises Anne's "most worthy mother" and insists that Anne has inherited all of Margaret's "faire parts" (59–60). The book begins with paired prefaces to Queen Anne (introduced in the second line as "most gratious mother of succeeding kings" [3]) and her eldest daughter, Elizabeth, and Lanyer uses Elizabeth's preface to memorialize Queen Elizabeth ("that deare Mother of our Common-weale" [11]), invoking a different kind of matriarchal lineage. The cumulative effect of maternal discourse is striking and, as Miller argues, the mothers of Lanyer's prefaces "nourish their daughters not simply with milk, but with the white ink (to paraphrase Hélène Cixous) of a woman's writing voice," a network of (re)productive associations reinforced by Lanyer's extended dedication to Mary Sidney Herbert as her own authorial foremother.[92]

The preface to Lady Katherine and her daughters takes the extra step of linking these reproductive themes (both biological reproduction and authorial production) to key figures and details from the Passion narrative, utterly collapsing prefaces and Passion poem. Lanyer makes oblique references to both Mary ("faire Gabriel . . . brought downe a message") and Eve (Lady Katherine is the "faire tree from which the fruit of Honor springs") before previewing, in shortened form, Christ's crucifixion. "Here may they see him in a flood of teares," Lanyer foreshadows, "crowned with thorns, and bathing in his blood" ("To the Lady Katherine," 61–62). This is not simply a foreshadowing of what is to come, but also a representation of Christ's crucifixion, a dilation that functions as a formal folding to bring together the *now* of the dedication and the *then* of the crucifixion, both happening in the indeterminate present tense of Lanyer's poetic telling. Lanyer takes four stanzas of the preface to describe the "blacke infernall booke" of sin, "cancel[led]" by Christ's "dying breath," along with the beauty of Christ's dying body (40–84). In the present tense, Christ's beauty "*showes* more clearer than before" through "the sable Cloudes of Shame and Death," and in the past tense, "fresh beauty *shin'd*" in "Deaths ashie pale discoloured face" (77–78, my emphasis). These stanzas rhetorically echo (and anticipate) Lanyer's longer depiction of the crucifixion in the central Passion poem, most evident in her use of blazon imagery to describe Christ's "rare parts" to Katherine and her daughters: "no dove, no swan, nor ivory could compare / with this fair corps, when twas by death imbrac'd / no rose, nor no vermillion halfe so faire / as was that pretious blood that interlac'd His body" (82–83).

When Lanyer later describes Christ's "body torn" on the cross, blurring his body with Mary's as he "bleed[s]" and "faint[s]," the text folds back on itself, returning us to the dedicatory poems. We have seen and heard this language before, we think, as we read Lanyer's blazon of the crucified Christ: "his alabaster breast, his bloody side, / his members torne, and on his head a Crowne / of sharpest Thorns" (ll. 1161–64). If the doubled language is not enough to prompt this connection between the pieces of her text (the description of the crucifixion in the central *Salve Deus* poem now acting as a dilation of the prefatory dedication), Lanyer offers an even more explicit cue by directly addressing one of her dedicatees immediately after the blazon in the Passion poem. "This with the eie of Faith thou maist behold, Deere Spouse of Christ" (1169–70), she writes to Lady Margaret, Dowager Countess of Cumberland, issuing a scopic invitation that aligns Margaret's pleasure in looking with the penetrative erotics of the book's prefaces. This direct address to Margaret in the Passion poem is one of many, including five margin notes that hail Margaret throughout the nonlinear crucifixion narrative next to lines 9, 145, 250, 1321, and 1673 (the first positioned alongside the line "to thee great Countess now I apply / my pen [9–10]). Each address to

Margaret seems to fold back onto the previous iterations, creating an indexical and spatial pleating across the pages of Lanyer's book. She creates this textual structure, with Margaret's body at its center, in order to finally leave Christ's "perfect picture" in Margaret's breast at the poem's end, "where it still shall stand, / deepely engraved" (1325–27).

These echoes, doublings, and foldings between the prefaces and central Passion poem are formally significant to Lanyer's incarnational project, and they help us better understand the womb-like structure of her book. Drawing on Rayna Kalas's work on the Renaissance frame as a very particular kind of poetic technology, I invite us to think about how Lanyer's prefaces (as the "framing" of *Salve Deus*) position her book as "a thing *in potentia* coming into presence as matter."[93] As Kalas demonstrates, "framing had a rich conceptual, lexical, and material history" before the word "frame" became associated with our contemporary connotation of "pictorial enclosure."[94] Renaissance framing described "a form of making that conforms itself to its object," a process of conjoining, converting, or transforming "one medium, or one means, to another."[95] Kalas's rich analysis helps us think about prefatory material—a pervasive form of framing in Renaissance texts—as a crafted part of the book rather than something that functions outside of or introductory to a central, self-contained text. It also prompts us to recognize the transformative potential of the prefatory space. For Lanyer, as I have been arguing, that transformative potential is an incarnational one—the transformation from word to flesh to word. When Lanyer describes "celestiall powres" guiding her "to *frame* this worke of grace" in her preface to Lady Katherine (7, my emphasis), or refers to her book as a "*frame* of Glory which I have erected" in her preface to Lady Anne (2, my emphasis), she underscores the "indivisibility of meaning and matter"—her "little booke" as "spotlesse lamb," as the "Lord Jesus himself" ("To all Vertuous Ladies," 84; "To the Lady Margaret," 7).[96] To borrow language from Cavendish's theory of form, Christ's fleshly presence manifests through Lanyer's inscriptive work of folding, "joyning," and admixing textual matter.[97]

Handling Eve's Womb

If Lanyer's "little booke" is a body—framed as fleshly matter (*materia*, *mater*)—that her readers are invited to hold, kiss, and embrace, but also consume, open, enter, "cherish, nourish, and increase," how does the *Salve Deus* respond to and converse with anatomists' renderings of the female reproductive body? Structurally, the womb of the early modern anatomical archive is a folded, pleated, fibrous organ. One of the most curious things about anatomical descriptions of the womb from this period is that the exact structure and composition of the womb seems to constantly elude anatomists' grasps. The womb is "divided into four parts, the *Fundus* or bottom, *Os internum* or Cervix, the *Vagina*, and the *Sinus pudoris* of outward Privity," Thomas Gibson confidently asserts in his *The Anatomy of Humane Bodies Epitomized* (1682).[98] But, actually, he later clarifies, it is the "Fundus" that "in a special manner is called the Womb, because all the rest [i.e., the other three parts] seem to be made for its sake."[99] Most anatomists agree that the womb has two membranes, an outer and an inner membrane (although they are divided on the exact purpose of these membranes), and that these membranes give the womb its folded structure. But, again, a precise or consistent account of that folded structure remains elusive. "Betwixt these Membranes," Gibson observes with a lot of vague handwaving, "there is a certain carnies and fibrous contexture."[100] Anyone hoping to reference a visual diagram of the womb for more information on its anatomical structure would likewise find themselves confronting a shape-shifting organ. Early modern illustrations of the womb depicted the organ as a pear-shaped, fluid space or, as we saw in Vesalius's title page (Figure 4.1), as a blossom opening from the center of a woman's torso. We find an example of the former in Helkiah Crooke's pages on the womb (those pages, remember, that could be "separate[d] . . . from the rest" of the book), which depict differently textured folds in the outer and inner membranes, both of which, Crooke clarifies, can be "extended or stretched, and againe contracted and gathered together" (Figure 4.3).[101]

The consistent detail across all of these representations is the fold. In Crooke's diagram, the fold manifests as both the "fleshy cake or liver of the wombe" and the "membrane called *Amnios*" (the "membranous, thight, & compact" substance of the womb).[102] In Vesalius's title page, the fold manifests in the skin, tissues, and muscles that have been dissected and pulled back to reveal the indeterminate space at the center of the woman's body. In Lanyer's *Salve Deus*, the fold manifests in the book's structural form, the organization and arrangement of its many parts. In the stanzas on Mary's womb, folded into the crucifixion narrative, we glimpsed the complex way in which Lanyer dilates the Passion poem. Like Lady Lucy's body in the prefaces, the poem opens itself to bring new bodies—and new poetic matter—into its folds, and the spatial and temporal compressions within Lanyer's prefaces extend this folded form to the *Salve Deus* volume as a whole. Along with the Marian stanzas, we find extended dilations throughout the Passion poem: in the stanzas encompassing "Eves Apologie"

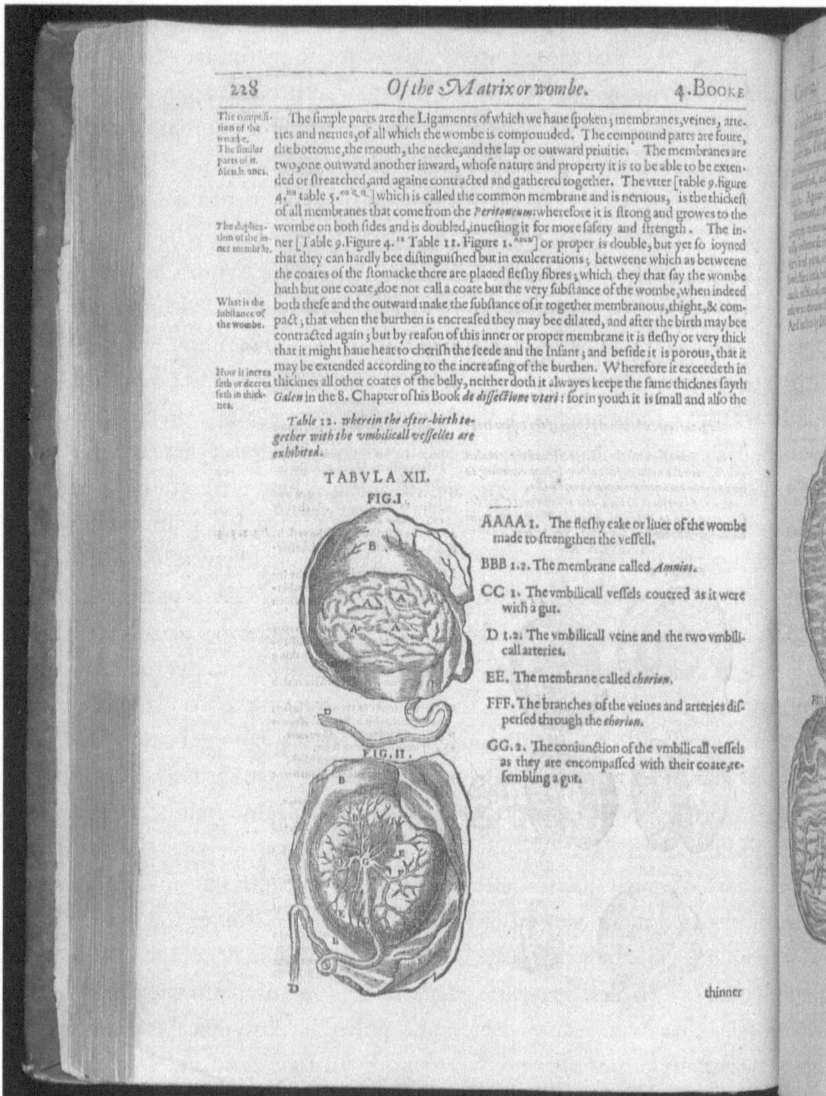

Figure 4.3. Illustrations of the womb. Helkiah Crooke, *Mikrokosmographia. A description of the Body of Man* (London, 1615), 228–29, Wellcome Collection, EPB/D/1683.

thinner, but after the courses begin it growes in thicknesse as in magnitude, as also it decreaseth both in magnitude and thicknesses when Women leaue to conceyue and theyr courses stay. For the beginning and end of their courses is, when they begin and cease to be fit for conception. So when the time of the courses is at hande, it growes thick and somwhat tumified, as being watered or dewed with the humidity of the bloode which floweth vnto it. Againe, the oftner a woman hath conceiued the thicker it is.

Moreouer, in the time of conception it is thicke and softer and growes thicker as the conception encreaseth, beeing thickest of all at the time of the birth. Yet *Galen* seemeth to say otherwise in his Booke *de dissectione vteri*, and the 8. chapter, and the 14. Booke of the vse of parts, and the 14. chapter (which opinion of his many follow) albeit it is against himselfe a little before where hee sayth, that it groweth thicke when the courses come because of blood; now we know that the greater the infant growes, the more bloode accrueth vnto the wombe.

And indeed by Dissection we are taught, that about the time of deliuerance, and in the last

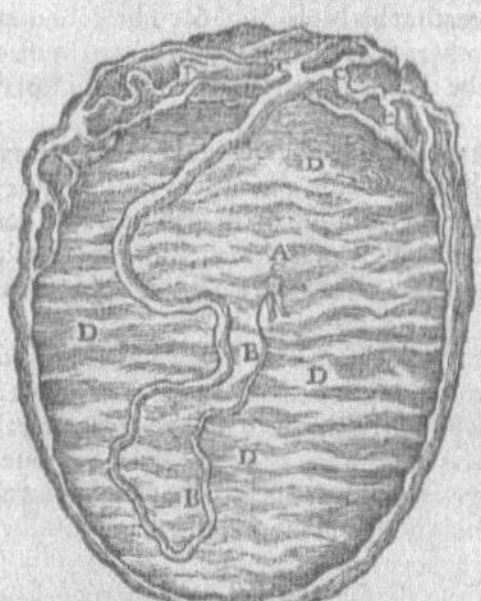

The first Figure sheweth an Infant of 14. dayes olde, in which all the parts are exactly delineated.

A. The Infant.

B B. The vmbilicall vessels meeting together.

C C. The diuarication of the vmbilicall Vessells, thorough the Chorion.

D D D. The Coate called *Amnios*.

FIG. II.

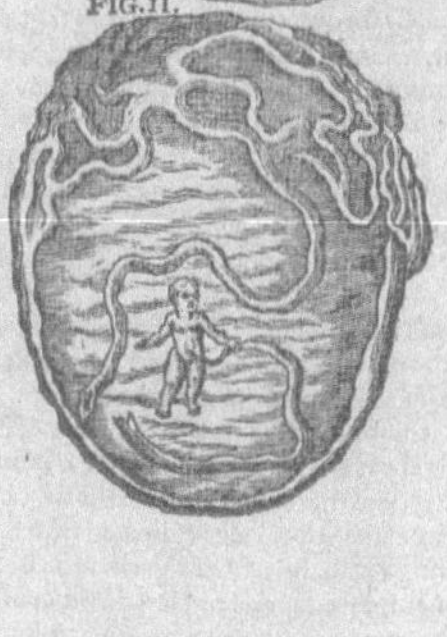

The second figure sheweth an abortiue Infant, which was auoided the xxv. day after conception, being depriued of blood to nourish it, because the vmbilical vessels were broken. The magnitude of that infant is perfectly described.

(ll. 753–840); in Lanyer's "invective against outward beauty" (ll. 185–247); in meditations on Cleopatra (ll. 1409–48), the Queen of Sheba (ll. 1569–1624), and other "famous women elder times have knowne" (ll. 1465–1544); and in the poem's opening and closing, both of which involve direct addresses to Lady Margaret. In Lanyer's text, we must notice, it is *women* who dilate. Or, more precisely, it is women who are hailed by and body forth within the text's narrative folds.

As Barbara Bowen's important work on Lanyer reminds us, the community of women gathered in *Salve Deus* is predominantly white and "faire," with the exception of Cleopatra, who becomes a particularly complex figure within the poem's gender and racial politics.[103] Lanyer herself was the daughter of an Italian Jew, Baptista Bassano, who was pressed into service as a musician in Henry VIII's court. She was writing "from the perspective of a racialized outsider," even as she held a privileged position as mistress to a powerful Protestant lord (Henry Cary, first cousin to Queen Elizabeth) and was baptized as a Christian.[104] *Salve Deus* testifies to Lanyer's own precariously raced and classed position within English court circles, "not so much explain[ing] as problematiz[ing] white womanhood," Bowen insists.[105] Lanyer simultaneously reflects on the "ways in which womanhood has been historically created" and excludes "some women from womanhood" on the grounds of race.[106] In Lanyer's poem, for instance, Cleopatra is the "blacke Egyptian" (1431) cast out of collective womanhood for "entertain[ing] disgrace, in stead of fame" (220), and Lanyer sides with "chaste Octavia" (1423), the Roman woman wronged by Antony in his pursuit of Cleopatra's foreign beauty. If the biblical, historical, and contemporary noblewomen in Lanyer's pages are focal points for the poem's "extraordinary meditation on the racialized process through which early modern women constituted themselves as a group" *and* the primary sites of textual dilation, then the poem's innovative formal structure is implicated in the boundaries it draws around communities based on gender, race, and class.[107]

A closer look at the poem's formal structure—the *folds* of Lanyer's poem—further illuminates the social, political, and epistemological projects in which it participates. Functioning like the indexical references in the margins of bibles, the marginal notes throughout *Salve Deus* (which include the five notes addressing Lady Margaret, along with "the sorrow of the Virgin Marie," "the salutation of the virgin Marie," and "Eves Apologie") invite a nonlinear, circular, and folded reading experience.[108] In her final address to Lady Margaret, Lanyer "fold[s] up" the devotional work of her book in Margaret's breast, and "humbly" praying to Christ before she begins the Passion narrative, Lanyer pleads for the "power and strength" needed to "unfold" the "manner of his Passion" (297–304).[109] The foldings and unfoldings that power Lanyer's formal work in *Salve Deus* enact her striking reclamation and repurposing of the anatomical and biblical hermeneutics concerning the womb. Hers is an epistemological project that takes the womb as its form—that is, the womb as it was conceived of in Renaissance anatomy, as a folded, pleated, multi-chambered part—and finds its most robust expression in the eleven stanzas marked as "Eves Apologie," undoubtedly the book's most famous section.[110] The stanzas on Eve establish women's reproductive bodies as knowledge repositories, challenge the history of extraction in relation to those bodies, and test the formal contours of Lanyer's incarnational devotional work—the enfolding of womb into "learned booke" (808).

Lanyer inserts her reading of the Fall during Pontius Pilate's judgment of Christ, so that "Eves Apologie in defence of Women" acts as another narrative dilation. These stanzas also fold back on Lanyer's references to Eve throughout the prefaces (most notably in the preface to Queen Anne, where Lanyer asks the Queen to "judge" if "faire Eve's Apologie . . . agree not with the Text," that is the biblical telling [6]). As Pilate deliberates on whether to crucify Christ or "Barrabas the Thiefe" (849), Pilate's wife interrupts with an extended comparison between Eve's "former fault" and Pilate's potential sin in crucifying Christ. "Her sinne was small, to what you doe commit," Pilate's wife argues, warning her husband that men will lose their "sov'raigntie" over women if he condemns Christ to death: "your fault beeng greater, why should you disdiane / our beeng your equals, free from tyranny?" (829–30). As Achsah Guibbory notes, these stanzas "offer a new understanding of Christ's crucifixion," one that redeems women and "liberate[s] them from their subjection to men under the Law."[111] Consequently, Lanyer's defense of Eve is often read as a proto-feminist stance, but we must consider the formal work of these stanzas—their structural function within the many folds of *Salve Deus*—in addition to their social and political import.[112]

"Eves Apologie" begins and ends with reminders that Eve is the ultimate progenitrix, the mother of all, the womb of all wombs. "Our Mother Eve, who tasted of the tree," Lanyer begins, "was simply good, / and had no powre to see" (763–65). These opening lines not only immediately position Eve as "mother" but also employ the first-person-plural pronoun, which, as John Garrison argues, "opens the narrative space for all women to collapse" into "an empowered, polyvocal subject position."[113] The collective voice of women linked by the "golden Chaine" of motherhood returns us to the maternal lineages in Lanyer's prefaces and propels us forward to Lanyer's reflection on virginal motherhood in the Marian stanzas ("The Authors Dreame to the Lady Marie," 21). In fact, the stanzas on "our Mother Eve" and the "Deare Mother of our Lord" mirror each other in the organization of Lanyer's poem. "Eves Apologie" is positioned just before the halfway mark of the poem, while the stanzas on Mary are positioned just after the halfway mark, bringing these points to touch around the narrative center of Christ's progression to his crucifixion. The polyvocal narrative voice bodies forth in both sections (recall Lanyer's "wonder" at the immaculate conception that could "repaire / our old decaies" [1101]), returning again at the end of "Eves Apologie" when Pilate's wife (and the poet? and all women? and all "vertuous" and "faire" women?) reminds him, "you came not in the world without our paine" (827). Lanyer's reference here to women's pain in child labor serves as a direct reminder of God's punishment for Eve's sin, worded thus in the King James Bible: "I will greatly multiply thy sorrow and thy conception; in sorrow thou shalt bring forth children; and thy desire shall be to thy husband, and he shall rule over thee."[114] If we recall John Pettus's language in his *History of Adam and Eve*, we remember that Mary is exempt from the "affliction and pains" that Eve's sin inflicts on all laboring women.[115] Consequently, the stakes of Eve's punishment become the foundation for the widespread speculation about the ontological and physical status of Mary's womb, the gynotheology that threads through premodern theological and anatomical archives.

If Eve is the womb of all wombs, for Lanyer she is also a source and site of knowledge that belongs to "faire" women. I find the most crucial epistemological hinge of "Eves Apologie" in the two lines that describe Adam taking knowledge from Eve's hand "as from a learned Booke." Lanyer frames these lines with a sharp distinction between Eve's desire for knowledge and Adam's attraction to the "faire" fruit—language that folds back on Lanyer's "invective against outward beautie" (185), and further evidence of Bowen's point that "Lanyer's poem finds that it cannot discuss womanhood without thinking about race."[116] "Greatest perills do attend the faire," Lanyer writes as she debates the lyric tradition that venerates "those matchlesse colours Red and White" (205; 193). Citing "great Cleopatraes Beautie" as one of her examples (215), Lanyer crafts another of her text's structural folds: "what fruit did yeeld that faire forbidden tree, / but blood, dishonour, infamie, and shame?" (217–18). Like Cleopatra ("pore blinded Queene, could'st thou no better see" [219]), Eve cannot see the effects of her actions ("had she knowne of what we were bereavid, / to his request she had not condiscended" [771–72]). Her "undiscerning Ignorance" falls prey to the serpent's "cunning" and "craft" (769; 773; 781). "If Eve did erre," Lanyer finally asserts, "it was for knowledge sake, / the fruit beeing faire perswaded [Adam] to fall" (797–98). The syntax here is important because it invites two possible readings. Read one way,

Eve wanted knowledge in the sense that she *lacked* knowledge. She was "ignorant" of God's command regarding the tree, whereas Adam "from God's mouth received that straight command" (787). Read another way, Eve wanted knowledge in the sense that she *desired* knowledge, a reading that juxtaposes Eve's sin more directly with Adam's but redeems her action because it was "for knowledge sake."

In this second reading, Lanyer claims the origin of human knowledge as the purview of women and vindicates Eve's action by connecting it to an epistemological birth—the moment when humanity gained both the desire and the means to know. Lanyer reinforces this interpretation in the subsequent stanza, by the end of which it is Adam's rather than Eve's act that deserves condemnation.

> Not Eve, whose fault was onely too much love,
> Which made her give this present to her Deare,
> That which shee tasted, he likewise might prove,
> Whereby his knowledge might become more cleare;
> He never sought her weakenesse to reprove,
> With those sharpe words wich he of God did heare:
> Yet Men will boast of Knowledge, which he tooke
> From Eves faire hand, as from a learned Booke. (ll. 801–8)

The stanza begins by aligning Eve's "fault" with "too much love." She brings the apple to Adam as a gift, giving him "this present" ("what she held most deare" [764]). Crucially, though, Eve's present to Adam is not the fruit, but the knowledge—"whereby his knowledge might become more cleare" (804)—reinforced by Adam's act of "prov[ing]" ("he likewise might prove"), here meaning "to test or assess the genuineness or qualities of."[117] By the end of the stanza, the tone of the knowledge exchange shifts dramatically from an act of gift giving to an act of forced extraction. Adam "took" knowledge "from Eve's fair hand," Lanyer insists, while criticizing, "yet men will boast of knowledge," a charge that anticipates Mary Astell's famous formulation (1694) that men "unjustly monopolize" knowledge and resent any woman "tast[ing] of that tree."[118]

Blaming Adam for taking knowledge "from Eves faire hand," Lanyer does several important things. She again conceptualizes knowledge as something tangible (something *to be* "took"), evoking the "key of knowledge" that unlocks Lady Lucy's breast in the prefaces. She infuses the history of knowledge and knowledge exchange with gender politics. And, most significantly for my argument, she compares Eve's hand to a knowledge repository: "as from a learned Booke." Eve's hand, standing in for the collective, polyvocal hands of women, becomes not only an original site of knowledge, but also a physical, material form from which knowledge can be extracted, not unlike how the womb as representative anatomical part, inscribed with powerful and mysterious knowledge, becomes subject to the probing hands of male anatomists.

As the mother of the human race in Christian theology, Eve was often the focus of anatomical interest in the womb. Several prominent anatomical illustrations from the period draw on Genesis imagery and tropes to cast Eve as potent paradigm for the "secrets of women." Thomas Geminus's *A Complete Delineation of the Entire Anatomy Engraved on Copper* (1545), which I discussed briefly last chapter, includes what historians of science refer to as an "Adam and Eve plate," a surface anatomy tableau featuring a man and woman (Figure 4.4).[119] Geminus, who began his publishing career in England by collaborating with Thomas Raynald on *The Byrth of Mankind, otherwise Named the Womans Book* (1545), revised this plate from Vesalius's *Epitome*, adding an apple held in the man's left hand and a serpent twisted through the skull at the figures' feet.[120] Geminus's surface anatomy of Adam and Eve merely teases (and withholds, à la John Banister) the "secrets of women" enclosed in Eve's womb, but other texts from the period offer a more comprehensive view. Adriaan van den Spieghel's *De Format Foetu* (*On the Form of the Foetus*, 1626), for instance, presents its readers with a fully dissected representation of Eve's womb in a plate depicting the gravid uterus (Figure 4.5).[121] Here, the woman holds an apple discreetly tucked into her left hand and the opening of her uterus is overpronounced, recalling the exaggerated womb on the *Fabrica*'s title page. The infant hovers above the folds of Eve's dissection wound (not so much wound as bloom), and the image draws on familiar imagery related to Eve's transgression in order to exhibit simultaneously the power of her womb (Eve's body as the origin of the human race) and the potential threat of her (and *the*) womb (Eve's body as the source of sin).

In these anatomical plates, Eve's reproductive body as representative female body is a knowledge repository shaped by and for men. In "Eves Apologie," Lanyer takes the anatomical and theological use of Eve's body and fractures its capacity as knowledge archive. That is, she splits the site holding the "secrets of women" between Eve's womb and her hand. Folded within the bookended references to "our mother Eve" and women's collective labor ("you came not in the world without our paine") is Eve's hand as a "learned booke," a comparison that defines Eve's hand (flesh) as a form of inscriptive media (word). For Lanyer's readers, this comparison might well have evoked the increasingly pervasive links between human hands and the lexicon of printed books. Early moderns often connected the popular term "handbook," for instance, to the Greek *enchiridion*, which lit-

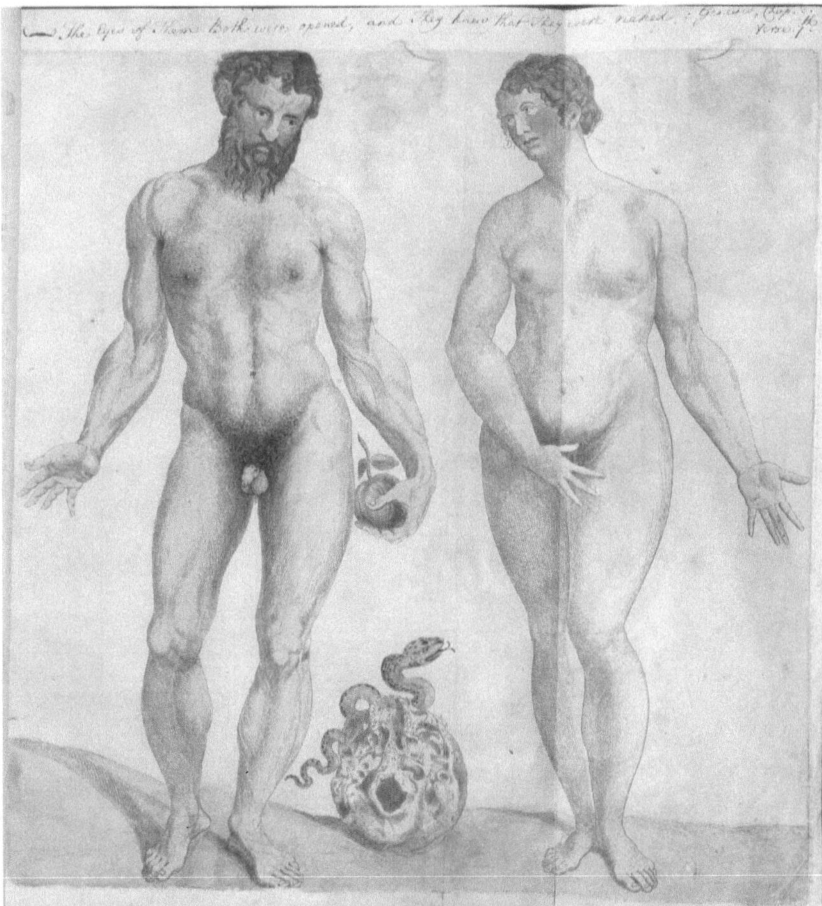

Figure 4.4. Surface anatomy "Adam and Eve" plate. Thomas Geminus, *Compendiosa totius anatomi[a]e delineatio* (London: J. Kingston, 1559), New York Academy of Medicine, VAULT 1. Courtesy of the New York Academy of Medicine Library.

erally translates as "in one's hand." In the fifteenth century, the English word "manual," previously used to refer to work relating to the hand, was increasingly applied to the form of a book ("of a book . . . intended to be kept at hand for reference").[122] Echoing the language of hands ("my unworthy hand writing" ["To the Ladie Margaret"]) and books ("desiring that this Booke Her hands may kisse" ["To the Queenes most Excellent Majestie"]) from the prefaces to *Salve Deus*,

Figure 4.5. Womb dissection plate. Adriaan van den Spieghel, *De format foetu liber singularis* (Padua, 1626), Table IIII, National Library of Medicine, WZ 250 S755dh 1627. Courtesy of the National Library of Medicine.

the stanzas on Eve reclaim and resituate the power of women's reproductive bodies to empower Lanyer's own writing act. Lanyer takes the mysterious, powerful, sacred knowledge of the womb, transfers it to women's hands, and claims the (folded) "learned booke" as its form.

"The First Fruits of a Womans Wit"

I have been arguing that Lanyer's *Salve Deus* is governed by gynecological and obstetrical thinking in the way it simultaneously confronts the epistemological exploitation of women's reproductive bodies and repurposes the anatomy of the womb as a poetic technology, an anatomical form that becomes a powerful site for incarnational experiments. This, however, cannot be the final word on Lanyer's intervention because, in her womb-oriented devotional poetics, she offers an important rejoinder to a Renaissance authorial commonplace—the womb as a metaphorical site for creative production. Perhaps the most famous employment of this metaphor comes at the opening of Philip Sidney's *Astrophil and Stella* when the frustrated poet, "great with child to speak, and helpless in [his] throws," is visited by his muse, who advises him, "look in thy heart, and write."[123] The successful production of the male poet manifests in the sonnet sequence that unfolds in the aftermath of this delivery. Edmund Spenser similarly figures the text of *The Fairie Queene* as his child in a dedication to Queen Elizabeth I (the product of "these his labors"), and John Donne woos his mistress in "Elegy 19" with the fertile craft of the pregnancy conceit. "Until I labor, I in labor lie," Donne insists, drawing on the erotic connotations of laboring ("to exert oneself physically").[124] We have also just heard this same language echo through the prefaces of anatomy texts—Crooke's book as his "first fruits," Vesalius's "rebirthing" of a scientific field. Throughout the sixteenth and seventeenth centuries, as Katharine Eisaman Maus and Elizabeth D. Harvey have demonstrated, writers employ such "analogies between mental creativity and bodily fecundity" in order to take advantage of exactly the secrets of women I have been tracing through the early modern anatomical archive in this chapter.[125] The womb's fecundity is so useful, Maus argues, because it "seems dependent upon hiddenness."[126] It is the mystery of the womb as a "container, itself concealed deep within the body," that makes it such a potent site for creative potential.[127]

In critical work on the "poetic use of birth metaphors," male poets take center stage and scholars tend to focus on explicit metaphorical language like the invocations of pregnancy and labor by Sidney, Spenser, and Donne.[128] Maus goes so far as to assert that "nobody in the English Renaissance argues that the analogy between poetic creation and childbirth means that women make better artists than men."[129] Lanyer's womb-oriented and woman-centric book must give us pause, though, because it is starting to sound as if Lanyer does, in fact, make that argument. It is harder to see, and relies on form rather than semantics, but now that we have unfolded her innovative formal work, we can study the language through which she frames her book with a keener eye and recognize that the anatomical interventions of her project are as much literary as they are scientific.

In the book's very first preface to Queen Anne, Lanyer acknowledges the entrenched humoral disparity that claims men (with dry and hot humors) are better suited to artistic endeavors than women (with moist and cold humors). Citing her "weake distempred braine and feeble spirits" ("distempred," as in, not properly tempered, disordered through excess or deficit), Lanyer states that she does not presume to "compare with any man," for "they are Scholers and by Art do write" (139; 148). "So Nature yeelds my Soule a sad delight" (149), she concludes, seemingly diminishing her poetic act within the humility topos framing that runs through this stanza ("pardon me [faire Queene] though I presume, / to doe that which so many better can" [145–46]). Lanyer begins the next stanza, though, by immediately recuperating the strength of her creative practice: "And since all arts at first from Nature came, / that goodly Creature, Mother of Perfection, / . . . why should not She now grace my barren Muse, / and in a Woman all defects excuse" (151–56). Connecting the logic of these two stanzas with the conjunction "and" instead of "but," Lanyer subtly overwrites—rather than directly refuting—the entrenched humoral paradigm cited in the previous stanza. If women are "distempred," moist, and cold, they are also children of nature, "mother" of "all arts." Lanyer reminds her readers of the maternal power of nature (in the previous stanza nature "yeelds," and here she is the "mother of perfection") before ending the stanza by linking her writing act to that fecund site of maternal productivity.

Or, rather, unproductivity. Lanyer's muse, unlike the fertile seedbeds of Sidney, Spenser, and Donne, is initially figured as "barren" (155). In fact, barrenness continues to feature in Lanyer's descriptions of her creative work. She refers to her "poore barren Braine" and her "weakling Muse" (276; 282), even as she describes her "Infant Verse" as "the first fruits of a womans wit" (279; "To the Ladie Elizabeth," 13). This paradox, however, becomes the site on which her formal innovation rests. Lanyer does not figure her text as a *product* of the womb but rather *as* a womb, infused with the mysterious, sacred, and transformational power inscribed onto that anatomical part. If the text itself is a womb, the barren poet needs only furnish it with the proper material for incarnational transformation—the power of virtuous women and the call of the divine.

At several points in the framing of her Passion poem, Lanyer proclaims her writing act as an act of immaculate conception, which folds our act of close reading back to where I began this chapter with the Virgin Mary's womb. The first of these declarations comes at the end of her retelling of Christ's Passion, where she asserts that she received a "charge" from heaven to write in praise of Lady Margaret: "and knowe, when first into this world I came, / this charge was giv'n me by th'Eternall powres" (1457–58). Echoing Gabriel's charge to Mary and her subsequent (re)productive act, Lanyer's heavenly charge also results in a productive act, the "learned booke" of *Salve Deus*. Another declaration comes in the form of the book's short epilogue "To the doubtfull Reader," a description of the poet's dream in which Lanyer performs one last act of folding, bringing the epilogue to touch with the book's title page. "Gentle Reader," she begins, "if thou desire to be resolved, why I give this Title, *Salve Deus Rex Judæorum*, know for certaine, that it was delivered unto me in sleepe many yeares before I had any intent to write in this maner" (137). Lanyer's dreamy prophecy "immediately" comes back "into [her] remembrance" after she pens the Passion poem, and she takes this confluence as "a significant token, that [she] was appointed to performe this worke" (137).

Twice over in the book's final pages, Lanyer frames her poetic labor not only in service of Christ but also as a direct result of a divine incarnational charge. She answers this charge by crafting a womb-oriented poetics given "membranous" and "fibrous" substance by the collective fertile and reproductive acts of the women (and mothers) folded into her book's pages. The "fruit" of her "wit" is thereby the product of this textual anatomical form and, finally, the poet's coupling with Christ as his "servant, mother, wife, and nurse" (1087). Drawing again on the language of infertility in the early lines of the Passion poem, Lanyer intimates that the task ahead is "a matter farre beyond [her] barren skill" in a plea not to her muse but to Christ himself. "Yet," she continues, "if he pleaseth t'illuminate my Spirit . . . that I may write part of his glorious Merit, / if he vouchsafe to guide my hand and quill, / to shew his Death . . . then will I tell of that sad blacke fac'd Night" (321–27). The poem that unfolds from this prayer, steeped in the incarnational language of the book (Christ as spirit and flesh touching Lanyer's flesh and pen, guiding her poetic act), is both anatomical-poetic site and the "fruit" of Lanyer's divine coupling.

In Lanyer's intimate, fleshly devotional work—propelled by her womb-oriented poetics—she confidently stakes a claim to early modern anatomical thinking, crafting her entire text as an anatomical form. If *Salve Deus* is "arguably now one of the most widely read religious poems of the seventeenth-century," what are the critical habits and paradigms that have kept us reading this text without recognizing Lanyer's active participation in Renaissance anatomical culture?[130] More than any body of work I have discussed thus far in *Anatomical Forms*, my reading of *Salve Deus* is a provocation to read women's writing differently, to continue to ask: What do we go to early modern women's writing for? What are the habitual moves we make when we encounter a new (or familiar) piece of writing by an early woman writer, and how do those habits regulate the conversations within which we situate (and *can* situate) women's writing? As a final, very direct engagement with these questions, I now turn to Hester Pulter, the most recently recovered poet of my book's archive, in order to think critically about how we do the work of literature and science studies.

CHAPTER 5

Situated Knowledges

Hester Pulter's Manuscript Laboratory and Our Archives of Scientific Forms

The decade from 1996 to 2006 was an extraordinary one for scholars of literature and science. In 1996, Mark Robson discovered a miscatalogued manuscript of poetry and prose in the University of Leeds archives. Attributed on various title pages to the "Noble Hadassah," lettering on the book's spine identified the manuscript as "Lady Hesther Pulter's Poems." This 5″ × 8″ volume of 167 leaves and several loose sheets contained a large collection of poems and one unfinished prose romance titled *The Unfortunate Florinda*. The poetry was split into two sections, one titled three times over—"Poems Breathed Forth by the Nobel Hadassas," "Hadassas Chast fances," and "Poem's Wrighten By ye R^t^ Honerable H.P."—which included what appeared to be political, devotional, and occasional poems. The second poetry section was titled "The sighes of a sad soule Emblimatically breath'd forth by the noble Hadassah," and included a highly unusual collection of emblem poems focused on different animals, insects, and plants. The story of Robson's discovery of Hester Pulter's (c. 1605–1678) manuscript is the stuff of dreams for premodern scholars—the recovery of a truly *lost*, and very substantial, collection of writing by a seventeenth-century English woman writer that had languished on the shelves of libraries and archives for more than two hundred fifty years.[1] I say "premodern scholars," but should clarify that Pulter's manuscript has been of particular interest to premodern *literary* scholars. Alice Eardley, scholar of early modern poetry and women's writing, edited the first complete critical edition of Pulter's poems in 2014, closely followed by Leah Knight and Wendy Wall's open-access digital critical edition, *The*

Pulter Project: Poet in the Making.[2] These efforts have made Pulter's writing accessible to a wide range of readers, and Pulter's poems are increasingly part of the standard early modern literature survey syllabus.

In 2006, ten years after the recovery of Pulter's manuscript, an auction-house representative was performing a routine valuation visit to a country house in Hampshire and was handed a "battered box of papers" that included a bulky manuscript of notes and meeting records.[3] On further examination, the manuscript's main hand was identified as Robert Hooke's, natural philosopher, scientist, and prominent founding member of the Royal Society. The manuscript's more than five hundred pages were split into two sections, the first comprised of Hooke's transcribed notes from the Royal Society records of Henry Oldenburg, who Hooke thought was actively sabotaging his career, and the second comprised of Hooke's own draft meeting notes during his time as Secretary of the Royal Society from 1678 to 1683. Historians of science have since used Hooke's notes to piece together and confirm important information about the early work of Society members. Howard Gest, microbiologist and historian of science, for instance, published a paper in 2009 arguing, "with the recovery of the Hooke Folio, we can now say with confidence that the origins of microbiology as a science have finally been clarified."[4] This story, too, is the stuff of dreams. As Robyn Adams and Lisa Jardine (both instrumental in the manuscript's recovery) chronicle it, the "Hooke folio," as it is now known, is "a veritable treasure-trove of freshly contextualized information about the day-to-day activities of the early Royal Society."[5]

I want to draw our attention to one particular difference between these stories of manuscript recovery. Hooke was already a well-known figure when his manuscript was discovered, and so the text (both the physical object and its contents) slotted very neatly into an established, received history of science narrative. In fact, the pages of the Hooke folio quite literally filled gaps in the Royal Society archives—empty stubs in the bound volumes of Society meeting records were waiting for Hooke's missing pages.[6] Pulter, on the other hand, was a completely unknown figure. As Eardley suggests, her manuscript may have remained in obscurity, miscatalogued for so many years "because no one was able to identify Pulter as an individual of historical interest."[7] At the time of my writing, scholars are still piecing together Pulter's biography, her familial and social networks, and the day-to-day conditions of her writing life. The difference here, I want to emphasize, is that Hooke's story is one of certainty and completion, while Pulter's story is one of uncertainty and partial knowledge. To put it another way, the difference between these two recovery stories is the difference *presence* makes, an archival condition that has resulted in the disparate treatment these two manuscripts have received in the aftermath of their recoveries. As I hold these two stories, these two manuscripts, and these two early modern thinkers together throughout this chapter, I also want to hold onto this tension in their recovery narratives, because it encapsulates one of the core inquiries driving *Anatomical Forms*: what counts as evidence of a certain kind of knowledge?

Although these two manuscripts have never been linked, it is probably clear by now that I think they ought to be. The story of this final chapter is a story about how Pulter's and Hooke's work overlaps in significant, complex, and fascinating ways. It is also a story about how we understand the interconnections between the histories of literature and science. As we will see, Pulter and Hooke are both keenly interested in acts of observation, embodied scientific practice, and modes of representation. Hooke's most famous work, *Micrographia: or Some Physiological Descriptions of Minute Bodies Made by Magnifying Glasses* (1665), was the first English treatise on microscopy, and Hooke used the book not only to exhibit hundreds of specimens observed through the microscope, but also to demonstrate the value of the microscope as a new tool for physiology and anatomy. *Micrographia* was also a landmark publication in the developing field of natural history, seeded in Francis Bacon's writings and particularly important to the founding of the Royal Society in 1660. Pulter's poetry, I argue throughout this chapter, is connected quite closely to the Society's early work, led by Hooke, Robert Boyle, Christopher Wren, Nehemiah Grew, and many other figures now firmly entrenched as pioneering scientists, natural philosophers, and natural historians. While Pulter is increasingly discussed as a scientific thinker, such arguments remain supplementary to critical claims about her value as a poet and literary writer. It has been the project of *Anatomical Forms* to animate the formal and material intersections of poetic and anatomical work. Returning to the moment of first encounter with Pulter's manuscript and building on developing critical work on Pulter's science, this chapter speculates: what if we had read Pulter as a scientific thinker *first*? Or, to ask a more expansive version of this question, what if we habitually approached early modern poetry as an indispensable site of scientific thinking?

If "form is both the enabling condition and the product of reading," as Ellen Rooney contends, then careful attention to our own processes of reading must be part of our formalist work.[8] In this chapter, I am especially interested in how we have read Pulter's and Hooke's writings, how entrenched paradigms of form have shaped our readings of their texts, and how the interdisciplinary methodologies I have developed and employed throughout this book might help us notice new things about these writers' formal experiments. This chapter is more of a thought experiment than the other chapters of *Anatomical Forms* have been thus far. It is the most speculative and, in writing this book, has been the most challenging because the pieces of evidence I bring together are laced with the uncertainty and partial knowledge that surrounds Pulter's manuscript itself and the circumstances of reading it. My aim here is to consider how the innovative interdisciplinary methods of early modern women poets can help us reorient our critical gaze. In the first part of this chapter, I connect Pulter's and Hooke's work through more conventional, familiar approaches to literature and science studies. I focus primarily on the content of their writings in order to demonstrate their shared interest in observation as a framework for understanding the world. In the second half of the chapter, I then explore new ways into this pairing through the methodological framework of *Anatomical Forms*. Margaret Cavendish, Mary Sidney Herbert, Mary Wroth, and Aemilia Lanyer have taught us to look at less obvious forms of scientific work, to trace material rather than conceptual links between poetry and science, links that have less to do with content and more to do with process, modes of knowing, and acts of forming. Attending closely to form and the acts of forming explored in Pulter and Hooke's texts, this chapter ultimately argues that anatomical thinking—a study of bodies practiced via careful inquiry about the "knowledge of the frame" and "use of each part"—functions as the mechanism powering both Hooke's and Pulter's innovative epistemological work.[9]

In the first half of this final chapter, it should be noted, I briefly turn away from this book's focus on anatomy and take up the more capacious, flexible, and unwieldy term "science" (*sciencia*, knowledge writ large) in order to use my analysis of Pulter's and Hooke's writings to raise broader questions about the critical field of literature and science studies. On the surface—that is, in its content and central themes—Pulter's poetry has little (almost nothing) to do with anatomy but a great deal to do with natural history, the prevailing field of increasingly institutionalized scientific inquiry in late-seventeenth-century England, largely a result of the nation's imperial project, ongoing colonial exploits, and burgeoning global trade economies.[10] As we will see, there is much to learn from a brief (re)consideration of an interdisciplinary literature and science studies approach that takes shared language, imagery, and ideas as its focus. And, as it will turn out, the work of natural history is not so far removed from the work of anatomy in early modern England. Nehemiah Grew "joyn[s]" these fields quite explicitly at the start of his *The Anatomy of Plants* (1682) when he insists that, in order to perform the thorough acts of observation required for natural history, "together with the Knife it will be necessary to joyn the Microscope" (what Al Coppola calls "Grew's anatomical method").[11] My shifting from anatomy to natural history and back again illuminates how anatomical thinking infused other forms of early modern English science, an ancillary but equally important claim of this chapter.

Specimen Poems and the Observational Stance

Hester Pulter was an exceptionally observant poet of things large and small, a quality on full display when, about halfway through her collection of poems, she takes a detour into the ant world. "Walking abroad, once in a Summers day," the poet narrates, looking for a place to rest her tired head and "panting Breast," she throws herself "upon the Grass" and leans her head "against a Siccamoore" (35.1–7).[12] The poet lies on the ground, "museing and looking on [her] Mother, Earth." After some time spent reflecting on her mortality and embodied connection to the earth ("to which I must; from whence I drew my breath" [10]), she spies an ant hill ("a hill of Pismires" [20]). She watches the ants at work, observing how some of them "lugger[ ] up and down their flatious issew," some transport "wheat, and other nibled grain" to store for winter, and others—the ants with "glittring wings"—seem to "keep perpetuall holy day" (21–26). In the second half of the poem, the poet connects her study of the ants to her metaphysical musings in the poem's opening lines. Reading the ants as "an embleme of the World" (28), the poet longs for the "glittring wings" of the playful ants that free them from the "Earthly Clog" (22; 36). "Grant mee the wings of som unspotted Dove," she finally pleads, shifting from entomology to ornithology, as she seeks a way to "ease [her] troubles" and "fly" to her "Eternall rest" (43–45).

Titled "The Pismire" (Poem 35), this poem, like many of Pulter's poems, uses the natural world as a catalyst for political and theological exploration. Pulter's representation of the ants as an "emblem of the world" seems especially important to the knowledge work of this poem. If we were to compare Pulter's depiction of the ants in "The Pismire" and Robert Hooke's depiction of the ant in *Micrographia*, for instance, we might conclude that Pulter's emblematic thinking is in direct contrast to Hooke's empirical thinking. The difference here, as William B. Ashworth Jr. outlines, has to do with the degree of associative thinking involved.[13] In Robert Hooke's "Observ. XLIX. Of an Ant or Pismire," he notes that the ant "is more than half the bigness of an earwig, of a dark brown, or reddish colour, with long legs" (203).[14] It has "a large head . . . at the upper end of which were two protuberant eyes, pearled like those of a fly, but smaller" and its body "seem[s] to consist of three joints or pieces" (204). For Hooke, every tiny detail about the ant's physical features, anatomy, and physiology becomes a possible data point, and his corresponding engraving of the ant as observed through a microscope reinforces our sense that this is a specimen studied in isolation (Figure 5.1). In contrast to this empirical knowledge work, the "emblematic world view," as Ashworth defines it, relies on a "web of correspondences," so that knowledge of any one thing "consists of an attempt to comprehend" as many of its "myriad hidden meanings" as possible.[15] Pulter's emblematic mode uses the natural world as analogical device to better describe or elucidate a social, cultural, or political concept. The ants represent the mortal coil of all beings "that moyle and labour in this dunghill Earth" (35.30), from kings to "the Naked, sunburnt female slave" (35.32).[16] "For who is free," Pulter finally exclaims, "untill his soul doth spring / from's Earthly Clog, and joy fully takes wing?" (35.35–36).

But we miss something about "The Pismire" if we shift too quickly into the "web of correspondences" and away from the poet's study of ants. Pulter's poet is an observer and student of nature—a quotidian natural historian. At the time of this poem's composition, insects had long been important emblematically, as allegories for social and political systems,[17] but they were also becoming increasingly central to the study of natural history, a field of scientific inquiry scaffolded by Francis Bacon in *Of the Proficience and Advancement of Learning, Divine and Human* (1605) and *Novum Organum* (1620), and advanced in practice by his *Sylva Sylvarum: Or, A Natural History* (1627), published posthumously by William Rawley. "The interpretation of nature," as Bacon outlined it, required a systematic "kind of logic" based on experiment and induction, and the field progressed immensely with seventeenth-century improvements to the microscope.[18] Some of the earliest published compound microscopy studies, dated between 1625 and 1630, were detailed examinations of insects (a bee and a weevil) circulated by Italian naturalists Frederico Cesi and Francesco Stelluti.[19] By 1665, Hooke was arguing in his *Micrographia* that insects were the ideal specimens for Bacon's natural history project. "We can with a microscope, in these smaller creatures," Hooke insisted, "quietly peep in at the windows, without frighting [Nature] out of her usual bias" (186). Hooke situated his microscopy project as an important corrective to dissection practice, which "endeavour[ed] to pry into [Nature's] secrets by breaking open the doors upon her, and dissecting and mangling creatures whil'st there is life yet within them" (186).[20] Instead, studying the "delicate and pellucid teguments of the bodies of insects" through the microscope allowed the scientist to leave nature "undisturbed" (186).

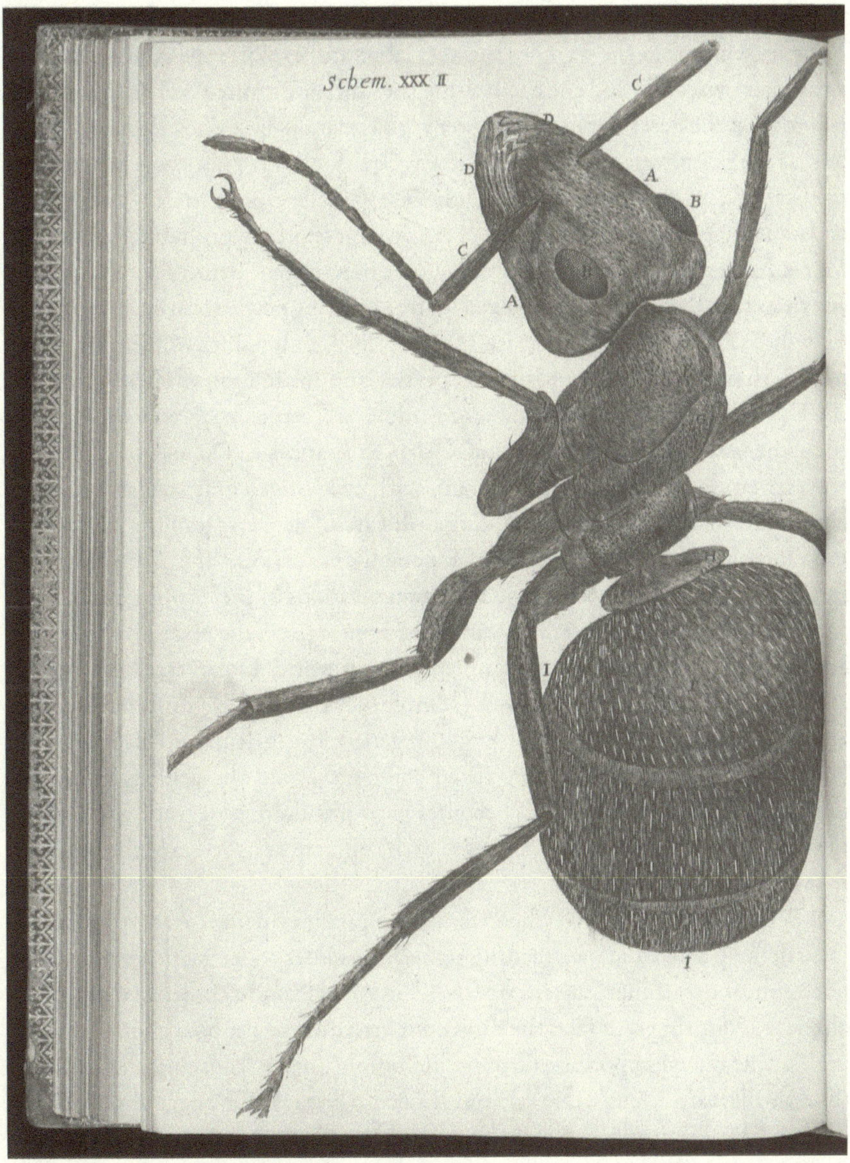

Figure 5.1. Robert Hooke's engraving of the ant. Hooke, *Micrographia: or some Physiological Descriptions of minute bodies made by magnifying glasses* (London, 1665), Schem. XXXII, Wellcome Collection, EPB/C/29309.

Pulter's account of the ants depicts a poet-observer "quietly peep[ing] in" at Nature's window as she watches the ants labor and explores the relationship between her own body and the earth using the "hill of Pismires" as her lens. Pulter records specific details of her encounter with the ants—not just a "summer's day" but, as the poet "well remember[s]," it was "in May," and she locates her study at the base of a sycamore tree (35.2; 7). Likewise, the ant Hooke studies in the pages of *Micrographia* was "one of many, of a very large kind, that inhabited under the roots of a tree" (203). Pulter's physiological observations identify three different activities that the ants were employed in: transporting new resources, storing food for winter, and playing. In contrast, Hooke records only a single, swarming movement: "they would sally out in great parties and make most grievous havoc of the flowers and fruits, in the ambient garden, and return back very expertly, by the same ways and paths they went" (203). As Frances E. Dolan notes, Pulter's representation of a hierarchical system, with the winged ants at play while the others "moyle and labour in [the] dunghill earth," departs from the more common early modern representation of ant colonies as collectives.[21] Pulter's anthill is more consistent with early modern representations of the beehive, often cited as models of ordered hierarchy. Even as she homes in on the pismire, the poem's lexicon signals a broader interest in the natural world. Dolan tracks how Pulter's language throughout the poem ("dunghill" and "moyle") links her study of ants to what we would now think of as regenerative agriculture.[22] The ants are not just laboring to store food for their colony; they are also actively maintaining the "dunghill," which itself becomes part of Pulter's observational study as she mingles her metaphysical musings "with concrete knowledge of the practices that build and maintain soil."[23]

What is "The Pismire" when we focus on such evidence? Are we still reading a literary text, or are we reading something closer to scientific jottings? It is not a minor detail that Pulter's poetic "I" in "The Pismire" is defined as a poet-observer. Four times, in fact, she draws our attention to the poet's active observing: "my heavy eyes upon the Ground did pore" (line 8); "museing and looking" (line 9); "I look'd about" (line 17); and "I spied" (line 19). The poet's observational work in "The Pismire" is bolstered by the insistent observational stance that drives Pulter's series of fifty-three emblem poems.[24] In these poems, the poet constantly calls on her readers to look at things, inviting us to turn our attentive eye to the natural world. "Behold this flying fish" (Poem 90); "Come look upon these pretty, innocent Doves" (Poem 101); "View but this Tulip" (Poem 105); "Seest thou this Horizentall Bird" (Poem 100). Pulter uses such language to extend the observational work we witnessed in "The Pismire" to the reader, calling on us to notice and study plants, animals, and insects as a way of learning something about ourselves and the events going on around us. "Behold how many Cobwebs doth invest / this ugly Spider," Pulter begins another emblem (Poem 102). Studying this "ugly spider" and how she draws "Flies, Wasps and Hornets" into her "dusty nest," Pulter asserts, we can learn something about man, "the greatest Beast of prey of all" (102.6–7; 27).

The act of observation, Lorraine Daston and Elizabeth Lunbeck assert, "is the most pervasive and fundamental practice of all the modern sciences, both natural and human."[25] And this fundamental scientific practice finds its most robust origins in the early work of the Royal Society, anchored in Bacon's proposal for a new scientific method based on observation and experiment. In Bacon's proposal for a "great renewal" of learning (outlined in his incomplete *The Great Instauration*, of which *Novum Organum* was the second of six parts), he laments that "men are wonderfully inattentive" and "observe nature only in a desultory and casual manner and after bodies are finished and complete, and not in its working."[26] Instead, Bacon argues, "one should keep a kind of round-the-clock watch on nature": "if you wanted to see the skills of a craftsman and observe his work, you would not only want to view the unformed materials of his craft, but also be there when he was at work himself and shaping his product. One must do something like that with regard to nature."[27] Above all, the knowledge maker must resist exactly the kind of conclusions that seem most important to emblematic thinking—conclusions (Bacon uses the phrase "highest generals") that are based on deductive reasoning and "some few particulars."[28] The practice of "solid and diligent observation[ ]," as Bacon defined it, will "cleave to the individual intrinsics of things."[29] The scientist must stay focused on particular details (extracted from "the very entrails of nature"), proceed systematically through a step-by-step experiment ("skillfully according to art invented and accommodated"), test and hone the epistemological veracity of the body's five senses ("sift and sound the information of sense itself; for the senses deceive"), and refrain from "seizing upon the highest generals, but in the last place."[30]

Bacon's utopian scientific method, as historians of science have argued, was just that—a utopian ideal, something to work toward but not necessarily (or always) attainable in practice.[31] Certainly, in the early days of the Royal Society, the methodological boundaries between Bacon's proposed empirical methods and the "vulgar logics" he decries (primarily, arguments based on syllogism) were still very slippery.[32] We see evidence of this in Thomas Sprat's *The History of the Royal-Society of London* (1667). Sprat frames the Royal Society's careful methods in contrast to the works of "Pliny, Aristotle, Salinus, and Aelian," which "abound[ ] more with pretty tales, and fine monstrous stories, than sober and fruitful relations."[33] And yet, in his later account of Thomas Henshaw's (noted alchemist) study of saltpeter, Sprat quotes Henshaw's explicit reliance on Pliny's *Natural History*, a vast encyclopedia on topics including zoology, astronomy, geology, and botany. "Concerning the production of nitre," Henshaw notes, Pliny's account "agrees very exactly with what I have been informed by a refiner of salt-peter."[34] The Royal Society Fellows were citing, testing, and refining Pliny's work—and so was Pulter. One might say she *instrumentalizes* Pliny's text as her emblems constantly reference, interpret, and extend Pliny's knowledge about plants and animals. The manuscript's main scribal hand has included indexical references to Pliny's *History* (likely Philemon Holland's English translation, first published in 1601) in the margins of several emblems, a testament to how thoroughly Pulter's poems engaged with the kind of encyclopedic knowledge building—instincts toward collection and study—that would become so central to early modern natural history. Marginal notes on Emblems 10, 11, and 12, for example, direct the reader to specific books and chapters in Pliny's *History* ("Plinie Natural Hist 10:12"; "Plinie 2nd booke; chap; 25th") and related classical texts ("Doctor Sanderson out of Aristottle his 6 Hist"). These emblems even include annotations on specific creatures referenced in the poem, directing readers to corresponding margin notes for further context (see Figure 5.2).[35]

Pulter's Emblem 53 (Poem 118) provides a particularly potent example of how different modes of knowing intersect in her explorations of nature. The poem initially reads as a kind of fable, steeped in classical imagery and primed for a purely symbolic rendering of the natural world. The poet describes the sun as "fair Aurora, drest with raidient Light," and introduces the poem's central character (a bee) as "the active Amizonian Maid" (118.1; 7). At first, this bee could be any bee—an abstract, emblematic figure buzzing about "extract[ing]" pollen from various plants: "out of every Gold Enamold Cup / her Mornings draught of Nectar shee did sup" (118.9–10). As Pulter's poem progresses, however, the bee's travels get more specific. The bee lands on "nightshade," "henbane," "acconite," "opium," and "hemlock," and the poet notes that the bee is able to "safely" extract pollen, even from these plants that are poisonous to other creatures (the "toad and spider") (118.11–14). The bee is about to fly away (back to her "sexanguler cell"), but then spies the poet's garden and "could not choose but stay" (118.16–18). The poet enters the emblem in the first-person voice and becomes simultaneously witness, narrator, and student of the bee's actions. She traces the bee's path to specific plants in her own garden—auriculas and tulips—even as she maintains the poem's emblematic tone, anthropomorphizing the bee ("this

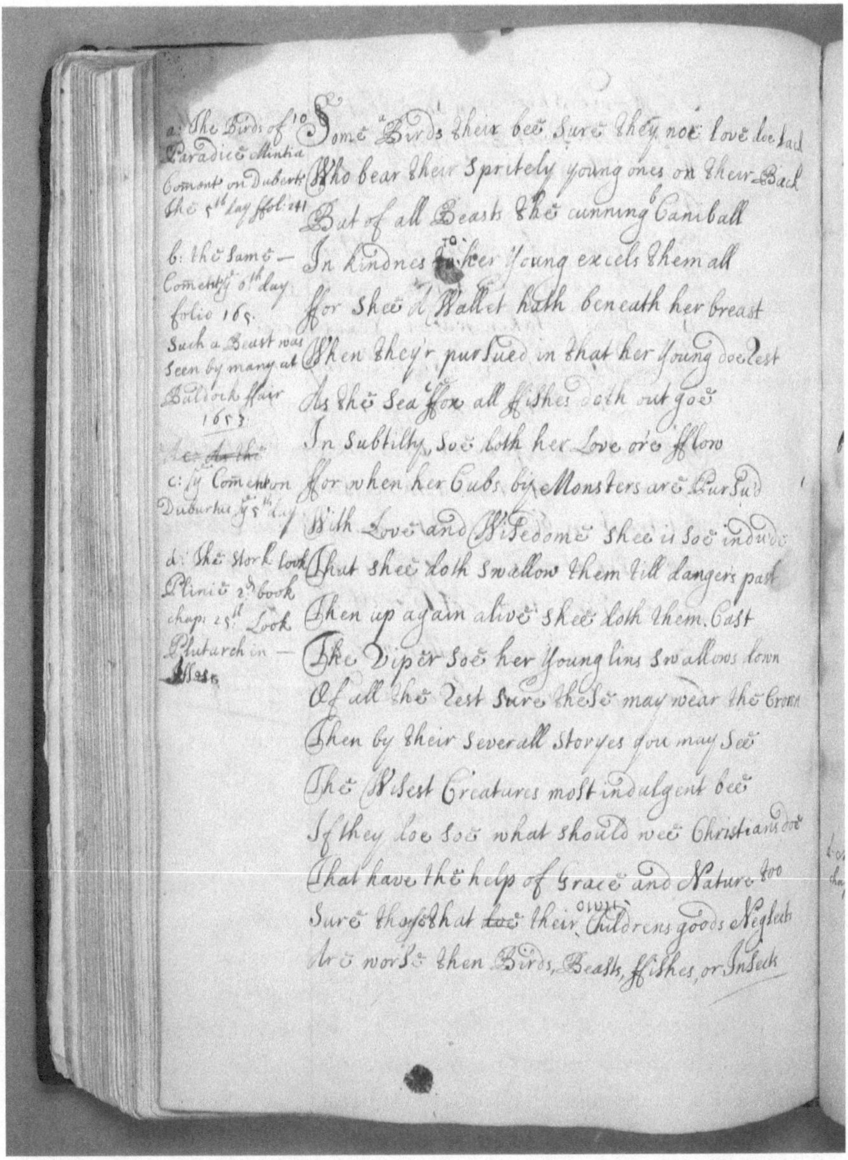

a: The Birds of Paradice Mintia Coment on Dubert the 5th day ffol: 141
b: the same — Coment the 6th day folio 165.
Such a Beast was seen by many at Bullock ffair 1653.
c: sy Coment on Duburtus the 5th day
d: the Stork look Plinie 2d book chap: 25th Look Plutarch in —

10 Some Birds their bee Sure they noe love doe lack
Who bear their Spritely young ones on their Back
But of all Beasts the cunning Caniball
In kindnes to her Young excels them all
For shee a Wallet hath beneath her breast
When they'r pursued in that her Young doe rest
As the Sea fox all fishes doth out goe
In subtilty, soe doth her Love o're fflow
For when her Cubs by Monsters are pursu'd
With Love and Wisedome shee is soe indu'd
That shee doth swallow them till dangers past
Then up again alive shee doth them Cast
The Viper soe her younglins swallows down
Of all the rest sure these may wear the Crown
Then by their severall storyes you may see
The Wisest Creatures most indulgent bee
If they doe soe what should wee Christians doe
That have the help of Grace and Nature too
Sure those that their owne Childrens goods Neglects
Are worse then Birds, Beasts, ffishes, or Insects

Figure 5.2. "Some Birds their bee Sure they noe love doe lack" (Emblem 10) with margin notes. Lady Hesther Pulter, *Poems breathed forth by the nobel Hadassas, and The Unfortunate Florinda*, Manuscript (c. 1645–1665), 97v, Leeds University Library, BC MS Lt q 32.

covetous Insect") and using the bee's actions to deliver moral lessons ("to carry home her wealth shee'd not the power / till shee had search'd the sweets of every flower") (118.22–24).

There is a wealth of emblematic precedent for this poem's central characters and plot points. Geoffrey Whitney's emblem titled "Vitae, aut morti" ("Life, or death") describes the "two contraries" that exist in "one flower"—poison and honey—and uses the spider and the bee as its evidence, as Pulter does in lines 11–14 of her poem.[36] The "toad and spider" of line 11 were frequent characters in early modern emblems and fables, and Pulter herself centers them in the action of another emblem poem where toad and spider duel to "trie the might / of eithers poyson, in a single fight" (Poem 88). This well-known fable circulated widely in English texts of the period, including in Edward Topsell's *The History of Four-Footed Beasts and Serpents* (1658), a popular and extensive bestiary that compiled fantastic legends about actual and mythical animals.[37] Similarly, the snail was a frequent emblematic figure, and it joins the bee at the end of Pulter's emblem.[38] Bee and snail are trapped together in a tulip as the plant closes its petals with the setting sun. The bee fights to free itself and dies struggling ("thus with Beating of her self did die"), while the snail calmly waits for morning and then safely exits the tulip's flowerhead (118.36). "Then let impatient spirits here but see," the poet concludes, "what 'tis to struggle with their destinie" (118.37–38).

As Rachel Dunn has noted, Pulter's poems "display[ ] a splintering of the emblem form itself," and we witness this splintering in Pulter's attention to nature in the emblem poems.[39] Nature is never simply symbolic or allegorical; it is active and alive, and the poet exists within and learns things about nature in real time. Pulter calls on her reader to look, and then puts herself in the poems ("my garden") as a firsthand observer of the insect and plant worlds. In Emblem 53, she cites specific times of the season: "then were my Tulips painted in there pride" (118.21). She watches as the setting sun affects her garden's plants: "the sun, from home, all, Influence receives, / bid them decline, the Tulip clos'd her Leaves" (118.25–26). And she documents her own embodied movement in the garden as the poem shifts temporally from sunset to sunrise. "Night being past," the poet records, "the Tulip then her gilded Leaves displais" (118.33–34). The present tense ("displais") marks the poet's return to the garden to observe, in real time, the outcome of the insects' night in the tulip. As the emblem form splinters, empirical study fills the gaps, and, by the end of the poem, Pulter's final motto reads less as another moral lesson and more as a prompt for further study: "think on the Snail, the Tulip, and the Bee" (118.51–52).

I say "final motto" because this is one of at least three possible mottos to Emblem 53, along with the poet's warning to "impatient spirits" (118.37–38) and her reflection that "'tis valianter by far to live then die" (118.50). Saturated with such maxims, we might consider that Pulter's emblem poems invite us to question exactly the kind of knowledge work emblems do. What makes a motto? What have I *learned* from this poem? Pulter's poetry, with the emblems as the most explicit example, participates in the expansive early modern epistemological project seeded by Bacon's argument for a new science, the ongoing theorization of how to make knowledge with and through nature in new ways. Robert Hooke's descriptions of his specimens in *Micrographia* similarly hail the reader's attentive eye—"look at this"—and focus its attention on specific examples of insects, animals, and plants. Like Pulter's poems, Hooke's descriptive prose zooms out for a telescopic view of the setting within which we encounter the specimen, and then zooms in for a microscopic view. Introducing "the tufted or brush-horned Gnat," for instance, Hooke first notes that "this little creature was one of the multitudes which fill our English air all the time that warm weather lasts," and then goes on to describe how he observed "several small swarms of them playing to and fro in little clouds in the sun" on a spring day (193). These general observations frame Hooke's particulars and, while Pulter does not go the extra step to pull out a microscope, the work of both writers is thoroughly informed by their observational stance.

Situated Knowledges and Histories of Science

Already, then, we must revisit Sarah Hutton's assessment in an early argument about Pulter's manuscript after its recovery in 1996: "Hester Pulter did not write 'scientific poetry' in the sense that the purpose of her poems was to convey scientific theory or scientific knowledge in poetic form."[40] Instead, Hutton goes on to argue, Pulter makes "allusions to contemporary science" in the service of "spiritual reflection," resulting in "unconventional adaptations of conventional spiritual motifs."[41] Hutton's assessment follows the contours of "scientific poetry" outlined by Robert M. Schuler, who defined it as "didactic poetry whose purpose is to instruct in science."[42] Schuler's argument, like much of the first wave of critical work in literature and science studies that I outlined in this book's introduction, took the model of modern-day science as its starting point and looked for early modern poetry that empirically and instructively explored the content of discrete scientific fields. Even within such a restrictive definition of Renaissance scientific poetry, however, Pulter's writing fulfills the remit—evident, most obviously, in her consistent treatment of the three dominant subjects Schuler ascribes to the genre: alchemy, medicine, and astronomy. On alchemy, one can reference Pulter's "The Circle" (Poem 21); on medicine and illness, "Made when I was not well. April 20. 1655" (Poem 51); and on astronomy, "The Revolution" (Poem 16).

More recent work on Pulter's manuscript has productively troubled Hutton's early assessment of Pulter's scientific "allusions," and scholars increasingly recognize the many forms that her science takes.[43] Frances Dolan positions Pulter as an active participant in seventeenth-century agricultural practices, "particularly the making of compost and amendment of soil," as we have already seen in Dolan's reading of "The Pismire."[44] Ruth Connolly's work on Pulter's "childbirth poetics" and Amanda Zoch's account of "felt mortality" in Pulter's poems, while not explicitly intervening in critical conversations on Pulter's science, both situate Pulter's poems about childbirth (all written from *within* the lying-in chamber if we are to believe the poems' titles) as provocative acts of observation based on Pulter's embodied experience of pregnancy and labor.[45] Lara Dodds, who also begins with Pulter's observational stance, urges us to attend to Pulter's "eclipse poetics," while Victoria E. Burke reads Pulter as a metaphysical poet employing astronomy and cosmology to traffic in alternate worldmaking.[46] Finally, Liza Blake finds a "deep interrelation" between Pulter's "physics and her poetics," but outlines a definition of *physics* that resists linking Pulter's work directly to "the systematic and sustained epistemological truth-making, or study of nature, comparable to that happening in the contemporary Royal Society."[47] For Blake, importantly, Pulter's physics exists alongside and proximal to, rather than within or in opposition to, the Society's work, a relational model I will draw on when I return to anatomy later in this chapter.

Pulter's natural history provides another way into this conversation, and it is a compelling anchor tethering Pulter's scientific work directly to the epistemological project of the Royal Society. Critical work on Pulter's engagements with a variety of scientific fields reminds us that women's writing is primed to destabilize entrenched paradigms about the relationship between literature and science in early modern England. Pulter's writing, as Blake demonstrates, offers particularly complex answers to literature and science studies questions in large part because her writing does not follow the familiar underlying warrants of such inquiries ("there is no 'fundamental' physics for Pulter"; "no one source . . . is fully sufficient to understand the vision of the universe that even one single poem conjures and plays with").[48] Up to this point, my attention to Pulter's writing has been focused on how early modern science can help us attend to different details of her poetry, what Howard Marchitello calls the "influence model" of literature and science studies in which "the issue most pressingly at stake is the labor of demonstrating the influence of science on literary texts."[49] In this next section, I follow the arc of literature and science studies history, staying focused on content, but attempting to craft a more "creative dialectic" between Pulter and Hooke's work.[50] What do we gain, this section asks, from an interdisciplinary approach that recognizes the complex, powerful dialogue between literature and science in early modern England? Thinking with second-wave science and literature studies methods, that is, I consider Pulter's poetry "as a knowledge practice" and use her observational methods to adjust our approach to Hooke's archive.[51] Both Hooke and Pulter are observational thinkers but, as we will see, both are also more interested in theorizing the *act* of observation than they are in the knowledge that results from their observational acts.

In order to trace Hooke and Pulter's attention to method over results, we need a better sense of how they structure their observational acts. As we glimpsed in Hooke's study of the ant, his approach in *Micrographia* is based on the static specimen studied primarily in isolation from its surrounding ecological network. Hooke divides his book into discrete acts of observation, with each case study titled "Observ.," along with corresponding number and descriptive subtitle, such as "Observ. LIV. Of a Louse" (211). Observation, therefore, is not only Hooke's primary method of knowledge making, but also the structuring apparatus of his text—what we might call the *specimen form* of *Micrographia*. One can imagine an entomology exhibit at a contemporary natural history museum, with individual insect specimens preserved and pinned to a display board for study, and the effect is similar to reading Hooke's insect observations in succession. He constructs his text as an accumulation of individual examples, each taken in turn, isolated from the others, and often accompanied by a visual illustration of the specimen. This structure enables Hooke to introduce and model (again and again) the merits of his microscopy methods, but this focus on the micro-details of individual specimens also neglects the interconnected systems of the insect, plant, and animal worlds.

Pulter's poetry, on the other hand, is highly attentive to such ecologies. Her emblem poems offer the reader focused studies of sunflowers, birds of paradise, moose, turtles, ravens, porcupines, elephants, turtledoves, raccoons, marmots, spiders, tulips, ostriches, and bees, but always as one part of a surrounding environment. In a rich discussion of maternal care in Emblem 10, for example, Pulter focuses on the "cunning Caniball"—an unidentified type of marsupial that has "a wallet . . . beneath her breast" where she carries her young (76.3–6). Pulter links the canibal to birds that "bear their spritely young ones on their Back," as well as the "sea fox" (a type of shark) and the "viper," both of which swallow their young when they are in danger and then spit them up alive when "danger's past" (76.1–2; 7–13). Zooming in with specific details of each creature and then zooming out to link them together as examples of "wisest creatures" from which "Christian" mothers can take a lesson, Pulter uses the poem to craft a zoological web centered on protective parenting. Her poems play with scale and illuminate, as Wendy Beth Hyman argues of John Donne's "The Flea," "how small forms tether the microcosmic to the cosmic through the collapsible lens of imagination."[52] Focused on natural systems rather than isolated specimens, the dynamic "scopic effect" of Pulter's poems is just as active, changeable, and evolving as the animals, plants, and insects she studies.[53]

In the different ways Hooke and Pulter structure their observations, we find different modes of relation to their objects of study. Through his specimen form, Hooke takes on the status of the authoritative, objective scientist, with "objective" here defined on two different but related registers. On a theoretical register, Hooke's objectivity "derives from (although it is not identical with) the ideal of 'representing things as they really are,'" what Allan Megill defines as "philosophical objectivity."[54] This is the utopian objectivity of Baconian science, an ideal scientific method intent on "disentangling facts from interpretations," but ultimately unrealistic in practice.[55] On a practical register, Hooke's objectivity leaves more room for the "subjectivity of the knower" even as it constitutes objects *as* objects "in the course of an interplay between subject and object," what Megill defines as "interactional objectivity."[56] Hooke's observational stance simultaneously aims at the objective ideal and positions the scientist at the center of its epistemological project. He endeavors to "represent[ ] things as they really are," but remains captivated by the process of his own scientific knowledge-making and, therefore, constantly inserts himself as subject into the representational work. In his study of the "Structure and motion of the Wings of Flies," for instance, Hooke begins not with facts about the wings, but rather his own desire to know more about them. "The Wings of all kinds of Insects, are, for the most part, very beautifull Objects," he declares before describing his "many trials" to discern "what the vibrative motion of the wings is, and after what manner they are moved" (173). Hooke's study of the fly's wings is framed through his own curiosity (about "beautifull *Objects*" of nature) and the resulting "trials" that allow him to represent knowledge to the reader, all of which reinforce the binary between the subject doing the knowledge making and the specimen from which that knowledge is extracted.

Pulter is instead an active part of the environment she studies (recall, for instance, her garden in Emblem 53), so her observational stance resists a stark binary between subject and object. The best evidence for this is the form of her emblem poems—or, more precisely, what she omits from her epistemological work in the emblem poems. Traditional emblems are composed of three parts—*inscriptio* (motto), *pictura* (visual image), and *subscriptio* (short epigrammatic verse). By removing the *pictura* element, Pulter employs what scholars call the "naked emblem" form, which draws heightened attention to the lack of a visual reference point.[57] A lack, that is, of the scientist's "faithful eye," so important in Hooke's formulation of the reformed scientific method ushered in by the Royal Society. "There is not so much requir'd" for this new "reformation in Philosophy," Hooke argues, but only "a sincere Hand, and a faithful Eye, to examine, and to record, the things themselves as they appear" (sig. a2^{v}). In *Micrographia*, Hooke's "faithful Eye" is transferred to the book's pages with his detailed engravings, aligning the reader's eye with Hooke's microscopic lens and situating a scientist-instrument hybrid as authority and expert. Conversely, the work of Pulter's emblem poems relies on the absence of the scientist's "faithful Eye." Reinforcing the ecological perspective of her emblem studies (no isolated specimens here), the absent visual also resists the universalizing impulse of Hooke's specimen form. Moreover, it invites speculation and individual, local acts of observation on the part of Pulter's reader. Absence, then, is a key component of Pulter's form of natural history. Hers is an epistemological project that constantly *situates* the subject of knowledge making.

Situated knowledge, a key tenet of feminist history of science studies, is based on "partial perspective"—the knowledge maker's "particular and specific embodiment," and their recognition that their perspective is one part of a much broader network of knowing.[58] Correspondingly, the object of situated knowledge is never displayed "as a screen or a ground or a resource," but rather always represented as "an actor and agent."[59] Not only does Pulter's blurring of emblematic and empirical thinking position objects of knowledge this way (the "bee" as "active Amizonian maid" is an actor and agent in Emblem 53 [118.7]); she also uses her poetry to explore—and often break down—the dichotomy between subjects and objects of knowledge making. In one of Pulter's longest poems, the flowers speak to the poet as she lays in her garden, asking her to serve as "umpire" to decide "which of them did theire fellowes all excell / in vertue, Coulour, Beuty, fashion, smell" (12.7–9).[60] This poem is, in part, a study of each plant (including the woodbine, lily, rose, and gillyflower), but crucially the poet ventriloquizes the flowers, so that each plant voices and describes their own characteristics, blurring the boundaries between subject and object of knowledge. And, lest we think that the poet as "umpire" retains a certain amount of authority in the poem, Pulter infinitely defers the poet's final judgment. The poem ends with the sun going down and all the participants consenting to "prorogue [adjourn] this Parliament" (12.522). Similarly, in "Universall Dissolution" (Poem 6), Pulter positions herself within a vast natural ecosystem, constructing an encyclopedic collection of animals, trees, and plants as she documents the inevitability of death (the "universal dissolution" of the poem's title). The "fraile fabrick" of the poet's own body is one of a long catalog of natural beings subject to death and decay (6.4).

"Universall Dissolution" is, in fact, worth closer attention as we begin to make the important shift in this chapter away from scientific content and back toward scientific forms. It is the *form* of "Universall Dissolution" that does the work of deconstructing the subject/object binary between poet and creatures of the natural world. The early lines of this 176-line poem maintain a consistent analogical pattern: four lines describing something in nature (always represented as actor) that is subject to dissolution, followed by a two-line simile linking the example to "Mortall Man's" inevitable death (6.9). Like the "triumphant Laural" whose "virgin Leaves all withered Fillimott to bee," reads one example, "e'ne soe, Mans youth and beauty doth decay" (6.20–21).[61] Pulter breaks this formal pattern, though, at line 23 with the poet's description of the "stately Cedar that aspires soe high." In the margin of this poem, Pulter inserted four additional lines to extend the description of the cedar and introduce a second tree, "the Sypriss" (6.29–30). For the next seventy-three lines, the descriptions of natural phenomena swell in length and "Mortall Man" is increasingly less present. The effect is that this poem, which purports to be a reflection on the poet's own mortality (signaled in the opening apostrophe to "My Soule"), transforms into an extended documentation of and reflection on the natural world, with the poet as one of many natural beings. Subjects and objects of knowledge-making blur rapidly as the poem unfolds to its conclusion. At line 133, Pulter abruptly drops the simile structure completely after turning her attention from the Earth's surface to the cosmos. The poem's original formal structure descends into chaos as Pulter trips and skips through planetary and celestial examples—Jupiter (141), Mars (143), Saturn (145), the Pleiades stars (153), the Hyades constellation (157), the Asoph comet (165), and the Dog Star (169)—all without returning to "Mortall Man" between the descriptions. Rather, in these lines, the poet starts to fold references to herself and her own experiences into the descriptions of the stars and planets. She describes Saturn, for example, as a "Malignant, Melancholy Star" that does "mischiefe" to "sweete Hibernie" (another name for Ireland) where the poet "first had life" (145–47). By the poem's end, the natural world and the poet's body cease to be tenor and vehicle of a simile and are instead woven together as the breakdown of the subject/object binary becomes just as inevitable as "universall dissolution."

The complex category of *subject* in "Universall Dissolution" begins with the poem's lengthy title, which represents Pulter as both subject and object of the poem's knowledge work: "Universal Dissolution, made when I was with Child of my 15th Child [my Son John] I being as everyone thought, in a consumption, 1648." Pulter is the poetic subject making poetry during her illness and pregnancy ("made when"), and this poem becomes, in part, a documentation of that experience through the autobiographical, first-person poetic "I" marked in the title. The poet is also, though, an object of study in the mist of her illness—a patient diagnosed ("as everyone thought") with "consumption." The poet's status as "object of study," observed and documented by both herself and others, is partly an effect of the diary-entry form of this title, and here we find an especially intriguing overlap in Pulter's and Hooke's observational methods. Hooke was also very interested in his own body, its actions, and its surrounding circumstances as a site for empirical study. Hooke's diary, which he kept throughout his career with the Royal Society, contains daily observations on the weather, directions of the wind, and planetary positions, right alongside notes on his own daily activities: medical experiments on his body, interactions with Royal Society members, movements around London, financial transactions, lending of books and other materials, and progress on new ideas or inventions. Pulter's observational stance works alongside and complicates the increasingly pervasive scientific methods of the Royal Society and, correspondingly, her situated knowledge project offers an alternative mode of knowing within the same epistemological orbit. Pulter's situated knowledge, that is, helps us return to Hooke with a different eye.

In Hooke's writings beyond *Micrographia*, including the pages of meeting notes recovered in 2006, we get a glimpse at scientific knowledge in formation, situated in the political and social contexts of Hooke's day-to-day dealings with the Royal Society. A key feature of Hooke's diary entries is his consistency in tone. His matter-of-fact recordings of data points related to the weather, for example, mimic the observations and recordings of his own movements. Historian of science Lotte Mulligan demonstrates the striking way in which the organization of Hooke's diary draws attention to his attempt to produce "a fully objective 'history' with himself as the datum."[62] The opening pages of the diary are divided into four columns, one for the date, another for observations on the weather, and the last two columns reserved for Hooke's record of his daily activities (one column) and memos for future activities (another column). One entry on the weather notes, "mercury stood at 185. a very cleer but cold morn. wind South but gentle. it grew thick & very cold about 11 [mercury] falling" (March 11, 1671/72).[63] The following day's personal entry reads, "I took last night Laudanum gr. 1½. Aloes [half a dram]. it wrought good effects on me. I carryd in the Account of the Bubble Experiment to Arundell house" (March 20, 1671/72).[64]

Over time, as Mulligan tracks, Hooke's organization schema breaks down: "after a year of careful entries he began to make errors, recording personal entries on the weather side or writing right across the page instead of in the personal memo column."[65] For Mulligan, this suggests a loss of interest in weather data and an increased interest in the exercise of "self-scrutiny," with Hooke positioning himself as "external object," a self "largely devoid of feelings, emotions, sensibility, even thought."[66] Read through the lens of Pulter's situated knowledge-making, though, we might consider how the dissolving organization of Hooke's diary pages enacts the same blurring of subject and object that we saw in Pulter's "Universall Dissolution," an instance in which the *situated-ness* of Hooke's science overpowers the objective observational stance he strives to inhabit. As the diary progresses, his science becomes increasingly embedded in the social, cultural, and political contexts of his daily life—and writing through (and then simply foregoing) the lines of his original grid evinces the fact that separating scientific work from the circumstances of daily life will always be a constructed discursive act. Hooke's resolutions to quit tobacco ("Began to take tobacco again" [September 25, 1674]; "Left off taking tobacco" [October 16, 1674]) shape and are shaped by the interpersonal conflicts between Royal Society members ("Brouncker a dog for belying me to the king" [April 29, 1675]), which shape and are shaped by his natural history work ("Discovered the fabrick of the muscles of a crabb" [May 24, 1675]).[67]

This returns us, briefly, to the lost Hooke folio—the document of Hooke's presence in the Royal Society meetings and his distillation of the discussions and activities therein. The manuscript situates the knowledge-making work of the early Royal Society in bodies around a room. One entry, for instance, records Hooke's reading of a letter from "Dr. Lister" (likely, Martin Lister, nephew to the Matthew Lister connected to the Sidney family from Chapter 2) about a man who was cured of parasites. Hooke documents the chain of associations that move the group's conversation from Lister's parasite case to a discussion of "the various and strange ways of the productions of insects" (April 5, 1682).[68] Another entry records the group's conversation on "the transformation of Creatures by means of the qualifications of the place wherein they are fostered" (April 12, 1682).[69] The "creatures" in question range from "redstrake fruit" and "wild asparagus" to human beings: "It was likewise urged that there were many of the Jews Black which yet are very strict in not mingling with other nations. And that Europeans by continuing to Inhabit in Africa have been found to turn black and that Blacks in England after a few generations do become white" (April 12, 1692).[70] Hooke's meeting notes *do* provide key insight into the daily workings of the early Royal Society, as Robyn Adams and Lisa Jardine emphasize, but what they reveal are the structures of epistemological power in formation, exposing that power as anecdotal, speculative, and (above all) situated.

Surface Anatomies in Robert Hooke's Laboratory

Thus far, in my account of the connections between Pulter and Hooke's work, I have stayed focused primarily on the content of their writings (with a dash of analysis on their specimen and ecological forms), and the resulting natural history narrative within which I have situated Pulter's poetry extends ongoing critical work on Pulter's science. We need, I think, further investigation of Pulter's interventions into the observational and experimental work of the early Royal Society that takes seriously what David Carroll Simon would call the "observational mood" of her poetry—a "careless" observational stance of "seeming inconsequence" that nonetheless "implies an awareness of the perceptual and cognitive advantages it affords."[71] Such investigation, however, is not the work of this book because it relies on science looking like we expect it to look. In the preceding pages, I have retained the familiar vocabulary, methods, and contexts of contemporary science (natural history, observation, experiment, even situated knowledge) and have simply absorbed Pulter's writing into that paradigm.

Reorienting our approach to the connection between these two figures, attending closely to the formal and material practices that shape their work, in the remainder of this chapter I connect Pulter and Hooke as conscientious students of an epistemology—a mode of knowing—compelled by anatomical forms. Hooke was particularly interested in the diverse textures available on the surfaces of bodies and, in *Micrographia*, he uses the book's pages as a laboratory space in which to conduct a complex study of surface. How do different kinds of marks on the page enable readers to feel different textures? How can the page extend the work of the microscope? How does one render an anatomy of surface? In Hooke's explorations of these questions, I find space for thinking in new ways about the graphic details of Pulter's manuscript—her marks of revision, her use of punctuation, and her sustained interest in the messy, in-progress, embodied work of knowledge making. The textual field of Pulter's manuscript is layered, multiple, and, above all, fragmented. And this is a result of both her poetic experiments (different poems on the same theme, for instance, that accumulate through the manuscript's pages) and the material traces of her forming work (the cuts, inserts, and changes she makes to the poetry). She, too, employs her text as a laboratory space, one in which and through which to mount a sustained study of poetic formation. Building on my claims from the first part of this chapter—that Hooke and Pulter are observational thinkers, more interested in theorizing the *act* of observation than they are in the resulting knowledge—I now excavate the anatomical thinking that underlies their modes of knowing.

The influence of anatomy is more explicit for Hooke, who did not think of himself as a natural historian in the way we currently define the term. By today's definition, natural history is the study of organisms and plants in their natural environments, with a methodological emphasis on observation over experiment (the definition, notice, that I applied to Hooke's and Pulter's work in the first half of this chapter). But Hooke was immensely interested in experimentation. He refers to his work as "experimental philosophy" and "physical inquiries" in the preface to *Micrographia* (sig. d2^{r}), and some of the longest sections of the text are explanations of experiments (see, for instance, his "Observ. VI. Of small Glass Canes," in which he spends eleven pages on an "Explication of this Experiment" [10–22]). *Micrographia* also betrays Hooke's curiosity about the insides of things as much as the outsides, along with what we might call his science of texture—sustained, systematic investigation into the different textures manifest visually and tactilely on the surface of specimens. We find evidence of this curiosity in his "anatomy or dissection of the head" of the gray drone-fly (177) and his study of a piece of cork, in which he cuts up the cork to better observe its "perforated," "porous," "honey-comb" texture ("I with the same sharp pen knife, cut off from the . . . smooth surface an exceedingly thin piece of it, and place[d] it on a black object plate" [113]). Facilitating Hooke's inquiries into the "frothy" surface of cork (112), the "grain" of the "kettering-stone" (93), and the "curious texture of seaweeds" (140), the microscope becomes, as Kristin M. Girten argues, "both a visual and tactile instrument."[72]

Attending to the textures and tactile work underlying Hooke's microscopic observations is to read *Micrographia* against the grain. Hooke's groundbreaking book is discussed almost exclusively as a text about the microscope, the increasing dominance of vision in scientific investigation, and the trend toward observation as the privileged methodological mode of the Royal Society.[73] If we take a closer not just at *what* Hooke represents but also *how* he represents it, however, we start to understand Hooke's project quite differently. The *how* of Hooke's epistemological work is particularly evident in his depiction of the flea, perhaps the most famous of the book's illustrations (Figure 5.3). Recalling the interactive handwork of the anatomical flap figures from Chapter 1, the flea plate folds out to more than double the size of the book's pages in the 1665 edition. As the largest foldout in the text, it presents an exaggerated visual image of one of the smallest creatures Hooke studies, inviting readers not only to "see the shape and structure" of the flea's parts, but also to "feel their texture."[74] In his brief description of the flea, Hooke directs us to notice the insect's "polish'd suit of sable armour . . . beset with multitudes of sharp pins" and the "small cavity" behind the flea's eyes that contains a "certain thin film beset with many small transparent hairs" (210). As he works to convey the highly textured body of the flea to his reader, he finally concedes that there are "many other particulars, which, being more obvious, and affording no great matter of information, I shall pass by, and refer the reader to the figure" (211). Hooke invites us to shuttle between the highly textured language of his prose descriptions and the intricate micro-details of his engravings, reminiscent of the active "bookwork" of Vesalius's *Fabrica*.[75] We begin to see feelingly, to see with our hands, as tactile and visual modes of knowing intersect to facilitate our interaction with Hooke's specimens.

In most copies of the 1665 *Micrographia*, the flea plate is bound as a double-page foldout, so the two pages unfold from the center binding and the book cradles the large foldout, functioning like a microscope slide holding a specimen (Figure 5.4). Because the engraving is so large and the specimen so exaggerated, we are able to see details on the surface of the flea beyond what a sixteenth-century microscope would allow us to discern. The highly detailed, textural work of Hooke's plates is partly a product of his illustration technique. Hooke used copperplate engraving and, for Meghan C. Doherty, it was Hooke's "understanding of line and particularly the engraved line" and his engagement with "mid-seventeenth-century visual culture" that allowed him to "see differently."[76] As we saw in Chapter 3, though, copperplate engraving was a tactile as much as a visual form, a technique primed for the attention to texture Hooke

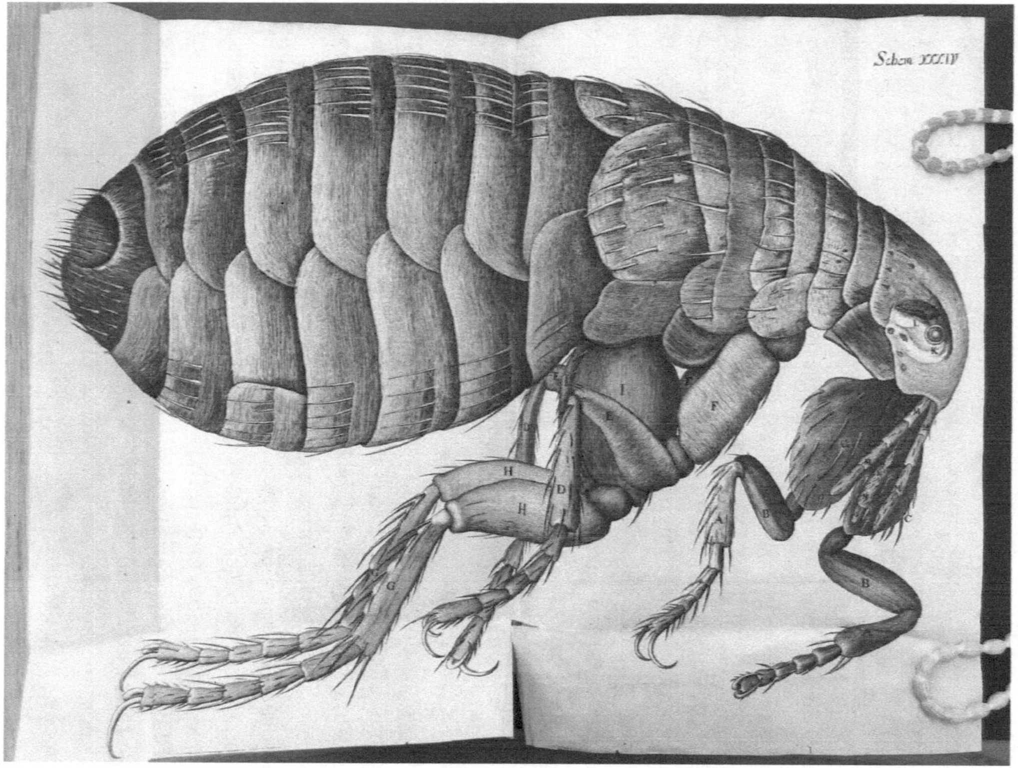

Figure 5.3. Robert Hooke's engraving of the flea. Hooke, *Micrographia: or some Physiological Descriptions of minute bodies made by magnifying glasses* (London, 1665), Schem. XXXIV. From the British Library Collection: 39.e.1.

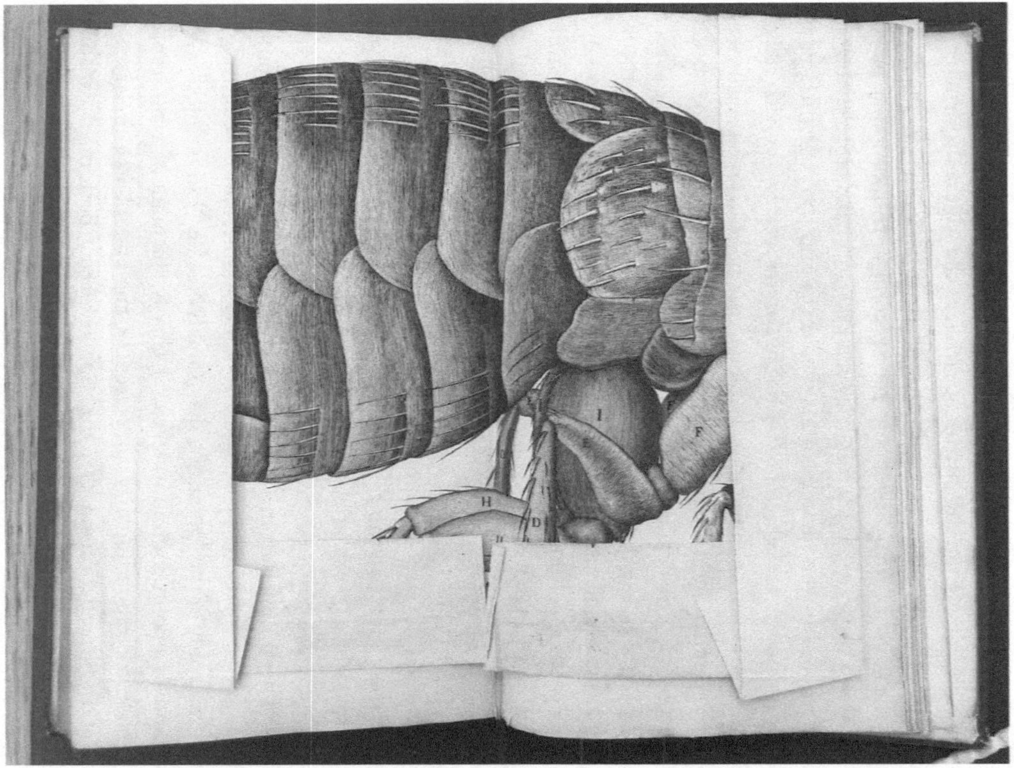

Figure 5.4. Robert Hooke's flea engraving, foldout detail. Hooke, *Micrographia: or some Physiological Descriptions of minute bodies made by magnifying glasses* (London, 1665), Schem. XXXIV. From the British Library Collection: 39.e.1.

demonstrates throughout *Micrographia*.[77] Hooke wields the medium of print, the art of engraving, and the materiality of his text's pages to craft a microworld—a textual curiosity cabinet of small things—and demonstrate the value of "experimental philosophy" and "physical inquiries" to his readers, promising them not only the "high rapture and delight of the mind," but also a range of "material and sensible pleasure[s]" (sig. d2^{r}). The book's pages become an even more useful technology than the microscope itself as the page transforms into a laboratory space that enables and invites privileged access to microscopic forms.

This is the *how* of *Micrographia* and the anatomical thinking that powers Hooke's work—a constant interaction between eyes and hands that enables an anatomization of both the surfaces of his specimens and the page surfaces on which those specimens are represented. I began *Anatomical Forms* with the assertion that Renaissance anatomy was a book-making project, and in each chapter we have explored different techniques anatomists and poets used to translate flesh to page. Hooke's flea plate reveals a thinker interested in the multiple surfaces of his specimens, the question of texture (both as it is rendered by the microscope and as it can be rendered in the pages of a book), and the dimensional capacity of paper as a technology that can be manipulated and handled to different effects. When we understand *Micrographia* as a study of polytextural and multidimensional surfaces, it activates the same questions about the relationship between bodies and texts that have threaded throughout *Anatomical Forms*, but reorients our attention to the micro-graphic (to riff on Hooke's title) details of the page and their effects on this relationship.[78] As a case in point, Hooke begins *Micrographia* with a specimen that literary and history of the book scholars know intimately: a dot of ink on a piece of paper. "As in Geometry, the most natural way of beginning is from a mathematical point; so is the same method in observations and natural history the most genuine, simple, and instructive" (1). After an initial study of the "point of a sharp small Needle," Hooke's second example is "more of a point commonly so call'd, that is, the mark of a full stop, or period" (3). The corresponding diagram presents a small period in a circle (scaled to the size of the punctuation marks in the book's typeface) positioned just above a large, fuzzy ball that turns out to be a microscopic view of the period in the circle (Figure 5.5)—a microscopic rendering of ink on a paper surface.

Hooke's corresponding description attempts to articulate the intricacies of the ink dot's texture, marveling at how even "the most . . . smoothly engraven strokes and points" are actually "misshapen," "rugged and deformed" with "many furrows and holes" (3). He suggests that one explanation for the "furrows and holes" of the ink dot is the surface of the paper itself ("which at best appears no smoother then a very course piece of shag'd cloth" [3]), bringing the reader's attention not only to the materiality of the page but also to the interactions between ink and paper that bring out the latter's mottled texture. In Hooke's study of the "rough daubings of . . . printing ink," both ink and paper are revealed to be incredibly textured materials, the ink's texture revealed through its interaction with the uneven texture of the page surface. Hooke's microscopic study of the dot/point/period presses the reader's "sincere eye and faithful hand" into the service of scientific inquiry (sig. a2^{v}). Opening his text with this study, Hooke immediately transforms his readers' relationship to the materiality of the book's pages and turns our reading and handling of the book into a scientific endeavor. The act of reading Hooke's book—any book—now becomes an occasion to pay closer attention to surface textures and the micro-graphic details of the printed page. We are invited to attend to the periods, commas, and colons not as punctuation marks supporting our reading of prose, but rather as specimens. We become hyper-attuned to the feel of the "furrows and holes" of the book's pages in our hands as we read, and that haptic engagement becomes part of the knowledge we take away from our act of reading.

Hooke studies both "printing ink" and the inked "points made with a pen" (which, he reports, are "much more rugged and deformed" [3]), and his close attention to graphic features and material components holds much promise for studying the range of textual technologies at work in early modern England. Increased scientific attention to the surfaces of things resonated within a culture already highly attuned to the physicality and materiality of textual forms. In Mary Thomas Crane's theorization of surface reading, she argues that Renaissance humanists imagined texts as "flat surfaces divided into discrete places that [could] be overseen by the reader and excerpted for reuse."[79] Crane's orienting framework is cognitive science, but her spatial conceptualization of early modern texts finds a foothold in the period's anatomical thinking, albeit with an important modification. Recall, for instance, the écorché figures of Chapter 3, which remove the body's surface only to reveal another surface underneath its skin. These were not so much "surfaces divided into discrete places" as they were surfaces divided into discrete *layers*, all still conceptualized as surface—an adjustment that takes us from horizontal to depth-oriented page space. Hooke's epistemological project, not unlike that of Govard Bidloo and Gerard de Lairesse twenty years later, treats the page like an anatomical écorché and extends our understanding of the myriad and complex ways early moderns conceived of their inscriptive media forms. For Hooke and (as we are about to see) Pulter, the textual page held multiple representational layers at once and functioned as a laboratory of surfaces. They

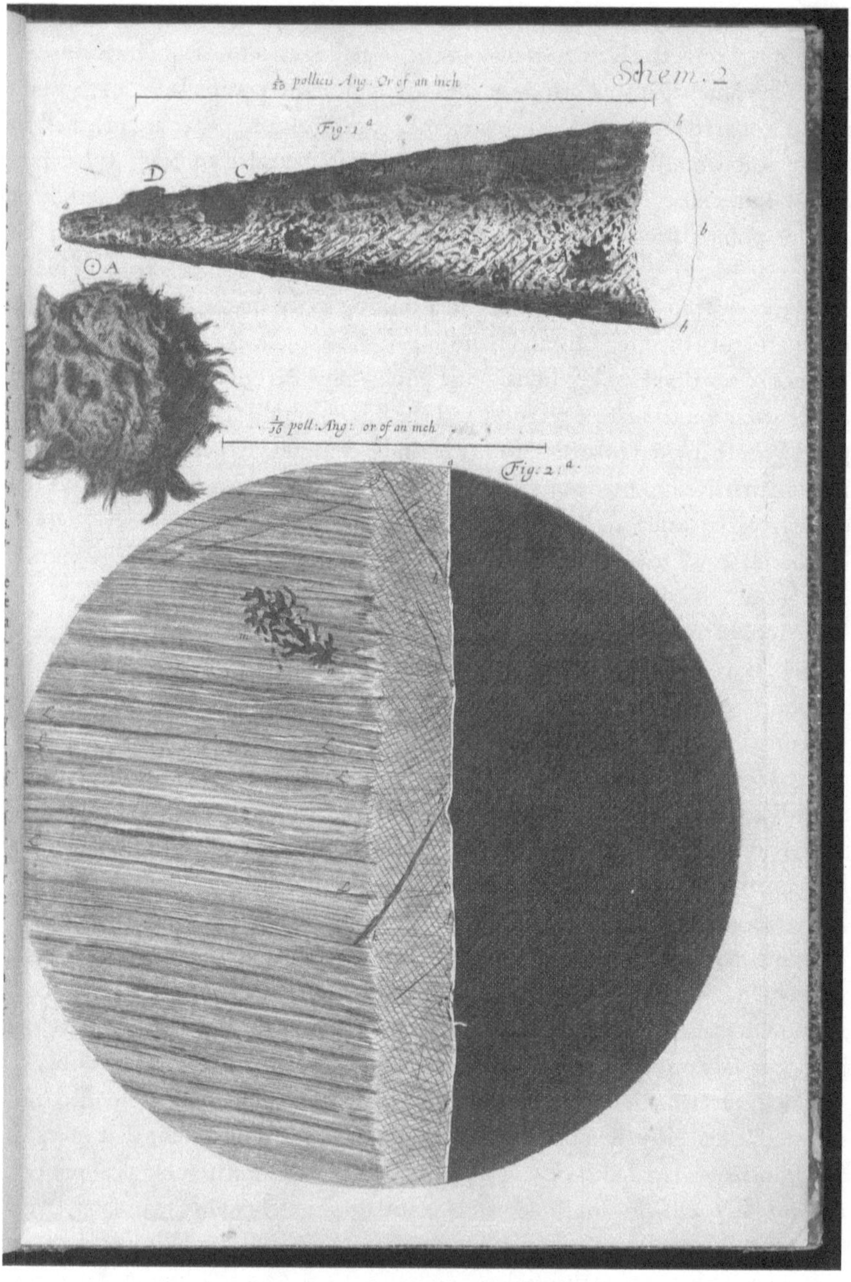

Figure 5.5. Robert Hooke's engraving of the ink dot. Hooke, *Micrographia: or some Physiological Descriptions of minute bodies made by magnifying glasses* (London, 1665), Schem. II, Wellcome Collection, EPB/C/29309.

uments
ore can
he Ru-
it view
whence
ely, we
s of In-
e *hairs*,
oks, or
iriæ or
any of
view'd
llion of
roscopes
pores,
-room,
bodies.
in in-
in the
whose
ove a
D
C
a
a
A
$\frac{1}{16}$ *poll:*
m

think with and through the materials of their practice, and their texts reveal deep curiosities about the relations among reader, writer, and textual representation—relations that take shape through surface contact.

Before turning back to Pulter, I want to offer one final example from *Micrographia* in which Hooke's experiments with bodies and pages become, quite literally, three-dimensional. "Reading one day in Septemb[er]," Hooke recounts in one of his specimen studies, "I chanced to observe a very small creature creep over the book I was reading, very slowly; having a microscope by me, I observ'd it to be a creature of a very unusual form" (207). Hooke goes on to describe the insect's various features, and includes a representation of it in yet another fold-out illustration plate, together with several other specimens (Figure 5.6). Hooke introduces this specimen (what he calls the "crab-like insect") and his encounter with it through a narrative of reading, outlining a scenario in which a familiar daily activity becomes the occasion for scientific inquiry, not unlike the poems from Pulter's manuscript we have read thus far, wherein walking, resting under a tree, and spending time in the garden prompt the poet's study of the natural world. In Hooke's book bug example, the page again becomes a laboratory environment, but here a very literal one—a space on and within which the bodies of specimens can be found, handled, and studied. The book page is not solely a conduit for representing and circulating Hooke's microscopy work, but also itself a site within which (and with which) to perform that work. In the book bug entry, the actual physical specimen crawls across the surface of the page, is transported by the scientist's hands from that page for study, and then returned to the page in the form of engraved, diagrammed illustration and prose description. The three-dimensional insect crawling over the page of Hooke's book becomes a lively, textured, inked representation in the pages of *Micrographia*, imbued with the *anima*, materiality, and dimensionality of Hooke's original encounter with and study of the bug. At the same time, Hooke draws our attention to the epistemological capacities of the page we hold in our interaction with the bug illustration. Part of his rendering of surface anatomy is to narrate the conditions (the same conditions we encounter while reading Hooke's text) through which he noticed the bug in the first place, crawling across the surface of the page.

Making Cuts

Hooke's fastidious narration of his representational process finds a corollary in the messy material conditions of Pulter's manuscript, replete with traces of the

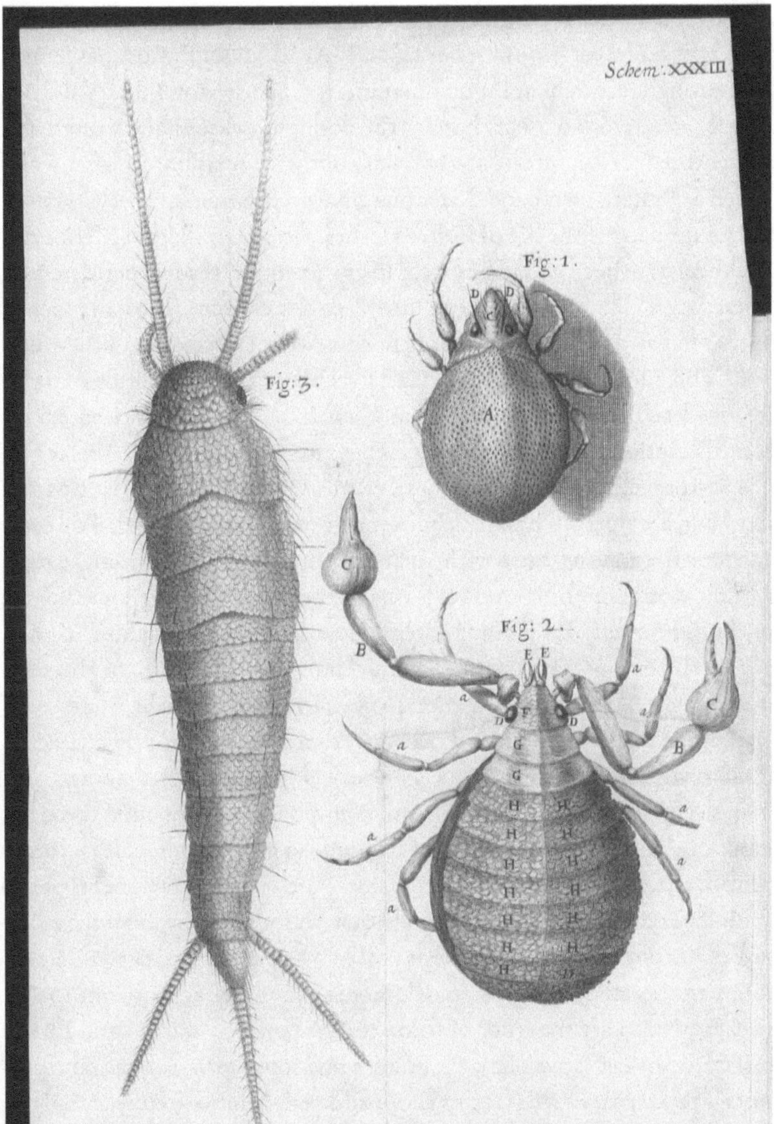

Figure 5.6. Robert Hook's engraving of the crab-like insect that crawled across his book page (bottom right of plate, Fig: 2). Hooke, *Micrographia: or some Physiological Descriptions of minute bodies made by magnifying glasses* (London, 1665), Schem. XXXIII, Wellcome Collection, EPB/C/29309.

representational act—the forming work of the poet's hands as she curates certain kinds of representations for her readers. As Sarah C. E. Ross has identified in her thorough research on Pulter's manuscript, Brotherton Library BC MS Lt q 32 involves the work of many hands. The dominant clear, italic hand is that of an unidentified scribe, interrupted throughout with revisions in what Ross has identified as Pulter's own hand. Later marginal annotations and new textual elements (a summary of the life of Pulter's father, James Ley, Earl of Marlborough, for instance) have been added by an eighteenth-century reader (perhaps, as Ross suggests, Angel Chauncy, rector of a nearby parish church). The hand identified as Pulter's has also recorded payments to three women servants in the volume's flyleaves, and another unidentified hand uses blank space on various title pages to practice letters and signatures. Finally, additional hands contribute a loose-sheet transcription of Judith Madan's eighteenth-century poem "Abelard to Eloisa," a pastoral elegy titled "A Pastorall on the Death of a young lady Mrs Ann Everett," and a stanza of a popular eighteenth-century song (scribbled three times throughout the volume, once with an altered final line). In addition to making significant word- and line-level revisions to the poetry throughout the manuscript, Pulter herself also penned several loose-leaf poems included in the volume, and the manuscript contains her authorial rough drafts of the second, unfinished part of the prose romance *The Unfortunate Florinda*.[80]

Reading Pulter's revisions and the messy draft pages of her romance activates what Carolyn Dinshaw calls a "queer historical impulse"—the strange, strong, and wonderful desire to touch and form communities across time through the materials of the archive.[81] Instances of revision—a crossed-out word, a corrected pronoun, a replacement verb, an inserted line—are very literal moments of touching, a direct encounter with a paper body that, in Pulter's case, is all the evidence we have of her physical being. Revisions pull us into meaningful encounters across time, in part, because they are so much more than minor adjustments to a text's content. Revisions are the work of form, to draw on Ben Burton and Elizabeth Scott-Baumann's evocative phrase.[82] Pulter's revisions are part of the fabric of her manuscript text, part of what forms the disordered, handwritten, material conditions of her existence in the archive. And, as such, they hold the manuscript in a state of formation, a state of active forming rather than static form. Pulter's revisions, we might say, pull us "backward" into the time, space, and conditions of writing. As Dianne Mitchell argues, early modern women's poetry in particular "work[s] against long-held ideals of linearity and progress, demanding a richer set of assumptions" and prompting us to "extend[ ] our understanding of form to accommodate effects of textual materiality, temporal delay, or authorial affect."[83]

Holding onto these deeply embodied accounts of our encounters with revision in the archives, I want to add revisions to our growing inventory of anatomical forms in this book. Revisions render visible the anatomy of the text itself. They function as a kind of dissective act that simultaneously constructs and reveals the textual page as a polytextural, multidimensional surface.

To understand manuscript revisions as an anatomical form, we can turn to the particularly striking example of Pulter's elegy to her daughter, Jane Pulter, a poem titled "Upon the Death of my deare and lovely Daughter J.P." (Poem 10). This poem has received much attention since the recovery of Pulter's manuscript, not only as a "heart-wrenching" early modern maternal elegy, but also as a remarkable case study of Pulter's revision process.[84] The poem begins with a plea that "all indulgent parents . . . lend one teare" to Pulter in her grief, and continues with thirty-two lines that reiterate on the same pattern: Pulter provides a figurative measure of time ("twice hath the Earth thrown Chloris mantle by" [10.7]) coupled with an assertion of her still-present grief ("yet still my heart is overwhelm'd with griefe" [13]). For the first two-thirds of the poem, this repeated oscillation between marks of time passing and the "yet still" of a seemingly unending present provides a formal enactment of the yawning time that engulfs the poet in her grief. It is also our first clue that the representational work of this poem relies on multiplicity—on a very literal *heaping* of figurative language as the poet works to convey her grief to the reader. When Pulter finally breaks the poem's opening pattern at line 37, the poem continues with an explanation of Jane's death that is saturated with blazon imagery, foreshadowed earlier in the poem with references to Jane's "milkey limbs" (11) and "sparkling Diamond eyes" (25). Pulter describes how the signs of fever appeared on Jane's "snowey skin" (37), comparing the "fevour spot[s]" to "drops of blood upon unsoiled snow" (44). Persistently returning to white and red imagery, Pulter also compares the spots of fever on Jane's skin to "white Lillies" scattered with "Damask Roses" (41–42), framing this metaphor with the observational language we are familiar with from her emblem poems: "soe have I som times seene" (39).

Notably, these lines describing Jane's illness and death are the lines that Pulter revises. After line 43, there is an asterisk in the right margin. Line 44 has been crossed out and there is a manicule drawn in the left margin next to it (Figure 5.7).[85] An annotation accompanying the manicule directs us to turn from page 29 to page 27, where we find a corresponding manicule and asterisk next to nine untitled lines of poetry, meant (we might hypothesize) to be inserted into the elegy (Figure 5.8). Following the manuscript's direction for this change, here is how Pulter's significant revision to "Upon the Death" reads:

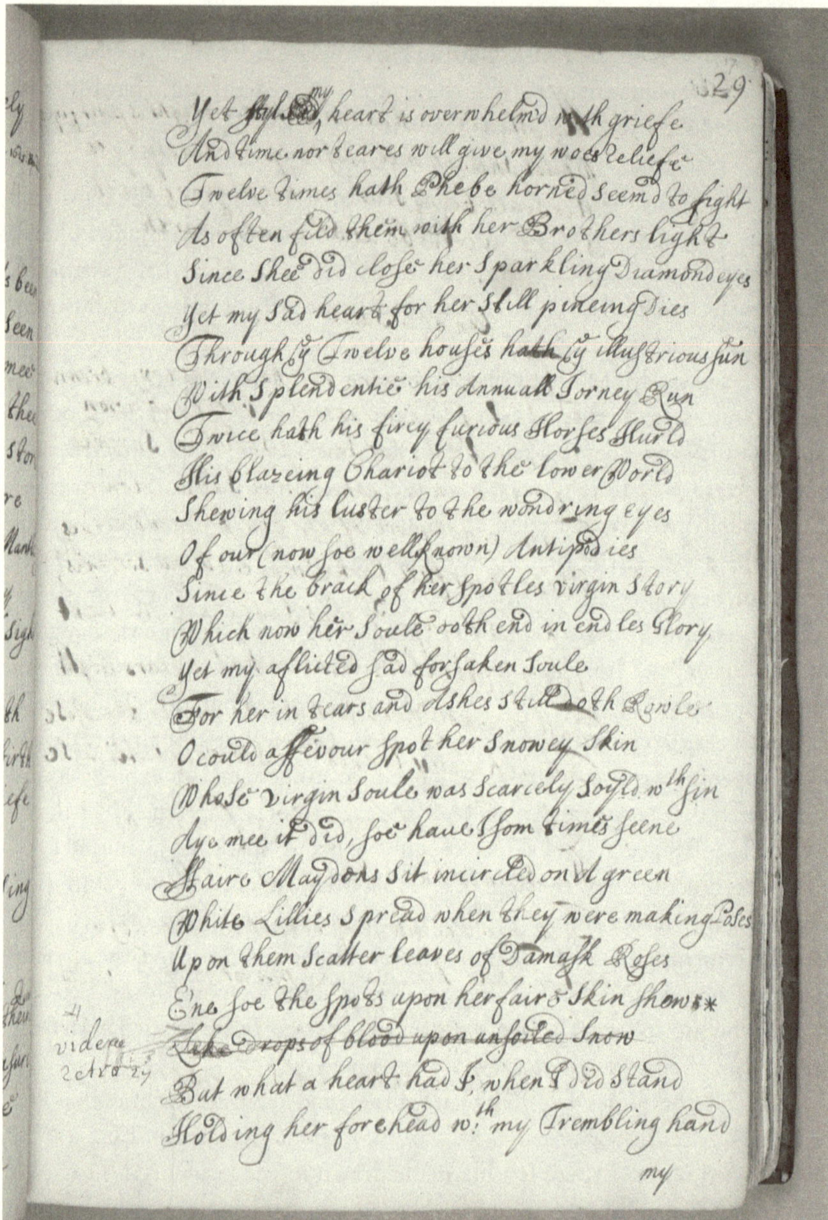
29

Yet ~~still~~ my heart is overwhelmd with griefe
And time nor teares will give my woes reliefe
Twelve times hath Phebe horned seemd to fight
As often fild them with her Brothers light
Since Shee did close her sparkling Diamond eyes
Yet my sad heart for her still pineing dies
Through ye Twelve houses hath ye illustrious sun
With splendentie his Annuall Torney Run
Twice hath his firey furious Horses Hurld
His blazeing Chariot to the lower World
Shewing his luster to the wondring eyes
Of our (now soe well known) Antipodes
Since the brack of her spotles virgin story
Which now her Soule doth end in endles Glory
Yet my aflicted sad forsaken Soule
For her in tears and Ashes still doth Rowle
O could a ffevour spot her snowey skin
Whose virgin Soule was scarcely soyld wth sin
Aye mee it did, soe have I som times seene
Faire Maydons sit incircled on A green
White Lillies spread when they were making Poses
Upon them scatter leaves of Damask Roses
Eve soe the spots upon her faire Skin show *
~~Like drops of blood upon unsoiled snow~~
But what a heart had I, when I did stand
Holding her forehead wth my Trembling hand
my

Figure 5.7. Manicule detail next to line 44 of "Upon the Death of my deare and lovely Daughter J.P." (Poem 10). Lady Hesther Pulter, *Poems breathed forth by the nobel Hadassas, and The Unfortunate Florinda*, Manuscript (c. 1645–1665), 29r, Leeds University Library, BC MS Lt q 32.

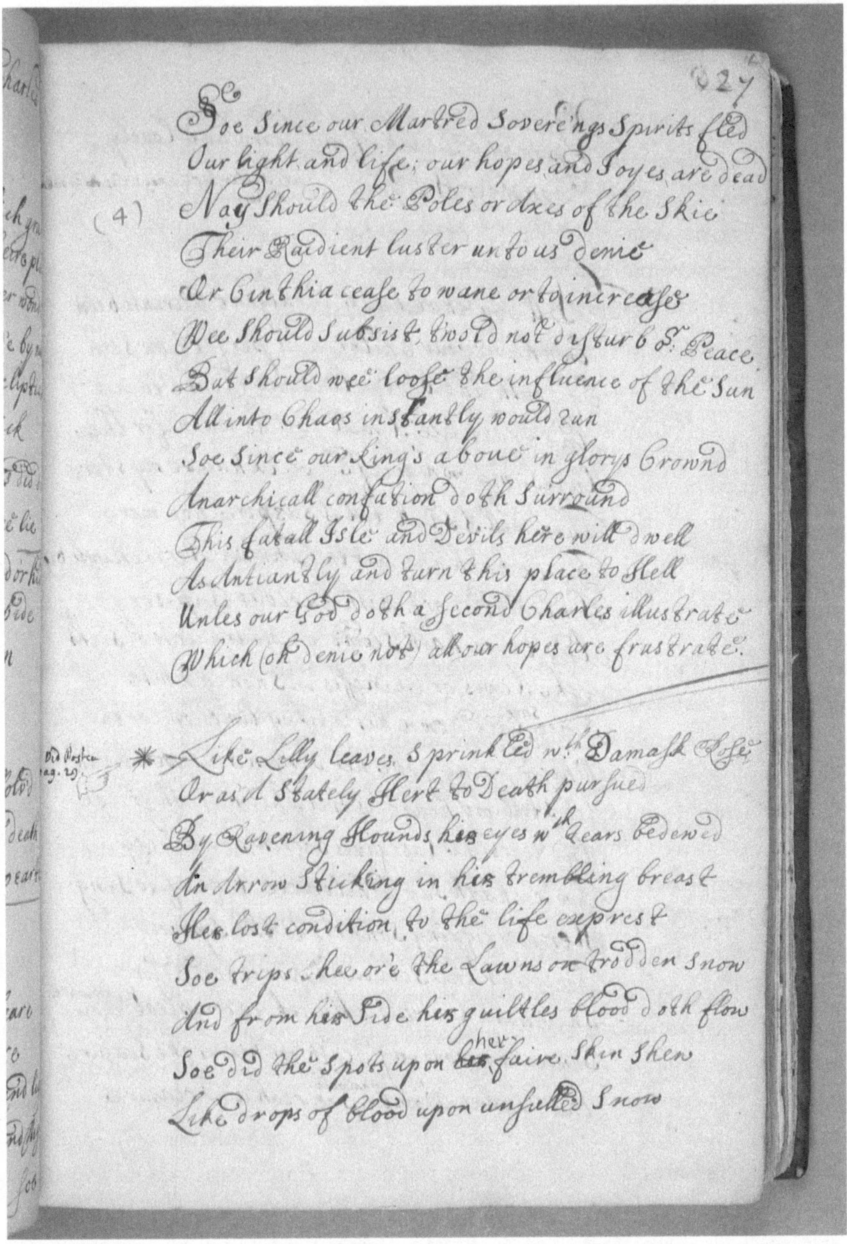

027

Soe Since our Martred Sovere'ngs Spirits fled
Our light and life; our hopes and Joyes are dead
(4) Nay Should the Poles or Axes of the Skie
Their Raidient luster unto us denie
Or Cinthia cease to wane or to increase
Wee Should Subsist &would not disturb ye Peace
But Should wee loose the influence of the Sun
All into Chaos instantly would run
Soe since our King's above in glorys Crown'd
Anarchicall confusion doth surround
This fatall Isle and Devils here will dwell
As Antiantly and turn this place to Hell
Unles our God doth a Second Charles illustrate
Which (oh denie not) all our hopes are frustrate.

* Like Lilly leaves sprinkled wth Damask Rose
Or as A Stately Hert to Death pursued
By Ravening Hounds his eyes wth tears bedewed
An Arrow Sticking in his trembling breast
His lost condition to the life exprest
Soe trips shee o're the Lawns on trodden snow
And from his Side his guiltles blood doth flow
Soe did the Spots upon her faire Skin shew
Like drops of blood upon unsullied Snow

Figure 5.8. Manicule detail next to untitled lines. Lady Hesther Pulter, *Poems breathed forth by the nobel Hadassas, and The Unfortunate Florinda*, Manuscript (c. 1645–1665), 27r, Leeds University Library, BC MS Lt q 32.

O could a fevour spot her snowey skin,
Whose virgin soule was scarcely soyld with sin
Aye mee it did, soe have I som times seene
Fair Maydons sit incircled on a green
White Lillies spread when they were making Poses,
Upon them scatter leaves of Damask Roses,
E'ne soe the spots upon her faire skin show
~~Like drops of blood upon unsoiled snow~~
⌈Like Lilly leaves sprinkled with Damask Rose
Or as a stately Hert to Death pursued
By Ravening Hounds his eyes with tears bedewed
An arrow sticking in his trembling breast
He[s] lost condition, to the life exprest[86]
Soe trips hee or'e the Lawns on trodden snow
And from his side his guiltles blood doth flow
Soe did the spots upon ~~his~~ her faire skin shew
Like drops of blood upon unsulled snow.⌉

In the inserted lines (marked by brackets), Pulter adds a second analogy (again employing white and red imagery) to her representation of Jane's illness. Jane is now also the "stately Hert to Death pursued / by Ravening Hounds," whose "guiltles blood" splattered on the "trodden snow" evokes the fever "spots upon [Jane's] faire skin." The added lines expand the poem's representational work from one dominant analogy to two, while the new line 44 reiterates the initial analogy—emphasizing the comparison between Jane's skin and the "lilly leaves." By replacing *and* adding to the original line 44, the new lines dilate (recalling my discussion of Aemilia Lanyer's *Salve Deus* in Chapter 4) the original comparison, folding a new representational component into the poem and functioning like the accumulative logic of "yet still" introduced in the poem's opening.

This particular dilation, though, also functions like a dissection. Attending to formal structures at the intersections of flesh and text enables us to perceive the crossed-out line as a *cut*, an opening or break in the surface of "Upon the Death" that creates the conditions for the poem's fragmentation and imagines the page as it is rendered in Hooke's *Micrographia*, as a textured, multi-layered surface. The new poetic part to be inserted into the cut exists separate from the poem, a dissected organ only reattached through collaborative acts of writing and reading—one act that resulted in the manicule additions, another that follows the instructions of the manicule to reference a different page, another (mine, above) that rewrites the poem with the added lines. Pulter makes her cut and another writer (or, perhaps Pulter—it is difficult to make out which hand drew the manicules) leaves a mark of their own body alongside that cut, using a drawing of a body part to mark a partitioned poem, a poem in parts, one crafted by cutting, overwriting, fragmenting, rewriting, inserting, and stitching. Structurally, the nine new lines exist both beside and beneath the old as they fill the poetic space opened by Pulter's cutting of the original line 44. Figuratively, the second comparison acts as another layer of metaphor, a layering of poetic figure that is pieced into and takes up space within, beneath, and between the cut line and its revised version—"Like drops of blood upon unsoiled snow" and "Like drops of blood upon unsulled snow." As readers, we are invited into the dissection moment and the surgical grafting of new lines onto and into old, evidence of knowledge making at work in the textual traces left behind.

The graphic detail of the manicule cues us into the *how* of Pulter's poetics—the anatomical thinking that undergirds her acts of poetic forming. This is the only pair of corresponding manicules in the manuscript, but Pulter's exploration of poetic practice as layered, multiple, textured—and, above all, *in parts*—is ubiquitous throughout the text.[87] We find another representation of Jane Pulter's death directly following "Upon the Death," in a poem titled "On the Same" (Poem 11). This time, Pulter employs the form of *apophasis*—a poetic mode of negation that references and describes the very thing the poet negates—to outline Jane's physical features and dominant characteristics. The apophasis form manifests in Pulter's echoing refrain, "tell me no more": "Tell mee noe more her haire was lovly brown"; Tell mee noe more of her black Diamond Eyes"; "Tell mee noe more her cheeks exceld the Rose," (11.1; 5; 9). Reading this poem within the manuscript's broader epistemological project, we could first situate "On the Same" as part of Pulter's sustained experiments with observation as a mode of knowing. The repeated refrain "tell me no more" responds directly to the impulse (whether that of an imagined interlocutor or a manifestation of the poet's own grief) to observe, record, and preserve specific details about Jane. Through apophasis, Pulter simultaneously marks the observed details (Jane's "curious curles," "snowey hand," "noble Mind") and resists the act of observation. As both blazon and anti-blazon, the poem questions the merits of that pervasive poetic form, and goes a step further to query the capacity of observational acts to preserve and document. Juxtaposing the observational stance of the blazon—marked by distance and fragmented description—with the embodied experiences of grief and loss, Pulter incorporates her sustained, systematic study of observation into the lyric form that is most informed by the observational mode. Simultaneously, she experiments with the possibility of a record formed through absence, through an active *not* knowing.[88] The details about Jane that body forth in this poem exist in the spaces between negations, and we are left wondering: what would it mean to *not* observe, to *not* document?

Reinforcing the work of the manicules in "Upon the Death," this second elegy on Jane's death is another kind of poem in parts. This poem is titled "On the Same," and it is one of two poems so titled in the manuscript, the other following a poem on the death of Charles I (Poems 14 and 15). In fact, eighteen out of roughly sixty-six poems in the first section of Pulter's manuscript (the section titled "Poems Breathed Forth by the Nobel Hadassas") have the same name as another poem in the section. In addition to the two poems titled "On the Same," Pulter writes two poems titled "Aurora" (Poems 3 and 37); four poems titled "The Circle" (Poems 17, 21, 25, and 36); two poems titled "The Welcome" (Poems 2 and 33); and three poems titled "to Aurora" (Poems 22, 26, and 34). To this catalog of repeated titles, we might add five additional poems in which the titles overlap significantly, a pair of poems titled "The Wish" (Poem 52) and "The Weepeing Wishe" (Poem 61) and a trio of poems titled "Made When I was Sick 1647" (Poem 31), "Made When I was Not Well. April 20. 1655" (Poem 51), and "Made When my Spirits were Sunk Very Low with Sickness and Sorrow. May 1667" (Poem 66). To this extended catalog of linked titles, we might further add the many lines in this section of the manuscript that are untitled, which include the nine lines that begin "like lilly leaves" already discussed above, along with a hundred-line stichic that begins "Why must I thus forever bee confin'd" (Poem 57), a collection of ten stanzas of alternate rhyme scheme that begin "My soul! why art thou full of trouble?" (Poem 40), and a collection of four stanzas of enclosed rhyme that begin "My God, I Thee (and onely Thee) adore" (Poem 50).

As Elizabeth Zeman Kolkovich notes, the poems with repeated titles "echo and expand upon each other . . . edit and revise one another," a kind of circular form that Kolkovich links to Pulter's interest in astronomy and Jayne Archer connects to Pulter's interest in alchemy.[89] These "tropes of turning and returning," though, are also a kind of fracturing, a formal pattern that reveals Pulter's keen interest in parts and wholes, an interest we know well from our study of Margaret Cavendish's poetry in Chapter 1. For Pulter, no representation can be reduced to one single, tidy form. Multiple versions of poems function as experiments in the multiplicity of representation—an anatomization of poetic formation itself. What happens, these poems ask, if I dissect this representation and put one part here and another part over here? Should we, as readers, stitch these poems back together? Are they parts of a whole meant to be *whole*, or are they meant to remain fragmented, dissected as separate parts for reading and study? Highly attuned to the capacity of poetry as an instrument for studying forms, Pulter—like Hooke—employs her text as a laboratory space in which to explore modes of knowing and experiment with different kinds of representational practice. Moreover, her textual experiments, too, carry the imprint of anatomical thinking. Pulter understands the poetic page as a multilayered space that can hold cumulative acts of forming. And, as we are about to see, she is also keenly interested in the textural possibilities of her inscriptive medium, in the "furrows and holes" of her page surfaces (*Micrographia*, 3).

Surface Anatomies in Hester Pulter's Laboratory

Reading one day in February, I chanced upon a collection of untitled lines in a manuscript. "Why must I thus forever bee confin'd / against the noble Freedome of my Mind," these lines began, continuing for one hundred lines, primarily in rhyming couplets of iambic pentameter with two rhyming triplets at lines 69–71 and 98–100 (57.1–2). The lines referenced a great number of real and mythical creatures, describing the different ways these creatures "freely range" and "injoy the Freedome of their Birth" while the poet, by comparison, lives in "sad confinement" and "shut up in a countrey grange" (57.17–18; 48–49). The writer of the lines seemed to possess an encyclopedic knowledge of insects, animals, and mythical creatures. The lines described the stages of life of the silkworm ("The Silkworm Feeds, then Works, then Shee involves [cocoons], / her Self, then Breeds, then Flies till Shee dissolves" [57.11–12]); referenced the "spontaneous breed[ing]" of reptiles (57.27); discussed the egg-laying habits of the manucode, or bird of paradise (57.41–43); and cited the writer's observations of swans nesting on the River Thames (57.53–54).

The lines stood apart from the poems around them in the manuscript. They began on a separate page after a fair amount of blank space at the top of the page, and continued for five pages. After the last of these hundred untitled, sequential lines ("to God alone my Suffrings Il'e deplore" [57.100]) the writer had, again, left blank space on the page—enough to fill with an additional four lines—before penning another series of six untitled lines. These six lines, all in rhyming couplet, began "For I noe Liberty expect to See" (58.1), ended "wee give a beeing to a nother World" (58.6), and included an ornamental flourish under the word "world." Having my close reading eye with me, and being a student of form, I also noted that these lines included almost no end-line punctuation. The only punctuation marks that seemed part of the original writing act were parentheses, a few stray commas, one colon, and a period at the end of the first set of a hundred lines and the second set of six lines. A few more commas and some end-line periods in the final lines seemed in darker ink, and I speculated that they were, perhaps, added by a different hand.

I begin the final act of this chapter channeling Hooke's observational stance because it is important to document what we notice when we encounter new materials in the archive. Records of first encounter help us defamiliarize things that have since been calcified into particular narratives through entrenched habits of reading. The lines I describe above have become some of the most famous lines from Pulter's manuscript and, as such, there are two specific, calcified habits of reading that I would like to consider here. First, the hundred untitled lines are always read as a discrete poem, assigned the title of the first line, "Why must I thus forever bee confin'd." In the two current *Pulter Project* editions of these lines, Leah Knight and Wendy Wall's elemental edition and Victoria E. Burke's amplified edition, the editors also combine the two separate sets of lines to form a single uninterrupted 106-line poem.[90] Knight and Wall do mention the fact that "this poem's conclusion is oddly separated from what precedes it by a sizeable gap" in the manuscript, but their edition does not include any break between the first hundred lines and concluding six lines. Second, the lines are read almost exclusively as an autobiographical study, one of Pulter's "defining poems of rural isolation."[91] The first line of the series, along with the poet's description of being "shut up in a countrey grange" (57.18) are frequently cited as autobiographical certainties, to better or worse effect. Alice Eardley admits, "it is not possible to gauge the extent or duration of Pulter's incarceration," while Maureen E. Mulvihill uses evidence from this poem and others to assert that "the long life of Hester Pulter . . . was a small life, a life pathetically played out in the dedicated space of Broadfield Hall manor."[92]

Since the recovery of her manuscript, Pulter's writing has been hailed as a galvanizing opportunity to break the mold of what we think we know about early modern women writers, to dismantle entrenched paradigms and interrogate, as Margaret J. M. Ezell urges, not just *what* we know but *how* we know what we know.[93] All the scholars cited in the previous paragraph have been at the forefront of Pulter scholarship and have had a guiding hand in ensuring that her writing is accessible to early modern scholars, students, and any interested readers. But old habits die hard. I want to use my last acts of close reading in this book to think otherwise about what we can learn from Pulter's manuscript and, perhaps in the process, undo some of the habits of critical reading that dominate early modern literary scholars' encounters with new writing—and particularly women's writing. I focus on these untitled lines because they have become so familiar to Pulter's modern-day readers, but it is worth remembering that a significant number of what we call "poems" in this manuscript are also clusters of untitled lines (Poems 47, 48, 49, and 50 in *The Pulter Project*, for instance). The provocations I offer in the following paragraphs, that is, can be much more broadly applied to Pulter's writing.

First, let us consider the interesting detail that there are two sets of untitled lines that share page space (for ease of reference, I will refer to the first set of a hundred lines as "Noble Freedom" and the second set of six lines as "No Liberty"). Are the lines of "No Liberty" part of "Noble Freedom"? If so, has the gap in the manuscript page been left to fill in additional lines at a later moment (the poetic act initiated by carving out blank page space, as we saw in Mary Wroth's *Urania* manuscripts)? Or are these, indeed, two separate sets of lines, meant to be read on their own terms? It is important, I want to insist, to maintain these multiple possibilities, for this is how Pulter teaches us to read and think throughout her manuscript. When read as part of "Noble Freedom," and the entire 106-line collection taken as a single poem, the additional six lines punctuate the poet's claim that she will only have freedom in death. In the first hundred lines, the poet insists, "I noe liberty expect to have / untill I find my freedome in my Grave (57.78–79), a sentiment echoed both thematically and formally in the series of six lines: "I noe liberty expect to see / until to Atoms I dispersed bee" (58.1–2). Lara Dodds, who also contemplates the status of these six untitled lines, notes that this fragment "shift[s] the register" of the preceding poem from a focus on "earthly freedom" to "a freedom that transcends such mortal bounds."[94] There also would be, though, some consistency between the "freedome" of the poem's opening and closing because the six lines bookend the poet's opening declaration that her mind is free while her body is "confin'd" (she is "confined / against the noble freedome of [her] Mind" [57.1–2]). Once dispersed to atoms, the poem's ending would read, the poet will be "free as her verse" (58.3), a conclusion that then prompts us to reread the entire poem as a reflection on poetry, perhaps a theorization of the relationship between poetry and death (hailing Mary Sidney Herbert).

If we link "No Liberty" to "Noble Freedom," we must at least maintain the white space between line 100 (57.100) and 101 (58.1), treating the six concluding lines as a separate stanza.[95] If one of Pulter's dominant formal modes is the *part*, the separation of these lines from the rest of the poem is particularly significant. The blank space after one hundred unbroken lines is imbued with a heightened sense of meaning and, through this white space, Pulter curates a pause, a shift, a rift, perhaps another kind of cut. This expansive blank space (which brings Wroth's manuscript poetry to mind) is pregnant with action, undisclosed in the poem's language but marked by absence (also recalling the "cauterizing" effect of Pembroke's white space in "Angell spirit"). Perhaps this is the action of the poet's prayer: "to God alone my Suffrings Il'e deplore" [57.100]. Perhaps that prayer goes unanswered, prompting the shift from devotional to natural philosophical discourse in the concluding six lines. The white space is similarly significant if we treat "No Liberty" as a discrete set of lines, an unattached part. These lines, like the poet, become cut off, removed, formally isolated from the content around them. As a result, the sequence "Noble Freedom" becomes less optimistic, stopped short by the act of prayer that continues the poet's lament privately, internally, and beyond the bounds of textual representation. And "No Liberty" becomes an even more potent, compressed philosophical meditation if taken in isolation.

Thinking about this isolated stanza on its own terms should remind us of the lines that begin "like lilly leaves," which are so frequently read (which I myself earlier read) as a revision to "Upon the Death of my deare and lovely Daughter J.P." (Poem 10) because the marginal manicules direct our reading in such a way. But our uncertainty about the addition of the manicules lingers: Who drew them? When were they added? Did Pulter authorize these marginal instructions? Does Pulter's authorization (or not) matter? The nine lines that begin "like lilly leaves" could be pieced into a different poem in the sequence, perhaps part of the elegy to Charles I, with which those lines share a page (such a reading is certainly bolstered by Emblem 25 [Poem 90], in which Pulter directly compares Charles to a hunted hart whose "guiltles blood still flows" [90.13]). What if "Noble Freedom" is *also* a part? This untitled sequence's opening line begins after a fair amount of white space at the top of the page, and the previous poem, a pastoral dialogue between two of Pulter's daughters, ends with one of them declaring, "let us cease in vain to make our moane, / and goe to our sad Mother shee's alone" (56.53–54). If we follow this thread, the untitled lines might be Pulter's addition to the dialogue with her daughters, her expression of loneliness in response to their care. As an extended pastoral dialogue, this new poem (Poems 56 and 57 in the *Pulter Project* edition) would present two very different perspectives on nature—first, a natural world that is sympathetic with the daughters' grief and, second, a natural world that retracts that sympathy and flaunts its freedom in opposition to the poet's confinement.

Giving space to these speculations, I think with Pulter about the manuscript as a laboratory space within which the poetic representational mode is the specimen under the microscope. Pulter's collection of poetry is, above all, an experiment in form and the act of poetic formation. As an anatomist of poetry, she offers readers a poetics of parts that is contingent on an understanding of the writing surface as layered, multiple, and textured. We are then invited to explore the different knowledges available depending on how we piece together the parts across the manuscript's pages. Attending closely to the parts of Pulter's text, there is one final detail of these hundred untitled lines that deserves our attention, or rather a series of micro-graphic details—the punctuation marks. In my opening observations of these lines, I noted that they contain almost no end-line punctuation and, of the intra-linear punctuation, the parentheses are the most numerous. In this regard, crucially, these lines are unexceptional. Most of Pulter's poems contain very little end-line punctuation, but parentheses abound throughout the manuscript.[96] In "This Flying Fish" (Poem 90), the poem I referenced above in which Pulter compares Charles I to a hunted hart, she uses two sets of parentheses in her description of Charles's plight: "Soe have I known (oh sad) the Best of kings / (Ay mee the thought of this such horrour brings / to my sad soul) his Princely spirit posed / in strange Delemas every where inclosed" (90.15–18). The parenthesis is, arguably, Pulter's dominant form of punctuation, and this fact deserves closer study because the parenthesis is a mark of a part (and a mark apart). That is, the parenthetical mark acts as another kind of cut, creating a fragment that stands apart from (even as it remains within) the rest of the line.

Parentheses hold a unique status among the punctuation marks used by early modern writers because they operate as both rhetorical figure and "material-textual mark."[97] The rhetorical figure *parenthesis* (which I will delineate by italicizing the word) is marked on the page by the parenthetical bracket (or *lunulae*, Latin for "little moon"). Perhaps because of this dual status, early modern rhetorical manuals struggle to pinpoint the value of *parenthesis*. They often describe the figure as both unnecessary and perfecting—a device that strengthens an expression, but can be removed without creating an inconsistency or conceptual gap in the sentence.[98] For instance, in John Smith's rhetorical treatise *The Mystery of Rhetorick Unveil'd* (1657), he defines *parenthesis* as "a form of speech or a clause comprehended within another sentence, which (though it give some strength) may very well be left out, and yet the speech perfect, or the sense sound."[99] As Jenny C. Mann points out, Smith uses his own parenthetical here to question the "strength" *parenthesis* adds to a sentence.[100] By relegating the figure's potential strength to the parentheses in his definition, Smith formally enacts the equivocal value he ascribes to this figure. George Puttenham is a bit more directly dismissive. In *The Art of English Poesy* (1589), he categorizes *parenthesis* as a "figure of . . . disorder," only to be used "when ye will seem, for larger information or some other purpose, to piece or graff in the midst of your tale an unnecessary parcel of speech, which nevertheless may be thence without any detriment to the rest."[101] Puttenham emphasizes the "unnecessary" rhetorical quality of *parenthesis*, while simultaneously invoking the material quality of these marks—by inserting lunulae, the writer adds a "piece" or "graff" to an otherwise complete sentence.

We have already seen how revision functions in Pulter's manuscript as an instrument of reiteration (repeating and reinforcing her expressive acts) and partition (fragmenting and separating representational acts across her pages). We have also learned from Hooke's experiments in *Micrographia* that surfaces themselves can be anatomized—dissected to render the microscopic details and textures of each surface part in order to study its function within the whole. As a micro-graphic and material feature, parentheses have much to add to these epistemological projects. Rhetorically, the *parentheses* in Pulter's poems reinforce and add to the content around them and, materially, the parenthetical marks imbue the surface of Pulter's poetic lines with texture. In another series of untitled lines that begin "My God, I Thee (and onely Thee) Adore," Pulter uses six sets of parentheses, employing this punctuation mark most often as an echoing figure. "Leave not (oh leave not) my dejected soul," the poet pleads, but "obliviate (oh obliviate) my story" (50.5; 13). The parenthetical phrases in these lines repeat the preceding words, and importantly add an "oh" at the start, emphasizing the parenthetical echo as a kind of hallelujah chorus. They also cut into the otherwise unbroken words of the poetic line, creating a new layer within the representational form, a visible furrow on the poem's surface.

The repeated "oh's" of these lines—"(oh leave not)" and "(oh obliviate)"—are representative of what we find in most of Pulter's parentheses. Within the furrows and holes of her poems, Pulter inserts sighs: "a sudden, prolonged, deep and more or less audible respiration, following on a deep-drawn breath," especially "indicating or expressing dejection, weariness, longing, pain, or relief."[102] The embodied act of sighing, remember, is the framework through which Pulter's poems are first introduced, recorded on two separate title pages as "Poems Breathed Forth by the Nobel Hadassas" and "The sighes of a sad soule Emblimatically breath'd forth by the noble Hadassah." Let us return to the hundred untitled lines with which I began this section and take a closer look at their parentheses. In those hundred lines, there are three sets of parentheses, which I italicize here for added effect.

Thus Inscects, Reptals that Spontaneus breed
From Such a Solitude as mine are Freed
And I *(oh my Sad heart)* and onely I
Must in this Sad confinement living Die (57.27–30)

But I that am more constant then this Dove
Unto my First and last and onely Love
Cannot from this Sad place *(Ay mee)* Remove. (57.69–71)

Thus Beasts, Birds, Fishes, Equivocall Worm & Fly
Injoy more liberty *(woes mee)* then I (57.90–91)

All three sets of parentheses juxtapose the poet's woe with one or many creatures of the natural world *and* situate her sorrow within a broader ecology. All three parentheses also cut into the surrounding poetic line, creating a new representational layer on the page surface. The first set contains the "oh" we have seen in other parentheses, here texturizing the first-person reference "I" with a more specific reference to an interior part ("my Sad heart") that sits beneath the surface of the poet's "I." Laden with the weight of the poet's melancholy, both the first and last sets of parentheses tug on the words around them. Finally, the middle set of parentheses contains the most common phrase that Pulter encloses in parentheses throughout the manuscript: "Ay mee." We find other instances of this specific parenthetical phrase, for example, in Poems 8, 14, 19, 22, 28, 30, 39, 43, 55, and 56. If we think about poetry as a voiced form, both "oh" and "ay" are sounds created with and through breath—they are particularly acute instances of pulmonic eggressive sounds. As parenthetical cuts, the sentiment of these sighs exist within a separate layer from the content of the lines around them. As breath, they seem to simultaneously float above and heave below the poem's surface.

Full of parenthetical marks that texturize and partition their poetic lines, Pulter's poems are inculcated with anatomical thinking even at the micro level. Parentheses, like the crossed-out line of "Upon the Death," are cuts, openings, insertions, breaks in the poem's surface. And, like the repeated titles throughout the manuscript, parentheses repeat, reinforce, expand, and splinter the act of poetic representation. Pulter's polytextural pages consistently accommodate multiple representations, multiple modes of knowing, multiple experimental acts. All of Pulter's parenthetical sighs exist as extraneous parts, as dissected parts excised by the material mark of the lunulae. They are both within and apart from the lines surrounding them, and they could all be removed without any syntactical disruption. But, of course, they are there. They are very deliberately there and, as part of a *poetic* line, the sighs are an integral part of the poem's meter, the sighs "breath'd forth" in the act of poetic writing. Wielding parenthesis as another instrument in her sustained experiments with poetic formation, Pulter pieces breath into the matter of her poetic lines, and her sighs are given substance—formal substance—by the parenthetical marks.

The still very recent discovery of Pulter's manuscript offers us a chance to turn a critical eye to the assumptions we bring to the archive, to dissect and study our own habits surrounding recovery, reclamation, and argument forming. Upon discovery (recovery) of Pulter's manuscript, this figure previously unknown to modern scholarship was defined as a poet and, only later, redefined as a poet with remarkably informed knowledge of contemporary science. In this chapter, I read Pulter and Hooke as early modern thinkers who smartly theorize acts of formation and use their texts to explore the material, textual, and embodied work involved in those acts. Poetry, in Pulter's hands, is a privileged tool for epistemological inquiry, one that has distinct value to developing modes of scientific practice in seventeenth-century England—most especially the shifting modes of anatomical inquiry (that science that "entailed the description of forms").[103] What, then, counts as evidence of certain kinds of knowledge? For me, Pulter's study of ants, her encyclopedic cataloging of plants, animals, and insects across her poems, her revisions, her "on the same's," her untitled lines, her blank spaces, and her punctuation marks count as evidence of a poet driven by anatomical thinking—by a sustained and deep curiosity about the way parts and wholes fit together, as well as the function and use of each part on its own terms. Through Pulter's anatomical forms, the surface of her manuscript pages come alive in new and different ways.

NEW HISTORIES OF LITERATURE AND SCIENCE (A CODA)

What do we go to early modern women's writing for? What are the habitual moves we make when we encounter a new (or familiar) piece of writing by an early woman writer, and how do those habits regulate the conversations within which we situate (and *can* situate) women's writing? I posed these questions at the end of Chapter 4, and I repeat them here again because they are urgent questions for literary scholars and, as this book has demonstrated, for those of us committed to cross-disciplinary thinking. Critical work on early modern women writers—as well as the broader perceptions of early women's writing within scholarly and popular imaginations—maintains implicit assumptions about women's participation in certain kinds of intellectual work, about women's access to certain kinds of knowledge and education, about the purposes for which women wrote or felt authorized to write, and about the homogeneity of the category of "early modern women." These assumptions determine the kinds of questions we ask about their work and, more importantly, the kinds of questions we do not ask.

Anatomical Forms has been a book driven by as yet unasked questions. I noticed fleeting references in poetic texts that otherwise seemed to have nothing to do with anatomy—the ink injection in Mary Sidney Herbert's elegy to Philip Sidney; Mary Wroth's reference to the tree bark as "skin" in *Urania*; and Aemilia Lanyer's reflection on the "strange" circumstances of the Virgin Mary's womb. What frame of mind, what burgeoning discourses, what epistemological capital cued these references? What was I seeing evidence of in these brief, but evocative, moments of poetic somatic potency? And if there is anatomical thinking here, as I suspect there might be, am I missing it elsewhere in this text? If we start from the premise that early modern poetry was a site of scientific thinking, what must we change about how we read poetry and how we read science? What if our narratives about the history of early modern literature and science *started* with women writers?

Addressing such questions, one orientation of this book's argument could have been through the proto-feminist framework so often brought to bear on early women's writing. In this version of the argument, I might have opened our inquiry not with Jane Barker's anatomical-poetic experiment ("we began our journey with the blood") but with Bathsua Makin's argument for women's education in *An Essay to Revive the Antient Education of Gentlewomen* (1673).[1] Makin's argument rests on the distinction between practical expertise (what she alternatively calls "labour" and "things") and formal learning ("understanding" and "words"). Women already have a great deal of the former, Makin asserts, because their daily tasks require it: "To buy wool and flax, to dye scarlet and purple, requires skill in natural philosophy. To consider a field, the quantity and quality, requires knowledge in geometry. To plant a vineyard, requires understanding in husbandry."[2] But in addition to this practical knowledge, Makin insists that women (specifically, white women of higher classes) must have access to the languages that enable a thorough understanding of their tasks. As she puts it, "words are the marks of things, and they are learnt better together than asunder."[3]

Much of Makin's argument hinges on language learning, and her own curriculum at the school she established in London focused on "gaining the Latin and French Tongues," followed by Greek, Hebrew, Italian, and Spanish for those students "that will bestow longer time."[4] Makin makes a strong case that women's exclusion from education—their inability to access the particular kinds of discourses that mark scientific expertise—diminishes the knowledge they do hold and restricts them from advancing that knowledge in meaningful ways. So, my argument might have gone, following Makin, if the language of science was not for women, *Anatomical Forms* will demonstrate that women radically seized access for themselves by turning from language to *form* in order to participate in science.

But that has not been the argument of this book. We do not need to apply a narrative of resistance and radical opposition in order to demonstrate early modern women's innovative thinking. Instead, I have aimed to show that the poets of this book always already thought about form as a material and tactile instrument. Their books and manuscripts are suffused with creative explorations of how poetic forming, along with the materials of paper and ink, can be animated with embodied practice—explorations also of great interest to Renaissance anatomists. Early modern textual forms were dimensional and physical, and the poets of this book demonstrate a deep curiosity about how bodies interact with that dimensional, physical textual matter, a curiosity that quite organically situates their work at the intersections of text and flesh. When we put the formal experiments of early modern poets and anatomists next to one another, we notice new things and we come to different conclusions about who was participating in whose field of expertise.

In building my argument about these poets, I also built an argument about what counts as evidence of scientific work, a critical investigation that requires us, as Margo Hendricks urges, to "read archival materials with the expectation that they will conceal as much as they reveal," and to recognize that "our task is to account for both."[5] The history of science archive (and therefore the history of literature and science) has been shaped by those with the privilege and institutional power to shape it. Taking in hand the contours of science as defined by that archive (already so productively destabilized by scholars such as Surekha Davies, Pablo F. Gómez, Evelyn Fox Keller, Katharine Park, and Banu Subramaniam) and applying it to early women's writing consistently results in singular examples of "exceptional women,"[6] the anomalous woman scientist story of Margaret Cavendish, or perhaps Lucy Hutchinson, increasingly recognized for her complex formulation of physics.[7] This history of science is one of exclusivity and isolated accomplishment and, as such, it also circumvents an important reckoning with how it perpetuates the continued gender-based gatekeeping within scientific fields in our contemporary moment.[8]

As this book has shown, our entrenched definitions of what counts as scientific work also stop our imaginations from thinking otherwise about forms of participation, engagement, and innovation. We sustain and maintain received ideas about what science is, instead of building a more dynamic and more historically robust account of what science was (and could be). In tracing unasked questions through women's literary history, *Anatomical Forms* has illuminated paths not taken in the history of science, and it turns out that those paths have much to do with formal dimensionality (Cavendish's paper bodies, Layner's womb-oriented poetics, Pulter's parentheses); embodied practice (Pembroke's "as if," Wroth's complex paper epistemology); work in progress (Cavendish's poet as Motion, Pulter's revisions); and affect (Pembroke's sustained study of grief, Wroth's of desire, and Lanyer's of devotion). Animated by interdisciplinary questions, my method has been to think through and with the formal experiments of these poets, and the result is a narrative about the intersecting histories of early modern science and literature that does not rely on a hierarchal relationship, evidence of direct influence, or a straightforward chronology of progress. I join Eve Kosofsky Sedgwick in proximal thinking—reading "beside" rather than "beneath" and "beyond"—in order to trace a ranging, variegated early modern science of the body.[9] And I take up Katherine Hayles's urge to understand disciplines as "heterogenous discursive field[s] through which multiple fissures run, with boundaries of fractal complexity."[10] In following the fissures between and through disciplines, I have tried to read, as Rayna Kalas articulates, "not from the outside in, but from the inside out," thinking with and learning from the writers I study in order to notice the "joints, mixtures, and temperings" that scaffold their thinking.[11]

In the end, this book's main contribution is not archival but perspectival. My intervention has been to bring together three familiar histories—poetics, anatomy, and women's writing—and see them all differently for the *proximity* curated in this book's pages. I am not interested in expanding what counts as science. I am interested in recognizing that science might not look like we expect it to look—a subtle but crucial difference. With the former approach, we start with our received, entrenched definition of science and then work to expand its contours, widen its reach, reshape something that already exists in our critical habits of mind. With the latter, we begin anew, we think creatively, and we build a definition of science based on what we encounter, are surprised by, and learn from the historical texts we study. Embedding our critical inquiries in the modes of knowing that animated sixteenth- and seventeenth-century texts, while keeping ourselves tethered to the situated knowledges of our own contemporary critical practice is no easy task. But we have guides to lead us through the "flesh, blood, skin, and bone" of premodern poetic and scientific forms. The poets and anatomists of this book teach us how to read form as a scientific instrument.[12] Taking in hand their lessons, we bring into focus histories of form centered on material practice and excavate the creative and poetic capacities of premodern science. What new and capacious pathways in literature and science are possible now?

NOTES

INTRODUCTION

1. Jane Barker, *Poetical recreations: Consisting of original poems, songs, odes &c. With several New translations* (London: Benjamin Crayle, 1688), sig. a2[v], British Library, 994.g.3.

2. Toni Bowers, for instance, describes Barker as "one of the most dyed-in-the-wool and unequivocal of Jacobites" ("Jacobite Difference and the Poetry of Jane Barker," *English Literary History* 64, no. 4 [1997]: 860). On Barker's politics, also see Rivka Swenson, "Representing Modernity in Jane Barker's Galesia Trilogy: Jacobite Allegory and the Patch-Work Aesthetic," *Studies in Eighteenth-Century Culture* 34 (2005): 55–80; Tonya Moutray McArthur, "Jane Barker and the Politics of Catholic Celibacy," *Studies in English Literature, 1500–1900* 47, no. 3 (2007): 595–618; Constance Lacroix, "Wicked Traders, Deserving Peddlers, and Virtuous Smugglers: The Counter-Economy of Jane Barker's Jacobite Novel," *Eighteenth-Century Fiction* 23, no. 2 (2010–11): 269–94; and Leigh A. Eicke, "Jane Barker's Jacobite Writings," in *Women's Writing and the Circulation of Ideas: Manuscript Publication in England, 1550–1800*, ed. George L. Justice and Nathan Tinker (Cambridge: Cambridge University Press, 2002), 137–57. For a notable exception that does consider Barker's engagement with contemporary science, see Rachel Mann, "Jane Barker, Manuscript Culture, and the Epistemology of the Microscope," *Eighteenth-Century Life* 43, no. 1 (2019): 50–75.

3. Here are the lines in full: "But may your Rhimes be still imploy'd to tell, / what satisfaction do's in Knowledge dwell; / and as you have begun, so yet go on, / to make coy Nature's secrets better known; / and may we learn in purest Verse, from thee, / the Art of Physick, and Anatomie; / while the much-pleas'd Apollo smiles to see / Medicine at once improv'd, and Poetrie" ("To the Incomparable Galæcia, On the Publication of Her Poems," in Barker, *Poetical recreations*, sig. a2[v]).

4. Jane Barker, "Poems on several occasions. in three parts," Magdalen Library, University of Oxford, Magdalen MS 343. All references to Barker's poem are from this version and will be cited by line number in the text. This poem is extant in two different versions, the manuscript version I cite here and an alternate version printed in *Poetical recreations* (see n1).

5. William Harvey outlined his theory of blood circulation in *Exercitatio anatomica de motu cordis et sanguinis in animalibus* (*Anatomical Exercise on the Motion of the Heart and Blood in Animals*), first published in 1628 and translated into English in 1653. In response to continued criticism on the topic, he then published a companion volume, *Exercitationes duæ Anatomicæ de circulatione sanguinis* (*Two Anatomical Exercises on the Circulation of the Blood*, 1649).

6. Barker's mention of "Bartholine" is likely a reference to Caspar Bartholin the Elder, who published a collected anatomical treatise in 1611, later augmented and illustrated by his son, Thomas Bartholine. Thomas updated the treatise, in part, to include William Harvey's theory of blood

circulation. Barker's reference to the "hipotheses" that Harvey denied refers not to debates surrounding blood circulation, but to theories on the womb and origins of life, another field of study in which Harvey's work was particularly influential. Lines 72–83 of Barker's poem treat different theories on the womb's form and purpose.

7. The one break in rhyming couplets in the first 101 lines corresponds with Barker's brief discussion of Eve, accompanied by a provocative lament about the status of women's knowledge: "And cause our sex precipitated first / to pains and ignorance we'r most accus'd / desire of knowledge cost so very dear, / that ignorance must be our only share, / and always move ith' low obedient sphere" (33–37).

8. Colleen Ruth Rosenfeld, *Indecorous Thinking: Figures of Speech in Early Modern Poetics* (New York: Fordham University Press, 2018), 10.

9. Domenico Laurenza, "Art and Anatomy in Renaissance Italy: Images from a Scientific Revolution," *Metropolitan Museum of Art Bulletin* 69, no. 3 (2012): 5.

10. Megan Heffernan, *Making the Miscellany: Poetry, Print, and the History of the Book in Early Modern England* (Philadelphia: University of Pennsylvania Press, 2021), 5. On the textures and ecosystems of writing materials, see Joshua Calhoun, *The Nature of the Page: Poetry, Papermaking, and the Ecology of Texts in Renaissance England* (Philadelphia: University of Pennsylvania Press, 2020).

11. Here, I have in mind the rich archive of critical work on the material culture of early modern poetry, which extends from ongoing work in book history as scholars continue to excavate the myriad ways in which early moderns interacted with paper and other media forms. For representative examples of this work, see Jennifer Andersen and Elizabeth Sauer, eds., *Books and Readers in Early Modern England: Material Studies* (Philadelphia: University of Pennsylvania Press, 2001); Claire M. L. Bourne, *Typographies of Performance in Early Modern England* (Oxford: Oxford University Press, 2020); Heidi Brayman, Jesse M. Lander, and Zachary Lesser, eds., *The Book in History, The Book as History: New Intersections of the Material Text; Essays in Honor of David Scott Kastan* (New Haven, CT: Yale University Press, 2016); Juliet Fleming, *Graffiti and the Writing Arts of Early Modern England* (Chicago: University of Chicago Press, 2001); Rayna Kalas, *Frame, Glass, Verse: The Technology of Poetic Invention in the English Renaissance* (Ithaca, NY: Cornell University Press, 2018); and Whitney Trettien, *Cut/Copy/Paste: Fragments from the History of Bookwork* (Minneapolis: University of Minnesota Press, 2021).

12. For representative examples of such arguments, see Amy Louise Erickson, *Women and Property in Early Modern England* (London: Routledge, 1995); Laura Gowing, *Common Bodies: Women, Touch and Power in Seventeenth-Century England* (New Haven, CT: Yale University Press, 2003); Mary E. Fissell, *Vernacular Bodies: The Politics of Reproduction in Early Modern England* (Oxford: Oxford University Press, 2004); and Gail Kern Paster, *The Body Embarrassed: Drama and the Disciplines of Shame in Early Modern England* (Ithaca, NY: Cornell University Press, 1993). The history of conduct books also offers rich insight into early modern women's lived experience. See, for instance, Patricia Akhimie, *Shakespeare and the Cultivation of Difference: Race and Conduct in the Early Modern World* (New York: Routledge, 2018); Ann Rosalind Jones, "Nets and Bridles: Early Modern Conduct Books and Sixteenth-Century Women's Lyrics," in *The Ideology of Conduct: Essays on Literature and the History of Sexuality*, ed. Nancy Armstrong and Leonard Tennenhouse (New York: Methuen, 1987), 39–72; Suzanne W. Hull, *Chaste, Silent and Obedient: English Books for Women, 1475–1640* (San Marino: Huntington Library, 1982); and Jessica C. Murphy, *Virtuous Necessity: Conduct Literature and the Making of the Virtuous Woman in Early Modern England* (Ann Arbor: University of Michigan Press, 2015).

13. Elizabeth D. Harvey, "Introduction: The 'Sense of All Senses,'" in *Sensible Flesh: On Touch in Early Modern Culture*, ed. Elizabeth D. Harvey (Philadelphia: University of Pennsylvania Press, 2003), 3.

14. Gowing, *Common Bodies*, 80. For more on the specific and embodied circumstances of non-white, poor, serving, and enslaved women in the early modern period, see Urvashi Chakravarty, *Fictions of Consent: Slavery, Servitude, and Free Service in Early Modern England* (Philadelphia: University of Pennsylvania Press, 2022), especially Chapter 3; Patricia Fumerton, *Unsettled: The Culture of Mobility and the Working Poor in Early Modern England* (Chicago: University of Chicago Press, 2006); Ann Kussmaul, *Servants in Husbandry in Early Modern England* (Cambridge: Cambridge University Press, 1981), especially Appendix 2; and Denys Van Renen, *The Other Exchange: Women, Servants, and the Urban Underclass in Early Modern English Literature* (Lincoln: University of Nebraska Press, 2017). Outside the scope of early modern England, Marisa J. Fuentes's *Dispossessed Lives: Enslaved Women, Violence, and the Archive* (Philadelphia: University of Pennsylvania Press, 2016) has been especially important on this topic.

15. On women's work within and among these various domains, see Carla Bittel, Elaine Leong, and Christine von Oertzen, eds., *Working with Paper: Gendered Practices in the History of Knowledge* (Pittsburgh: University of Pittsburgh Press, 2019); Bianca F. C. Calabresi, "'You Sow, Ile Read': Letters and Literacies in Early Modern Samplers," in *Reading Women: Literacy, Authorship, and Culture in the Atlantic World, 1500–1800*, ed. Heidi Brayman Hackel and Catherine E. Kelly (Philadelphia: University of Pensylvania Press, 2008), 79–104; Fleming, *Graffiti and the Writing Arts*; Susan Frye, *Pens and Needles: Women's Textualities in Early Modern England* (Philadelphia: University of Pennsylvania Press, 2010); Elaine Leong, *Recipes and Everyday Knowledge: Medicine, Science, and the Household in Early Modern England* (Chicago: University of Chicago Press, 2018); Patricia Pender and Rosalind Smith, eds., *Material Cultures of Early Modern Women's Writing* (London: Palgrave Macmillan, 2014); Helen Smith, *"Grossly Material Things": Women and Book Production in Early Modern England* (Oxford: Oxford University Press, 2012); Trettien, *Cut/Copy/Paste*, especially Chapter 1; and "Forum: Early Modern Women and Material Culture," *Early Modern Women: An Interdisciplinary Journal* 4 (2009): 135–246.

16. Trettien, *Cut/Copy/Paste*, 36.

17. Smith, *Grossly Material Things*, 8; and Calhoun, *Nature of the Page*, 62.

18. Pamela H. Smith, *The Body of the Artisan: Art and Experience in the Scientific Revolution* (Chicago: University of Chicago Press, 2004), 8.

19. Wendy Wall, *Recipes for Thought: Knowledge and Taste in the Early Modern English Kitchen* (Philadelphia: University of Pennsylvania Press, 2016), 246. For more on alternative scientific communities in early modern England, also see Deborah E. Harkness, "A View from the Streets: Women and Medical Work in Elizabethan London," *Bulletin of the History of Medicine* 82, no. 1 (2008): 52–85; Harkness, *The Jewel House: Elizabethan London and the Scientific Revolution* (New Haven, CT: Yale University Press, 2008); Lynette Hunter and Sarah Hutton, eds., *Women, Science and Medicine, 1500–1700: Mothers and Sisters of the Royal Society* (Gloucestershire: Sutton Publishing, 1997); and Rebecca Laroche, *Medical Authority and Englishwomen's Herbal Texts, 1550–1650* (New York: Routledge, 2009).

20. For some representative work, see Peter Dear, ed., *The Literary Structure of Scientific Argument: Historical Studies* (Philadelphia: University of Pennsylvania Press, 1991); Marjorie Hope Nicolson, *The Breaking of the Circle: Studies in the Effect of the "New Science" upon Seventeenth-Century Poetry* (Evanston: Northwestern University Press, 1950); and Robert M. Schuler, "Three Renaissance Scientific Poems," *Studies in Philology, Texts and Studies* 75, no. 5 (1978). On "sci-

entism," see Howard Marchitello, "Science Studies and English Renaissance Literature," *Literature Compass* 3, no. 3 (2006): 341–65; and the debate between Stephen Pinker and Leon Weiseltier in the *New Republic*: Pinker, "Science Is Not Your Enemy: An Impassioned Plea to Neglected Novelists, Embattled Professors, and Tenure-less Historians," *New Republic*, August 6, 2013, http://www.newrepublic.com/article/114127/science-not-enemy-humanities, accessed May 8, 2024; Weiseltier, "Crimes Against Humanities: Now Science Wants to Invade the Liberal Arts. Don't Let It Happen," *New Republic*, September 3, 2013, http://www.newrepublic.com/article/114548/leon-wieseltier-responds-steven-pinkers-scientism, accessed May 8, 2024; and Pinker and Weiseltier, "Science vs. the Humanities, Round III," *New Republic*, September 26, 2013, http://www.newrepublic.com/article/114754/steven-pinker-leon-wieseltier-debatescience-vs-humanities, accessed May 8, 2024.

21. Valerie Traub, "The Nature of Norms in Early Modern England: Anatomy, Cartography, *King Lear*," *South Central Review* 26, no. 1/2 (Spring–Summer 2009): 44. On "modes of knowing" or "ways of knowing," also see John V. Pickstone, *Ways of Knowing: A New History of Science, Technology and Medicine* (Chicago: University of Chicago Press, 2000). For representative work in the second wave of literature and science studies, see Mary Thomas Crane, "Form and Pressure in Shakespeare," *Philological Quarterly* 98, no. 1–2 (2019): 23–45; Jenny C. Mann and Debapriya Sarkar, "Introduction: 'Capturing Proteus,'" *Philological Quarterly* 98, no. 1–2 (2019): 1–22; Sarkar, *Possible Knowledge: The Literary Forms of Early Modern Science* (Philadelphia: University of Pennsylvania Press, 2023); Carla Mazzio, "Introduction: Shakespeare and Science, c. 1600," *South Central Review* 26, no. 1/2 (2009): 1–23; Henry S. Turner, "Lessons from Literature for the Historian of Science (and Vice Versa): Reflections on 'Form,'" *Isis* 101, no. 3 (2010): 578–89; and Turner, *Shakespeare's Double Helix* (London: Continuum Books, 2007). Also see Mary Baine Campbell, *Wonder and Science: Imagining Worlds in Early Modern Europe* (Ithaca, NY: Cornell University Press, 2002); Howard Marchitello, *The Machine in the Text: Science and Literature in the Age of Shakespeare and Galileo* (Oxford: Oxford University Press, 2011); John Rogers, *The Matter of Revolution: Science, Poetry, and Politics in the Age of Milton* (Ithaca, NY: Cornell University Press, 1996); and Elizabeth Spiller, *Science, Reading, and Renaissance Literature: The Art of Making Knowledge, 1580–1670* (Cambridge: Cambridge University Press, 2004). For more recent examples of just how this second-wave thinking is galvanizing and expanding the field of early modern literature and science studies, see Liza Blake, "The Physics of Poetic Form in Arthur Golding's Translation of Ovid's *Metamorphoses*," *English Literary Renaissance* 51, no. 3 (2021): 331–55; Blake, "The Grounds of Literature and Science: Margaret Cavendish's Creature Manifesto," in *The Palgrave Handbook of Early Modern Literature and Science*, ed. Evelyn Tribble and Howard Marchitello (New York: Palgrave Macmillan, 2017), 3–26; Timothy M. Harrison, *Coming To: Consciousness and Natality in Early Modern England* (Chicago: University of Chicago Press, 2020); Jennifer Park, "Discandying Cleopatra: Preserving Cleopatra's Infinite Variety in Shakespeare's Antony and Cleopatra," *Studies in Philology* 113, no. 3 (2016): 595–633; Jessica Rosenberg, *Botanical Poetics: Early Modern Plant Books and the Husbandry of Print* (Philadelphia: University of Pennsylvania Press, 2022); David Carroll Simon, *Light Without Heat: The Observational Mood from Bacon to Milton* (Ithaca, NY: Cornell University Press, 2018); and Justin E. H. Smith, *Nature, Human Nature, and Human Difference: Race in Early Modern Philosophy* (Princeton, NJ: Princeton University Press, 2015). For some sterling examples of new approaches to literature and science beyond the field of early modern English studies, see Ian Duncan, *Human Forms: The Novel in the Age of Evolution* (Princeton, NJ: Princeton University Press, 2019); Amanda Jo Goldstein, *Sweet Science: Romantic Materialism and the New Logics of Life* (Chicago: University of Chicago Press, 2017); Natania Meeker and Antónia Szabari, *Radical Botany: Plants and Speculative Fic-*

tion (New York: Fordham University Press, 2019); and Britt Rusert, *Fugitive Science: Empiricism and Freedom in Early African American Culture* (New York: New York University Press, 2017).

22. Mann and Sarkar, "Introduction," 3.

23. Mann and Sarkar, "Introduction," 3.

24. Mann and Sarkar, "Introduction," 3, original emphasis.

25. In literature and science studies, the primary cast of characters tends to be William Shakespeare, John Donne, John Milton, Edmund Spenser, George Herbert, Andreas Vesalius, Francis Bacon, Robert Boyle, and later members of the Royal Society. Foundational studies in the field of early modern literature and science studies set this trend. In Jonathan Sawday's influential *The Body Emblazoned: Dissection and the Human Body in Renaissance Culture* (London: Routledge, 1995), for instance, Margaret Cavendish is the only woman writer who warrants attention (250–56). In Michael C. Schoenfeldt's otherwise excellent *Bodies and Selves in Early Modern England* (Cambridge: Cambridge University Press, 1999), he advances a pervasive (but increasingly less common) refrain: the "cultural fact" that "women were barred from education in . . . medical discourse" (35). If they had not been so barred, Schoenfeldt goes on to speculate, they might also have produced the "highly theorized form of physiological interiority" that he finds in the writings of Spenser, Shakespeare, Herbert, and Milton (35). Ultimately, Schoenfeldt confesses that he "could not find an example of a woman writer practicing the philosophically rigorous and literarily intense engagement with physiological inwardness that marks the four canonical male writers" of his study. Undoubtedly, the readers of *Anatomical Forms* will be well-equipped to refute such claims by this book's end.

26. Work on early modern women and the history of science has worked to correct this trend, but scholars have focused almost exclusively on genres of writing other than literature. See, for instance, Hunter and Hutton, *Women, Science and Medicine, 1500–1700*; and Laroche, *Medical Authority and Englishwomen's Herbal Texts*. One notable exception to this is Liza Blake's work on early modern atomism and physics in the work of Margaret Cavendish, Lucy Hutchinson, and Hester Pulter. See Blake, "Hester Pulter's Particle Physics and the Poetics of Involution," *Journal for Early Modern Cultural Studies* 20, no. 2 (2020): 71–98; "Margaret Cavendish's Forms: Literary Formalism and the Figures of Cavendish's Atom Poems," in *Feminist Formalism and Early Modern Women's Writing: Readings, Conversations, Pedagogies*, ed. Lara Dodds and Michelle M. Dowd (Lincoln: University of Nebraska Press, 2022), 38–55; and "Non-Atomic Atomisms and Atomic Epistemologies in the Poetry of Margaret Cavendish and Lucy Hutchinson," *English Literary Renaissance* 53, no. 3 (2023): 401–24.

27. Mann and Sarkar, "Introduction," 2.

28. Danielle Clarke, "Aemilia Lanyer," in *Routledge Companion to Renaissance Literature*, ed. Catherine Bates (forthcoming).

29. Leah Knight and Wendy Wall, eds., *The Pulter Project: Poet in the Making* (Chicago: Northwestern University, 2018), http://pulterproject.northwestern.edu.

30. On this point, I think with Debapriya Sarkar's important question: "how might fresh scrutiny of familiar archives and aesthetic objects—rather than a pursuit to 'discover' new texts—enable us to see things we have missed?" (Sarkar, "The Arts of English Poesie: Making Worlds and Making Race, *RaceB4Race: Poetics*, Arizona Center for Medieval and Renaissance Studies, January 27, 2023).

31. Margaret J. M. Ezell, *Writing Women's Literary History* (Baltimore: Johns Hopkins University Press, 1993), 6–7.

32. Margaret J. M. Ezell, "The Laughing Tortoise: Speculations on Manuscript Sources and Women's Book History," *English Literary Renaissance* 38, no. 2 (2008): 335.

33. Ania Loomba, "The Color of Patriarchy: Critical Difference, Cultural Difference, and Renaissance Drama," in *Women, "Race," and Writing in the Early Modern Period*, ed. Margo Hendricks and Patricia Parker (London: Routledge, 1994), 33. On this point, also see Margo Hendricks, "Feminist Historiography," in *A Companion to Early Modern Women's Writing*, ed. Anita Pacheco (Oxford: Blackwell Publishing, 2002), 361–76; and Ian Smith, *Black Shakespeare: Reading and Misreading Race* (Cambridge: Cambridge University Press, 2022).

34. See Bernadette Andrea, "Pamphilia's Cabinet: Gendered Authorship and Empire in Lady Mary Wroth's *Urania*," *ELH* 68, no. 2 (2001): 335–58; Barbara Bowen, "Aemilia Lanyer and the Invention of White Womanhood," in *Maids and Mistresses, Cousins and Queens*, ed. Susan Frye and Karen Robertson (New York: Oxford University Press, 1999), 274–303; Kim F. Hall, *Things of Darkness: Economies of Race and Gender in Early Modern England* (Ithaca, NY: Cornell University Press, 1995), especially Chapters 2 and 4; Sujata Iyengar, "Royalist, Romancist, Racialist: Rank, Gender, and Race in the Science and Fiction of Margaret Cavendish," *ELH* 69, no. 3 (2002): 649–72; Margo Hendricks and Patricia Parker, eds., *Women, "Race," and Writing in the Early Modern Period* (London: Routledge, 1994); and Melissa E. Sanchez, "What Were Women Writers?," *Criticism* 63, no. 1–2 (2021): 63–73.

35. Laurenza, "Art and Anatomy," 5.

36. Sawday, *The Body Emblazoned*, 2.

37. See Mary E. Fissell, "Introduction: Women, Health, and Healing in Early Modern Europe," *Bulletin of the History of Medicine* 82, no. 1 (2008): 1–17; Harkness, "A View from the Streets"; Eve Keller, *Generating Bodies and Gendered Selves: The Rhetoric of Reproduction in Early Modern England* (Seattle: University of Washington Press, 2011); Laroche, *Medical Authority and Englishwomen's Herbal Texts*; and Katharine Park, *Secrets of Women: Gender, Generation, and the Origins of Human Dissection* (Baltimore: MIT Press, 2006).

38. Park, *Secrets of Women*, 26.

39. The rhetoric of "expanding what counts as science" has been especially pervasive in work on histories of science that take up women's writing and women's work. Representative of this rhetorical framing, Nina Rattner Gelbart opens her introductory essay for a forum on "Women and Early Modern Science" with the assertion, "if we go about it differently, if we define science with more elasticity to take into account the many unconventional places and spaces where women did research into the workings of the natural world, if we bring margins to the center, then numerous scientific women come into focus." See Nina Rattner Gelbart, "Adjusting the Lens: Locating Early Modern Women of Science," *Early Modern Women: An Interdisciplinary Journal* 11, no. 1 (Fall 2016): 116. This framing also scaffolds much of the work on early modern women's recipe cultures. See, for instance, Leong, *Recipes and Everyday Knowledge*; Sara Pennell, "Perfecting Practice? Women, Manuscript Recipes and Knowledge in Early Modern England," in *Early Modern Women's Manuscript Writing: Selected Papers from the Trinity/Trent Colloquium*, ed. Victoria E. Burke and Jonathan Gibson (Hampshire: Ashgate Publishing Limited, 2004), 237–58; Edith Snook, "'The Women Know': Children's Diseases, Recipes and Women's Knowledge in Early Modern Medical Publications," *Social History of Medicine* 30, no. 1 (2017): 1–21; and Wall, *Recipes for Thought*. My research is deeply indebted to this excellent body of critical work, even when I interrogate the frameworks through which women's writing is most often studied.

40. Germain Warkentin, Joseph L. Black, and William R. Bowen, *The Library of the Sidneys of Penshurst Place circa 1665* (Toronto: University of Toronto Press, 2013).

41. Banu Subramaniam, *Ghost Stories for Darwin: The Science of Variation and the Politics of Diversity* (Urbana: University of Illinois Press, 2014), 9; and Subramaniam, "Moored Metamor-

phoses: A Retrospective Essay on Feminist Science Studies," *Signs* 34, no. 4 (2009): 951–80. Representative work in this field includes, but is certainly not limited to, Donna Haraway, *Primate Visions: Gender, Race, and Nature in the World of Modern Science* (London: Routledge, 1989); Sandra Harding, *The Science Question in Feminism* (Ithaca, NY: Cornell University Press, 1986); Evelyn Fox Keller, *Reflections on Gender and Science* (New Haven, CT: Yale University Press, 1996); Carolyn Merchant, *The Death of Nature: Women, Ecology, and the Scientific Revolution* (San Francisco: HarperCollins, 1980); and Londa Schiebinger, *The Mind Has No Sex?: Women in the Origins of Modern Science* (Cambridge, MA: Harvard University Press, 1991).

42. On standpoint epistemology theory, see Sandra Harding, ed., *The Feminist Standpoint Theory Reader: Intellectual and Political Controversies* (New York: Routledge, 2003).

43. Hendricks, "Feminist Historiography," 366–67.

44. Bringing this question to bear on formalist methodologies, this is the crucial difference between applying a formalist paradigm constructed through (and by) male poets' work to women's writing and constructing new formalist paradigms with women's writing and feminist literary histories at the center. The latter approach is central to Blake, "Margaret Cavendish's Forms"; Lara Dodds and Michelle M. Dowd, eds., *Feminist Formalism and Early Modern Women's Writing: Readings, Conversations, Pedagogies* (Lincoln: University of Nebraska Press, 2022); and Sasha Roberts, "Feminist Criticism and the New Formalism: Early Modern Women and Literary Engagement," in *The Impact of Feminism in English Renaissance Studies*, ed. Dympna Callaghan (New York: Palgrave Macmillan, 2007), 67–93.

45. Subramaniam, *Ghost Stories for Darwin*, 9.

46. For foundational work on the concept of "situated knowledge," see Donna Haraway, "Situated Knowledges: The Science Question in Feminism and the Privilege of Partial Perspective," *Feminist Studies* 14, no. 3 (1988): 575–99.

47. See Joel Fineman, *Shakespeare's Perjured Eye: The Invention of Poetic Subjectivity in the Sonnets* (Berkeley: University of California Press, 1986); Constance Furey, *Poetic Relations: Intimacy and Faith in the English Reformation* (Chicago: University of Chicago Press, 2017); and Roland Greene, *Post-Petrarchism: Origins and Innovations of the Western Lyric Sequence* (Princeton, NJ: Princeton University Press, 1991).

48. Haraway, "Situated Knowledges," 582.

49. Subramaniam, *Ghost Stories for Darwin*, 9.

50. Sawday, *The Body Emblazoned*, 40.

51. In a firsthand account of a series of demonstrations Andreas Vesalius gave at the University of Bologna (1540), Baldasar Heseler, one of the students present at the lecture, records that Vesalius focused on "the testicles and the spermatic vessels of the male only. For we shall leave those of the female with the uterus, as we have no female body." Several days later, Heseler reports that Vesalius dissected the reproductive organs of a pregnant dog in order to show the students an example of the female reproductive parts. See Baldasar Heseler, *Andreas Vesalius' First Public Anatomy at Bologna, 1540: An Eyewitness Report by Baldasar Heseler, Together with His Notes on Matthaeus Curtius' Lectures on Anatomia Mundini*, ed. Ruben Eriksson (Uppsala: Almquist & Wiksells Boktryckeri, 1959), 175–209, Wellcome Collection, M24323.

52. Here I draw on Amanda Jo Goldstein's articulation of the "interanimation" and "cooperation" of poetic and scientific knowing in the Romantic period (Goldstein, *Sweet Science*, 2).

53. Margaret Cavendish, "Wherein Poetry chiefly consists," *Poems, or Several Fancies in Verse: with the Animal Parliament, in Prose* (London: A. Maxwell, 1668), sig. Aa4^{r}, British Library, C.118.g.4.

54. Laurenza, "Art and Anatomy," 5.

55. Ellen Rooney, "Form and Contentment," *Modern Language Quarterly* 61, no. 1 (2000): 36.

56. Alexander Read, *The manuall of the Anatomy or dissection of the body of Man, Which usually are shewed in the publike Anatomicall exercises. Methodically digested into 6 Books* (London: J. H., 1638), 1–2, Wellcome Library, EPB/A/5355.

57. Helkiah Crooke, for instance, lists anatomy as one node in a network of fields necessary for understanding and practicing "the art of chyrurgery," along with philosophical debate on the constitution of man's body; the "natures, differences, signs, and prognosticks" of diseases; and the nature of drugs, herbs, and minerals that surgeons need use of (Crooke, *Mikrokosmographia. A description of the Body of Man. Together with the controversies thereto belonging. Collected and Translated out of the Best Authors of Anatomy, Especially out of Gasper Bauhinus and Andreas Laurentius* [London: William Jaggard, 1615], 1–2, Wellcome Library, EPB/D/1683). Employing the phrase "science of the body" in this book, I join ongoing conversations among historians and philosophers of science on the use of the term "science" to describe any field of Renaissance practice. Many kinds of knowledges and methods could have been categorized as science in early modern England, but "science" was not a widely used term in the period. Rather, practitioners and thinkers would have referred to their work in a particular field of study (anatomy, chirurgery [surgery], astronomy, chymistry, or optics), or under the broader categories of natural philosophy (the study of the material world as experienced by the senses), natural history (the study of particular properties of natural entities), or mechanical arts (the practical application of mathematical and engineering knowledges). When I use the term "science" in this book, it is an expansive term, not always aligned with the more recent understanding of science as a regimen of systematized study following a set of given expectations and aligned with discrete disciplines. On this point, see Katharine Park and Lorraine Daston, "Introduction: The Age of the New," in *The Cambridge History of Science: Early Modern Science*, vol. 3 (New York: Cambridge University Press, 2006), 1–17.

58. Turner, "Lessons from Literature," 580.

59. Turner, "Lessons from Literature," 586, original emphasis; and Bruno Latour, *Reassembling the Social: An Introduction to Actor-Network Theory* (Oxford: Oxford University Press, 2005), 223. On the action of form, also see Elizabeth Scott-Baumann and Ben Burton, eds., *The Work of Form: Poetics & Materiality in Early Modern Culture* (Oxford: Oxford University Press, 2014).

60. Fleming, *Graffiti and the Writing Arts*, 13.

61. Turner, "Lessons from Literature"; Fleming, *Graffiti and the Writing Arts*; Trettien, *Cut/Copy/Paste*; Mitchell, "Lyric Backwardness"; Heffernan, *Making the Miscellany*; William H. Sherman, *Used Books: Marking Readers in Renaissance England* (Philadelphia: University of Pennsylvania Press, 2007); Adam Smyth, *Material Texts in Early Modern England* (Cambridge: Cambridge University Press, 2018); and Juliet Fleming, William H. Sherman, and Adam Smyth, eds., "The Renaissance Collage: Toward a New History of Reading," *Journal of Medieval and Early Modern Studies* 45, no. 3 (2015). My theorization of form throughout this book is also indebted to recent work by Katherine Schaap Williams and Adhaar Noor Desai, both of whom argue that to think with early modern forms is always already to think with bodies. See Schaap Williams, *Unfixable Forms: Disability, Performance, and the Early Modern English Theater* (Ithaca, NY: Cornell University Press, 2021); and Desai, *Blotted Lines: Early Modern English Literature and the Poetics of Discomposition* (Ithaca, NY: Cornell University Press, 2023).

62. Sachiko Kusukawa, *Picturing the Book of Nature: Image, Text, and Argument in Sixteenth-Century Human Anatomy and Medical Botany* (Chicago: University of Chicago Press, 2012), 3; Laurenza, "Art and Anatomy"; James Elkins, *Pictures of the Body: Pain and Metamorphosis* (Stan-

ford, CA: Stanford University Press, 1999); and Mechthild Fend, *Fleshing Out Surfaces: Skin in French Art and Medicine, 1650–1850* (Manchester: Manchester University Press, 2017).

63. Crooke, *Mikrokosmographia*, 18.

64. Crooke, *Mikrokosmographia*, 29. In his language of structure, action, and use, Crooke draws on Galen's instructions regarding anatomical study: "the Anatomist must first search out the structure or composition, then the action, and lastly the use" of each part (20). An anonymous 1655 anatomical treatise also uses the term "structure" to define anatomy as "the Fabrick or structure of the body of man revealed in dissection." *Anthropologie abstracted: or The idea of humane nature reflected in briefe philosophicall, and anatomicall collections* (London, 1655), 1, Early English Books Online Text Creation Partnership, 2011, https://quod.lib.umich.edu/e/eebo2/A75461.0001.001, accessed October 25, 2022. Read and Cooke both also use the term "frame"—as in, the "frame of man"—in their definitions of anatomy, a term that finds significant traction in early modern poetics (see Kalas, *Frame, Glass, Verse*, especially Chapters 1–2).

65. Katherine Rowe, *Dead Hands: Fictions of Agency, Renaissance to Modern* (Stanford: Stanford University Press, 1999), 28. Also see Rowe, "'God's handy worke,'" in *The Body in Parts: Fantasies of Corporeality in Early Modern Europe*, ed. David Hillman and Carla Mazzio (London and New York: Routledge, 1997), 285–312.

66. Crooke, *Mikrokosmographia*, 2; and John Banister, *The Historie of Man, sucked from the sappe of the most approved Anathomistes, in this present age, compiled in most compendious fourme, and now published in English* (London: John Daye, 1578), fol. 60v, Wellcome Library, EPB/D/657.

67. Alexander Read, *The manuall of the anatomy, or, dissection of the body of man* (London: R. Bishop, 1642), A4, Wellcome Library, EPB/A/43248.

68. Read, *The manuall of the anatomy* (1642), A4.

69. Rowe, *Dead Hands*, 29.

70. Crooke, *Mikrokosmographia*, 296.

71. For more on this history, see Owsei Temkin, *Galenism: Rise and Decline of a Medical Philosophy* (Ithaca, NY: Cornell University Press, 1973); Vivian Nutton, "Renaissance Galenism, 1540–1640: Flexibility or an Increasing Irrelevance?," in *Brill's Companion to the Reception of Galen*, ed. Petros Bouras-Vallianatos and Barbara Zipser (Leiden: Brill, 2019), 472–86. On the "patrilineal diagram of understanding" for the history of medicine, see Sawday, *The Body Emblazoned*, especially pages 39–43; and Park, *Secrets of Women*, especially 14–17.

72. Andreas Vesalius, *The Fabric of the Human Body: An Annotated Translation of the 1543 and 1555 Editions of* De Humani Corporis Fabrica Libri Septem, ed. Daniel H. Garrison and Malcolm H. Hast, 2 vols. (Basel: Karger, 2014), 3; and Vesalius, *De humani corporis fabrica libri septem* (Basel: Johannes Oporinus, 1543), II.

73. Vesalius, *The Fabric of the Human Body*, 6; *De humani corporis fabrica*, IV.

74. Andrea Carlino hails both of Vesalius's 1543 publications, *De humani corporis fabrica libri septem* and *De humani corporis fabrica librorum epitome*, as events "of extraordinary importance both for the history of anatomy and for the history of the book" (Carlino, *Books of the Body: Anatomical Ritual and Renaissance Learning*, trans. John Tedeschi and Anne C. Tedeschi [Chicago: University of Chicago Press, 1999], 39). On the intersections between the history of anatomy and the history of print, also see Sawday, *The Body Emblazoned*, especially pages 129–40; and Elizabeth L. Eisenstein, *The Printing Press as an Agent of Change: Communications and Cultural Transformations in Early-Modern Europe* (New York: Cambridge University Press, 1979), especially 267–69.

75. Turner, "Lessons from Literature," 584.

76. See especially Turner, "Lessons from Literature"; and Mann and Sarkar, "Introduction."

77. Mann and Sarkar, "Introduction," 3.

78. In the introduction to *Possible Knowledge*, Debapriya Sarkar urges us to consider "how close readings of imaginative methods and formal techniques can help us recover the kinds of knowledges that early modern literary forms were themselves producing" (7). For exemplary work in this vein, see Blake, "Hester Pulter's Particle Physics"; "Blake, "Margaret Cavendish's Forms"; Dodds and Dowd, *Feminist Formalism*; Michelle M. Dowd, "Breaking Form in Early Modern Literary Studies," *English Literary Renaissance* 50, no. 1 (2019): 40–46; Dianne Mitchell, "Lyric Backwardness," in *The Oxford Handbook of Early Modern Women's Writing in English, 1540–1700*, ed. Elizabeth Scott-Baumann, Danielle Clarke, and Sarah C. E. Ross (Oxford: Oxford University Press, 2022), 189–202; and Roberts, "Feminist Criticism and the New Formalism."

79. Margaret Cavendish, "Letter CXLIII," *CCXI sociable letters written by the Thrice Noble, Illustrious, and Excellent princess, The Lady marchioness of newcastle* (London, 1664), 296.

80. Cavendish, *Poems, or Several Fancies in Verse*, sig. Aa4^{r}.

81. Jacqueline Pearson, "Women Writers and Women Readers: The Case of Aemilia Lanyer," in *Voicing Women: Gender and Sexuality in Early Modern Writing*, ed. Kate Chedgzoy, Melanie Hanson, and Suzanne Trill (Edinburgh: Edinburgh University Press, 1996), 47.

82. Gillian Beer, "Translation or Transformation? The Relations of Literature and Science," *Notes and Records of the Royal Society of London* 44, no. 1 (1990): 92.

CHAPTER 1

1. Margaret Cavendish, "Letter CXLIII," *CCXI sociable letters written by the Thrice Noble, Illustrious, and Excellent princess, The Lady marchioness of newcastle* (London: William Wilson, 1664), 295.

2. Cavendish, *CCXI sociable letters*, 296.

3. Cavendish, *CCXI sociable letters*, 296.

4. Sylvia Bowerbank and Sara Mendelson, eds., *Paper Bodies: A Margaret Cavendish Reader* (Peterborough: Broadview Press, 2000), 11.

5. Bowerbank and Mendelson, *Paper Bodies*, 11.

6. Andreas Vesalius, *The Fabric of the Human Body: An Annotated Translation of the 1543 and 1555 Editions of* De Humani Corporis Fabrica Libri Septem, ed. and trans. Daniel H. Garrison and Malcolm H. Hast, 2 vols. (Basel: Karger, 2014), 8; and Vesalius, *De humani corporis fabrica libri septem* (Basel: Johannes Oporinus, 1543), 4r. All English translations of Vesalius's *Fabrica* are from Malcolm and Hast's edition and will be cited in the footnotes by page number along with corresponding pages in the original 1543 Latin text.

7. Alexander Read, *The manuall of the anatomy, or, dissection of the body of man* (London: R. Bishop, 1642). The 1638 edition of Read's text includes a shorter version of this title page advertisement, along with a woodcut of a dissection scene.

8. Margaret Cavendish, "To Poets," *Poems, and Fancies: written By the Right honourable, the Lady Margaret. Countess of newcastle* (London: T. R., 1653), sig. R1^{r}, British Library, C.39.h.27.(1.). Unless otherwise noted, Cavendish's poems are quoted from this edition. I include poem or preface titles and page signatures in the endnotes, along with poem line numbers where appropriate. In my work with *Poems* for this chapter, I am also greatly indebted to Liza Blake and the undergraduate research assistants responsible for the robust digital critical edition, which offers a fully

searchable text and tracks all emendations between the various editions. See Margaret Cavendish, *Margaret Cavendish's* Poems and Fancies*: A Digital Critical Edition*, ed. Liza Blake, May 2019, http://library2.utm.utoronto.ca/poemsandfancies/, accessed May 8, 2024.

9. For a complete list of Cavendish's publications, see Liza Blake, "Locating Margaret Cavendish's Books: Database, Map, and Analysis," *Digital Cavendish Project*, November 2018, http://digitalcavendish.org/locating-margaret-cavendish.org, accessed November 19, 2024.

10. Margaret Cavendish, "An Epistle to the Reader, for my Book of Philosophy," *The Philosophical and Physical Opinions* (London, 1655), Early English Books Online Text Creation Partnership, 2011, https://name.umdl.umich.edu/A53055.0001.001, accessed October 24, 2022. On this point, also see Liza Blake's argument that, particularly in Cavendish's later writings, she was "preoccupied with ideas of *grounds* and whether poetry could serve as the grounds of natural knowledge" (Blake, "The Grounds of Literature and Science: Margaret Cavendish's Creature Manifesto," in *The Palgrave Handbook of Early Modern Literature and Science*, ed. Howard Marchitello and Evelyn Tribble [London: Palgrave Macmillan, 2017], 7).

11. Cavendish, "To Poets," *Poems, and Fancies*, sig. R2^{r}; and "Phantasmes Masque," *Poems, and Fancies*, sig. X2^{r}.

12. For scholarship on *Poems* as a work of natural philosophy, see Deborah A. Boyle, *The Well-Ordered Universe: The Philosophy of Margaret Cavendish* (Oxford: Oxford University Press, 2018), Chapter 2; Stephen Hequembourg, "The Poetics of Materialism in Cavendish and Milton," *SEL: Studies in English Literature, 1500–1900* 54, no. 1 (2014): 173–92; and Jay Stevenson, "The Mechanist-Vitalist Soul of Margaret Cavendish," *SEL: Studies in English Literature, 1500–1900* 36, no. 3 (1996): 527–43. For scholarship that focuses on Cavendish's poetry, see Katharine Capshaw Smith, "'Bisket of Love, Which Crumbles All Away': The Failure of Domestic Metaphor in Margaret Cavendish's Poetic Fancies," in *Domestic Arrangements in Early Modern England*, ed. Kari Boyd McBride (Pittsburgh: Duquesne University Press, 2002), 48–73; Lara Dodds, "Bawds and Housewives: Margaret Cavendish and the Work of 'Bad Writing,'" *Early Modern Studies Journal* 6 (2014): 29–65; Megan J. Fung, "Art, Authority, and Domesticity in Margaret Cavendish's *Poems and Fancies*," *Early Modern Women* 10, no. 1 (Fall 2015): 27–47; and Elizabeth Scott-Baumann, "'Bak'd in the Oven of Applause': The Blazon and the Body in Margaret Cavendish's Fancies," *Women's Writing* 15, no. 1 (2008): 86–106. Important exceptions to this generic split are Liza Blake, "Margaret Cavendish's Forms: Literary Formalism and the Figures of Cavendish's Atom Poems," in *Feminist Formalism and Early Modern Women's Writing: Readings, Conversations, Pedagogies*, ed. Lara Dodds and Michelle M. Dowd (Lincoln: University of Nebraska Press, 2022), 38–55; and Jessie Hock, "Fanciful Poetics and Skeptical Epistemology in Margaret Cavendish's *Poems and Fancies*," *Studies in Philology* 115, no. 4 (2018): 766–802. It is also worth noting Lisa T. Sarasohn's argument that Cavendish's "materialism" and "fantasy" are "part of the same universe," but I do protest Sarasohn's suggestion that *Poems* divides easily "into two sections," the first containing the "atomistic poems, and the second, meditations or 'fancies.'" See Sarasohn, *The Natural Philosophy of Margaret Cavendish: Reason and Fancy During the Scientific Revolution* (Baltimore: Johns Hopkins University Press, 2010), 53, 36.

13. Cavendish herself emphasizes this natural philosophical content in her prefaces (the epistle "To Naturall Philosophers," for example) and by publishing a natural philosophy treatise, titled *Philosophicall Fancies* (1653), as a companion volume to *Poems* (it was, she laments, only "a week too short of the press" to be bound together with *Poems* ["To the Reader," *Philosophicall Fancies. Written by the Right Honourable, the Lady Newcastle*" (London: Tho: Roycroft, 1653)]). In one extant copy of Cavendish's writings, held at the British Library, *Poems* is bound together with *Philosophical and Physical Opinions* (1655). Both titles are advertised on the book's spine with the

date 1653. There is no separate title page to introduce *Philosophical and Physical Opinions*. After the closing paratextual material for *Poems* (the last of which is a short note: "Reader, I have a little Tract of *Philosophicall Fancies* in Prose, which will not be long before it appear in the world"), the opening paratextual material for *Philosophical and Physical Opinions* begins. It is worth noting two additional features about this combined volume. First, the opening prefatory material for *Philosophical Opinions* is reordered so that several of the epistles addressed to the reader (four total) come first (sig. a1^{r}–a3^{r}), rather than William Newcastle's preface ("To the Lady Marquesse of Newcastle, on her Book intitled her Philosophicall, and Physicall Opinions"), which is customary for other extant printed editions of the text. Second, this particular volume includes a number of handwritten notes and corrections, possibly in two distinct hands, one identified as Cavendish's and the other unidentified.

14. Cavendish, "Letter CXCIX," *CCXI sociable letters*, 417. The terms "poetaster" and "poetastress" were used to describe inferior poets ("a mere versifier") in early modern parlance (*OED*).

15. Cavendish, *CCXI sociable letters*, 418.

16. Cavendish, *CCXI sociable letters*, 418.

17. Cavendish, *CCXI sociable letters*, 418.

18. Cavendish, *CCXI sociable letters*, 418–19.

19. Cavendish, *CCXI sociable letters*, 420–21.

20. Cavendish, "The Epistle Dedicatory," *Poems, and Fancies*, sig. A2^{r}.

21. Cavendish, "To Poets," *Poems, and Fancies*, sig. R1^{r}.

22. Cavendish, *Poems, and Fancies*, sig. Kk3^{v}.

23. Cavendish, *Poems, and Fancies*, sig. Kk3^{v}.

24. For more on women writers and the material conditions of needlework, weaving, and spinning, see Susan Frye, *Pens and Needles: Women's Textualities in Early Modern England* (Philadelphia: University of Pennsylvania Press, 2010); Michele Osherow, "Mary Sidney's Embroidered Psalms," *Renaissance Studies* 29, no. 4 (2015): 650–70; Emma L. E. Rees, "A Well-Spun Yarn: Margaret Cavendish, and Homer's Penelope," in *A Princely Brave Woman: Essays on Margaret Cavendish, Duchess of Newcastle*, ed. Stephen Clucas (Farnham: Ashgate, 2003), 171–81; and Ann Rosalind Jones and Peter Stallybrass, *Renaissance Clothing and the Materials of Memory* (Cambridge: Cambridge University Press, 2000), especially pages 134–71. My reading here offers an important corrective to arguments on Cavendish's textile metaphors that read them as Cavendish's attempt to "feminize and naturalize her writing." See, for instance, Janet Todd, *The Sign of Angellica: Women, Writing, and Fiction 1660–1800* (London: Virago Press, 1989), 62.

25. Philip Sidney, "The Defence of Poesy," in *Sir Philip Sidney: The Major Works*, ed. Katherine Duncan-Jones (Oxford: Oxford University Press, 2002), 216.

26. Cavendish, "The Epistle Dedicatory," *Poems, and Fancies*, sig. A2^{r}.

27. Cavendish, "An Epistle to Mistress Toppe," *Poems, and Fancies*, sig. A5^{r}.

28. Cavendish, "To All Noble, and Worthy Ladies," *Poems, and Fancies*, sig. A3^{r}.

29. Cavendish, *Poems, and Fancies*, sig. Kk4^{r}.

30. Raymond Williams, *Keywords: A Vocabulary of Culture and Society* (London: Croom Helm, 1976), 138.

31. Sidney, "The Defence of Poesy," 216.

32. Hequembourg, "Poetics of Materialism," 174.

33. Hequembourg, "Poetics of Materialism," 176. Even those who are more attuned to Cavendish's generic hybridity often construct a hierarchy between the philosophical and literary components of her writing. Line Cottegnies, for example, in an article on Cavendish's "generic bricolage," claims that Cavendish's "choice of fiction as vehicle for science undermine[d] her claim

to authority." For Jay Stevenson, Cavendish's fancy is purely a gendered tactic, one that "makes her scientific writing seem less presumptuously assertive." See Line Cottegnies, "Generic Bricolage and Epicureanism in Margaret Cavendish's Imaginative Works," in *A Companion to the Cavendishes*, ed. Lisa Hopkins and Tom Rutter (Leeds: Arc Humanities Press, 2020), 330; and Stevenson, "Mechanist-Vitalist Soul," 532. For responses and important corrections to this critical tendency, see Blake, "The Grounds of Literature and Science"; Blake, "Margaret Cavendish's Forms"; and Katherine Eggert, *Disknowledge: Literature, Alchemy, and the End of Humanism in Renaissance England* (Philadelphia: University of Pennsylvania Press, 2015), especially Chapter 5.

34. Cavendish, "Of Poets, and their Theft," *Poems, and Fancies*, sig. R2^r.

35. Cavendish, "Of Poets, and their Theft," *Poems, and Fancies*, sig. R2^r.

36. Sidney, "The Defence of Poesy," 217.

37. Sidney, "The Defence of Poesy," 217–18.

38. In Maura Smyth's account of "fancy" as a vibrant, endlessly variable, and pleasure-seeking force in late seventeenth- and early eighteenth-century English literature, she argues that Cavendish's texts "replicate Fancy's prolific and varied motion." For Smyth, Cavendish's textual worlds are created *from* fancy, which then becomes "something that can exist and have independent agency in the external world" (Smyth, *Women Writing Fancy: Authorship and Autonomy from 1611 to 1812* [New York: Palgrave Macmillan, 2017], 120).

39. "work, *v.*," and "wrought, *adj.*" (*OED*).

40. Cavendish, "Of Poets, and their Theft," *Poems, and Fancies*, sig. R2^v.

41. Cavendish, "To Poets," *Poems, and Fancies*, sig. R1^v.

42. Cavendish and her husband, William, first Duke of Newcastle, were in exile for many years as Royalist supporters during the English Civil Wars. In both *The Life of the Thrice Noble, High and Puissant Prince William Cavendish* (1667) and *A True Relation of My Birth, Breeding, and Life* (1656), Cavendish recounts periods of time when the two were apart, particularly during two extended trips to London that she took to petition Parliament for the return of William's lands. She also alludes to this time apart in one of the prefaces to *Poems*: "For my Rest being broke with discontented Thoughts, because I was from my Lord, and Husband, knowing him to be in great Wants, and myself in the same Condition; to divert them, I strove to turn the Stream, yet shunning the muddy, and foul ways of Vice, I went to the Well of Helicon, and by the Wells side, I have sat, and wrote this Work" ("To the Reader," *Poems, and Fancies*, sig. A7^r–A7^v). For more on Cavendish's biography, also see Douglas Grant, *Margaret the First: A Biography of Margaret Cavendish, Duchess of Newcastle, 1623–1673* (Toronto: University of Toronto Press, 1957); James Fitzmaurice, "The Life and the Literary Reputation of Margaret Cavendish," *Quidditas* 20 (1999): 55–74; and Katie Whitaker, *Mad Madge: Margaret Cavendish, Duchess of Newcastle, Royalist, Writer and Romantic* (New York: Vintage Books, 2011).

43. Cavendish, "Similizing Thoughts," *Poems, and Fancies*, sig. U1^r.

44. Cavendish, "Similizing the Head of Man to a Hive of Bees," *Poems, and Fancies*, sig. U3^r.

45. Cavendish, "Similizing Fancy to a Gnat," *Poems, and Fancies*, sig. U4^r.

46. Cavendish, "Similizing Fancy to a Gnat," *Poems, and Fancies*, sig. U4^r.

47. Hequembourg, "The Poetics of Materialism."

48. Margaret Cavendish, "Wherein Poetry chiefly consists," *Poems, or, several fancies In verse: with the Animal Parliament, In prose* (London: A. Maxwell, 1668), sig. Aa4^r, British Library, C.118.g.4. Here I quote the 1668 edition (for these lines, the same as the 1664 edition) because the relationship between "fancy" and "form" is more precise than in the 1653 version. The 1653 version reads: "Fancy is the Form, Flesh, Blood, Bone, Skin; Words are but shadows; have no Substance in."

49. Joad Raymond, "Introduction: The Origins of Popular Print Culture," in *The Oxford History of Popular Print Culture*, ed. Joad Raymond, vol. 1 (Oxford: Oxford University Press, 2011), 12.

50. Andrea Carlino, "Knowe Thyself: Anatomical Figures in Early Modern Europe," *RES: Anthropology and Aesthetics* 27 (1995): 53. Carlino makes the important point that the same anatomical fugitive sheets often circulated for many years with little to no revisions based on developing anatomical information. "They were a game as well as a source of information, with colored figures that opened and closed," Carlino notes, "and it did not really matter that some anatomical details were vague, mistaken, or out-of-date" ("Knowe Thyself," 63).

51. Andrea Carlino, *Paper Bodies: A Catalogue of Anatomical Fugitive Sheets 1538–1687*, trans. Noga Arikha (London: Wellcome Institute for the History of Medicine, 1999), 46.

52. For more details on the history of anatomical fugitive sheets, see Carlino, *Paper Bodies*, especially pages 46–73; Gianfranco Crupi, "'Mirabili Visioni': From Moveable Books to Moveable Texts," *Italian Journal of Library and Information Science* 7, no. 1 (2016): 25–87; Suzanne Karr Schmidt, *Interactive and Sculptural Printmaking in the Renaissance* (Leiden: Brill, 2018), especially pages 107–202; and L. H. Wells, "Anatomical Fugitive Sheets with Superimposed Flaps, 1538–1540," *Medical History* 12, no. 4 (1968): 403–7.

53. Suzanne Karr Schmidt, "Printed Bodies and the Materiality of Early Modern Prints," *Art in Print* 1, no. 1 (June 2011): n.p. Also see L. H. Wells, "The 'Sabio' and 'Sylvester' Families of Anatomical Fugitive Sheets: Note on a Pair of Sheets in the National Library of Medicine," *Bulletin of the History of Medicine* 40, no. 5 (1966): 467–75. For the most up-to-date catalog of extant anatomical fugitive sheets from the period, including their archival locations, see Carlino, *Paper Bodies*.

54. Edward William Tuson was the House Surgeon at Middlesex Hospital, London from 1836 to 1847. His *Myology* (1825) and *A Supplement to Myology* (1828) were large books of anatomical flap figures focused on the muscles. Each flap figure was comprised of twelve or more hand-colored flaps, intricately pasted and requiring an exceptional amount of dexterity to study. Frederick Hollick was an American physician and sex educator in the mid- to late nineteenth century. He published several manuals on sexual health that included flap figures, with particular emphasis on the female sex organs and female pleasure, while maintaining that practices of masturbation, homosexuality, celibacy, and interracial relations were "unhealthful." See Tuson, *Myology, Illustrated by Plates. In Four Parts* (London: Callow and Wilson, 1825); and Hollick, *The Origin of Life and Process of Reproduction in Plants and Animals* (London: Greening & Co., 1878). For more on Tuson and Hollick's flap figures, see Meg Brown, "Flip, Flap, and Crack: The Conservation and Exhibition of 400+ Years of Flap Anatomies," *The Book and Paper Group Annual* 32, no. 6 (2013): 6–14; and Whitney Trettien, Rachel Ingold, and Meg Brown, "The Origin of Life and Process of Reproduction in Plants and Animals, Frederick Hollick," in *Animated Anatomies* (Duke University Library, 2011), https://youtu.be/w1KsqVAa-TA, accessed March 23, 2023.

55. "A Supplement to Myology," book review, *The Lancet, 1828–1829*, ed. Thomas Wakley, vol. 1 (London: Mills, Jowett, and Mills, 1829), 468.

56. Carla Bittel, Elaine Leong, and Christine Von Oertzen, eds., *Working with Paper: Gendered Practices in the History of Knowledge* (Pittsburgh: University of Pittsburgh Press, 2019), 8–10.

57. Henry S. Turner, "Lessons from Literature for the Historian of Science (and Vice Versa): Reflections on 'Form,'" *Isis* 101, no. 3 (2010): 586.

58. Elizabeth Spiller, *Science, Reading, and Renaissance Literature: The Art of Making Knowledge, 1580–1670* (Cambridge: Cambridge University Press, 2004), 3.

59. On records of reading and use in early modern books, see Bradin Cormack and Carla Mazzio, *Book Use, Book Theory: 1500–1700* (Chicago: The University of Chicago Library, 2005);

Lisa Jardine and Anthony Grafton, "'Studied for Action': How Gabriel Harvey Read His Livy," *Past & Present*, no. 129 (1990): 30–78; Elaine Leong, "'Herbals She Peruseth': Reading Medicine in Early Modern England," *Renaissance Studies* 28 (2014): 556–78; William H. Sherman, *Used Books: Marking Readers in Renaissance England* (Philadelphia: University of Pennsylvania Press, 2007); and Helen Smith, *"Grossly Material Things": Women and Book Production in Early Modern England* (Oxford: Oxford University Press, 2012), especially Chapter 5.

60. Cavendish, *sociable letters*, 296.

61. Cavendish, "Wherein Poetry chiefly consists," *Poems, or, several fancies* (1668), sig. Aa4ʳ.

62. Cavendish, "Similizing Fancy to a Gnat," *Poems, and Fancies*, sig. U4ʳ.

63. There is debate among historians of science about the pre-1619 editions, which seem to have been printed by Michelspacher without Remmelin and Kilian's consent. I treat the flap figures as a collaboration primarily between Remmelin and Kilian. On this debate, see Massey, "Alchemical Womb," 209n4; and W. B. McDaniel, "Affair of the '1613' Printing of Johannes Rümelin's *Catoptron*," *Transactions & Studies of the College of Physicians of Philadelphia* 6 (1938): 60–72.

64. Given the evidence of circulation and translations based on extant copies of Remmelin and Kilian's text, it is clear that the *Catoptrum* was widely popular throughout the seventeenth and eighteenth centuries. For more on the book's history, see Karr Schmidt, "Printed Bodies"; Lyle Massey, "The Alchemical Womb: Johann Remmelin's *Catoptrum Microcosmicum*," in *Visual Cultures of Secrecy in Early Modern Europe*, ed. Timothy McCall and Sean Roberts (University Park: Penn State University Press, 2013), 208–28; and Kenneth F. Russell, *A Bibliography of Johann Remmelin the Anatomist* (Australia: J. F. Russell, 1991), especially pages 55–90.

65. Karr Schmidt, *Interactive and Sculptural Printmaking*.

66. On the theological, alchemical, and eschatological themes running through the *Catoptrum* prints, see Massey, "Alchemical Womb."

67. Johann Remmelin, *A survey of the microcosme: or, The anatomy of the bodies of man and woman. Corrected by Clopton Havers* (London: James Moxon, 1695), British Library, C.107.k.13.

68. On this possibility, see Suzanne Karr Schmidt, with Kimberly Nichols, *Altered and Adorned: Using Renaissance Prints in Daily Life* (Chicago: Art Institute of Chicago, 2011); and Karr Schmidt, *Interactive and Sculptural Printmaking*. There are extant uncut impressions of Kilian and Remmelin's plates, housed at the Feinberg School of Medicine at Northwestern University. Karr Schmidt uses these plates to retrace the most likely process of assembly, speculating that the engraved top sheets would have "incorporated the upper level of the flaps, most of which show the figure's skin." "These were then partially cut out to reveal the cavities below, leaving hinges on one side to keep the flaps attached to the page." The flaps and organs were then printed on slightly thicker paper and cut out, "glued together in bundles of layers in descending order," and finally "affixed to the verso of the main sheet" (*Altered and Adorned*, 85–87).

69. Vesalius includes a diagram of anatomical instruments in his *Fabrica* and *Epitome* (titled, "Anatomicorum Instrumentorum Delineatio" ["Illustration of Anatomical Instruments"]), along with a short description of the tools necessary for dissection. His catalogue includes "small knives of the type with which pens are sharpened"; "curved needles with heavy thread, with which we sew printed fascicles"; and "smaller needles, which we use to sew up wounds." Rather than simply listing knives, needles, and thread, Vesalius's descriptive details drag the practices of writing, bookmaking, and sewing into the dissection room. See Vesalius, *The Fabric of the Human Body*, 478; *De humani corporis fabrica*, 237.

70. Katharine A. Craik, "Poetry, Anatomy, Presence," *Renaissance Studies* 32, no. 5 (2017): 777.

71. David Hillman and Carla Mazzio take Kilian and Remmelin's *Catoptrum* as representative of how "parts were imagined as dominant vehicles for the articulation of culture" in the early modern period. See David Hillman and Carla Mazzio, eds., *The Body in Parts: Fantasies of Corporeality in Early Modern Europe* (New York: Routledge, 1997), xiii–xiv.

72. For more detail on Cavendish's natural philosophy, see Blake, "The Grounds of Literature and Science"; Boyle, *The Well-Ordered Universe*; Sarasohn, *The Natural Philosophy of Margaret Cavendish*; and Brandie R. Siegfried, "The Natural Philosophy of Margaret Cavendish, Duchess of Newcastle: Nature, Self-Knowing Matter, and the Dialogic Universe," in *The Palgrave Handbook of Women and Science since 1660*, ed. Claire G. Jones, Alison E. Martin, and Alexis Wolf (London: Palgrave Macmillan, 2022), 27–45.

73. Cavendish, "The foure principall Figur'd Atomes make the foure Elements, as Square, Round, Long, and Sharpe," *Poems, and Fancies*, sig. B4^{r}.

74. Cavendish, "The foure principall Figur'd Atomes make the foure Elements, as Square, Round, Long, and Sharpe," *Poems, and Fancies*, sig. B4^{r}.

75. Cavendish's terms, however, are a bit slippery throughout the atom poems, particularly "figure" and "forme." As Liza Blake demonstrates, "figure" becomes a "close synonym for form" in Cavendish's later natural philosophy, and in several poems Cavendish does use "forme" to refer to the atoms themselves (Blake, "Margaret Cavendish's Forms," 41). We can see this slippage, for instance, when Cavendish switches to prose to clarify her vitalist, monist theory about halfway through the atom poems. She writes: "Thus the Fancy of my Atomes is that the foure Principall *Figures*, as Sharp, Long, Round, and Square, make the foure Elements; not that they are of several matters, but are all of one matter, onely their severall *Figures* do give them severall Proprieties; so likewise do the *mixt Figures* give them mixt Proprieties, and their several composures give them other Proprieties, according to their *Formes* they put themselves into, by their severall Motions" ("A World made by foure Atomes," my emphasis). "Mixt Figures" is the key phrase at the center of this quote. It stands in for "Formes" to describe a combinatory system, the "several composures" of atoms, which Cavendish then refers to as "Formes" later in the same sentence.

76. Cavendish, "Of Aire," *Poems, or, several fancies* (1668), sig. C2^{r}, ll. 5–8. Here I cite the 1668 edition of *Poems* because these lines are a bit sharper than the 1653 version, which reads: "For howsoere it moves, to what Forme joyne, / yet still that Figure lies in every line. / For Atomes long, their Formes are like a Thread, / which interweaves like to a Spiders Web."

77. Cavendish, "Of the Sympathy of Atomes," *Poems, and Fancies*, sig. C2^{v}.

78. Blake, "Margaret Cavendish's Forms," 51.

79. Blake, "Margaret Cavendish's Forms," 40, 41.

80. Blake, "Margaret Cavendish's Forms," 45.

81. Cavendish, "The joyning of severall Figur'd Atomes make other Figures," *Poems, and Fancies*, sig. C1^{r}.

82. Blake, "Margaret Cavendish's Forms," 41.

83. Cavendish, "Wherein Poetry chiefly consists," *Poems, or, several fancies* (1668), sig. Aa4^{r}.

84. For more on the "corporeality of intellectual particles" in Cavendish's philosophy, see Stevenson, "Mechanist-Vitalist Soul," 535.

85. Cavendish, "Of Loose Atomes," *Poems, and Fancies*, sig. C1^{v}, ll. 1–4.

86. Cavendish, "The foure principall Figur'd Atomes make the foure Elements, as Square, Round, Long, and Sharpe," *Poems, and Fancies*, sig. B3^{v}, ll. 7; 19.

87. Cavendish, "What Atomes make Fire to burne, and what Flame," *Poems, and Fancies*, sig. C3^{r}; and "Quenching, and Smothering out of Heat, and Light, doth not change the Property, nor Shape of sharpe Atomes," *Poems, and Fancies*, sig. E1^{r}.

88. Cavendish, "What Atomes make Fire to burne, and what Flame," *Poems, and Fancies*, sig. C3^{r}; and "The Power of Fire," *Poems, and Fancies*, sig. E2^{r}.

89. Cavendish, "The foure principall Figur'd Atomes make the foure Elements, as Square, Round, Long, and Sharpe," *Poems, and Fancies*, sig. B3^{v}, l. 17; "Of Burning," *Poems, and Fancies*, sig. E2^{r}; and "What Atomes make the Sun, and the Sea, go round," *Poems, and Fancies*, sig. C3^{v}.

90. Cavendish, "To Naturall Philosophers," *Poems, and Fancies*, sig. A6^{r}.

91. Cavendish, "To Naturall Philosophers," *Poems, and Fancies*, sig. A6^{r}.

92. "error, *n*." (*OED*).

93. Cavendish, "To Naturall Philosophers," *Poems, and Fancies*, sig. A6^{r}.

94. Cavendish, *Observations upon Experimental Philosophy*, 242.

95. Cavendish, "An Epistle to Mistris Toppe," *Poems, and Fancies*, sig. A5^{r}.

96. Cavendish, "To Naturall Philosophers," *Poems, and Fancies*, sig. A6^{r}.

97. Boyle, "Treading the Digital Turn," 86.

98. Cavendish, "Motion makes Atomes a Bawd for Figure," *Poems, and Fancies*, sig. D1^{r}–D1^{v}.

99. Cavendish, "Motion makes Atomes a Bawd for Figure," *Poems and Fancies*, sig. D1^{r}, ll. 1–4.

100. Cavendish, "Motion makes Atomes a Bawd for Figure," *Poems and Fancies*, sig. D1^{v}, ll. 5–8.

101. The 1664 edition revises all of the pronouns in this poem to the masculine pronouns "he/his," also consistent in the 1668 edition. For more on Cavendish's pronoun revisions, see Elizabeth Scott-Baumann, *Forms of Engagement: Women, Poetry and Culture 1640–1680* (Oxford: Oxford University Press, 2013), 65–67.

102. Cavendish, "The Purchase of Poets, or a Dialogue betwixt the Poets, and Fame, and Homers Marriage," *Poems, and Fancies*, sig. H5^{r}.

103. Cavendish, "Wherein Poetry chiefly consists," *Poems, or, several fancies* (1668), sig. Aa4^{r}.

104. On this point also see Debapriya Sarkar, *Possible Knowledge: The Literary Forms of Early Modern Science* (Philadelphia: University of Pennsylvania Press, 2023). Sarkar argues that Cavendish's poet-creator is "a key component of an evolving literary theory that sees the creator as both instrument and agent of a changeable and pluralist natural world" (126).

105. Cavendish, "Motion makes Atomes a Bawd for Figure," *Poems, and Fancies* (1664); and *Poems, or, several fancies* (1668), sig. E4^{r}. For more on Cavendish's revisions between the different editions of her texts, see Blake, "Textual and Editorial Introduction," *Margaret Cavendish's* Poems and Fancies; James Fitzmaurice, "Margaret Cavendish on Her Own Writing: Evidence from Revision and Handmade Correction," *The Papers of the Bibliographical Society of America* 85, no. 3 (1991): 297–307; and Scott-Baumann, *Forms of Engagement*, Chapter 2.

106. Cavendish, "The Epistle Dedicatory: to Sir Charles Cavendish, My Noble Brother-in-Law," *Poems, and Fancies*, sig. A2^{r}, my emphasis.

107. Cavendish, "Of a wrought Carpet, presented to the view of working Ladies," *Poems, and Fancies*, sig. Dd1^{r}.

108. Cavendish, "Of a wrought Carpet, presented to the view of working Ladies," *Poems, and Fancies*, sig. Dd1^{r}.

109. Cavendish, *Poems, and Fancies*, sig. A5^{v}.

110. Vesalius, *The Fabric of the Human Body*, 334; *De humani corporis fabrica*, 169.

111. Vesalius, *The Fabric of the Human Body*, 334; *De humani corporis fabrica*, 169.

112. Vesalius, *The Fabric of the Human Body*, 828; *De humani corporis fabrica*, 314. In some copies of Vesalius's *Fabrica*, the pages containing the foldout diagram are bound horizontally rather than vertically, so the figure spans two bound pages. This binding anomaly would have made it

especially tricky to assemble the flap figure without carefully extracting the improperly bound leaf. See, for instance, one of the University College London copies (Folio 1555 V 2).

113. In the *Epitome* version of this flap figure, the full body depicting veins and arteries is printed on the same page as the small flap woodcuts, so the reader would likely have extracted the entire page from the text and assembled it as an anatomical fugitive sheet. See, for instance, the Wellcome Library copy, which includes the uncut flap figure page: Vesalius, *De humani corporis fabrica librorum epitome* (Basileae: J. Oporini, 1543), Wellcome Collection, EPB/F/6565. In the *Fabrica*, the separate page of flap woodcuts allows the reader to assemble the anatomical flap figure on a bound page, preserving it within the text. Very few copies of the *Epitome* survive, and only six include constructed flap figures—those held at Cambridge University, Dartmouth College, New York Academy of Medicine, Royal Library of Sweden, Huntington Library, and Karolinska Institute Library.

114. Vesalius, *The Fabric of the Human Body*, 828; *De humani corporis fabrica*, 314.

115. Vesalius, *The Fabric of the Human Body*, 830; *De humani corporis fabrica*, 314.

116. Vesalius, *The Fabric of the Human Body*, 834; *De humani corporis fabrica*, 314.

117. For the most up-to-date information about the status of flap figure constructions in extant *Fabrica* copies at the time of my writing, see Dániel Margócsy, Mark Somos, and Stephen N. Joffe, *The Fabrica of Andreas Vesalius: A Worldwide Descriptive Census, Ownership, and Annotations of the 1543 and 1555 Editions* (Leiden: Brill, 2018).

118. Cavendish, *Poems, and Fancies*, sig. A8^{v}, British Library, BL C.39.h.27.(1.).

119. Cavendish, "To Poets," *Poems, and Fancies*, sig. R2^{r}. These instructions are included at the end of the preface "To Poets" in the 1653 edition, but then are extracted and retitled "To the Reader" in the 1664 and 1668 editions. On Cavendish's instructions for reading *Poems*, also see Randall Ingram, "First Words and Second Thoughts: Margaret Cavendish, Humphrey Moseley, and 'the Book,'" *Journal of Medieval and Early Modern Studies* 30, no. 1 (2000): 101–24; and Denise Tillery, "'English Them in the Easiest Manner You Can': Margaret Cavendish on the Discourse and Practice of Natural Philosophy," *Rhetoric Review* 26, no. 3 (2007): 268–85.

120. Cavendish, "To Poets," *Poems, and Fancies*, sig. R1^{v}–R2^{r}.

121. Cavendish, "To Poets," *Poems, and Fancies*, sig. R1^{v}.

122. Cavendish, "To Poets," *Poems, and Fancies*, sig. R2^{r}.

123. Amy Scott-Douglass, "Self-Crowned Laureatess: Towards a Critical Revaluation of Margaret Cavendish's Prefaces," *Pretexts: Literary and Cultural Studies* 9, no. 1 (2000): 30.

124. Scott-Douglass, "Self-Crowned Laureatess," 40. This line, "the first who ever wrote this way" comes from the opening of a preface in *Poems* attributed to "E. Toppe." This unknown interlocutor directly addresses Cavendish and praises her poetry collection: "You are not onely the first English Poet of your Sex, but the first that ever wrote this way: therefore whosoever that writes afterwards, must own you for their Pattern, from whence they take their Sample; and a Line by which they measure their Conceits and Fancies" (sig. A5^{v}). This preface is untitled in the 1653 edition, then titled "To Her Excellence The Lady Marchioness of Newcastle" in the later editions.

125. My discussion of deforming poetic matter throughout this section has been richly informed by Katherine Schaap Williams's argument that "deformity is not form's opposite but its multiplicity" and Adhaar Noor Desai's understanding of revision in early modern poetic practice. For Desai, "revision senses that underneath the architecture of discourse throbs a restless volatility, the potential energy of doing things in a different way." See Schaap Williams, *Unfixable Forms: Disability, Performance, and the Early Modern English Theater* (Ithaca, NY: Cornell University Press, 2021), 17; and Adhaar Noor Desai, *Blotted Lines: Early Modern English Literature and the Poetics of Discomposition* (Ithaca, NY: Cornell University Press, 2023), 120.

126. Blake, "Reading Poems (and Fancies): An Introduction to Margaret Cavendish's *Poems and Fancies*," in *Margaret Cavendish's* Poems and Fancies, my emphasis.

127. Cavendish, "Wherein Poetry chiefly consists," *Poems, or, several fancies* (1668), sig. Aa4[r].

128. Cavendish, "The several Keys of Nature, which unlock the several Boxes of her Cabinet," *Poems, or, several fancies* (1668), sig. Aa4[v]. I cite the 1668 edition here because Cavendish has made the language of "unlocking" consistent across the poem, while the 1653 edition uses "open" and "unlock" interchangeably. Cavendish and/or her printer changed the layout of this page for the 1668 edition, formatting "The several Keys" title like the other titles in the section and removing the illuminated "A."

129. Cavendish, "Natures Cabinet," *Poems, and Fancies*, sig. R3[v]; "Meat Drest for Natures Dinner; an Olio for Nature," *Poems, and Fancies*, sig. S1[r]; "A Dessert," *Poems, and Fancies*, sig. S2[v]; "A Heart dressed," *Poems, and Fancies*, sig. S2[r]; and "A Hodge-Podge for Nature's Table," *Poems, and Fancies*, sig. S2[r].

130. Mihoko Suzuki, "Margaret Cavendish and the Female Satirist," *Studies in English Literature, 1500–1900* 37, no. 3 (Summer 1997): 484. On Cavendish's recipe poems, see Fung, "Art, Authority, and Domesticity"; Theodora A. Jankowski, "Good Enough to Eat: The Domestic Economy of Woman-Woman Eroticism in Margaret Cavendish and Andrew Marvell," in *Privacy, Domesticity, and Women in Early Modern England*, ed. Corinne Abate (Aldershot: Ashgate, 2003), 83–109; Scott-Baumann, "'Bak'd in the Oven of Applause'"; and Edith Snook, "Margaret Cavendish and the Recipe Form in *Poems and Fancies*," in *Feminist Formalism and Early Modern Women's Writing: Readings, Conversations, Pedagogies*, ed. Lara Dodds and Michelle M. Dowd (Lincoln: University of Nebraska Press, 2022), 56–74.

131. Cavendish, "A Bisk for Natures Table," *Poems, and Fancies*, sig. S1[r].

132. Cavendish, "A Bisk for Natures Table," *Poems, and Fancies*, sig. S1[v].

133. Cavendish, "Of Poets, and their Theft," *Poems, and Fancies*, sig. R2[r].

134. Cavendish, "A Tart," *Poems, and Fancies*, sig. S2[r]–S2[v], ll. 1–12.

135. Scott-Baumann, "'Bak'd in the Oven of Applause,'" 90.

136. This is another instance where Cavendish's revisions between editions are especially interesting. When she revised "A Tart" for the 1664 edition of *Poems, and Fancies*, she changed the syntax of several lines, further muddling the structure of these metaphors. She changed line 5, for instance, from "then did she take some Cherry Lips that's red" to "then did she take some Lips, like Cherries red." In this revised version, the ingredient is not "cherry lips" but rather "lips, *like* cherries." She uncouples the body part from the actual food item, and "cherries" becomes purely analogical rather than part of the ingredient list.

137. Harry Berger Jr., "The System of Early Modern Painting," *Representations* 62 (1998): 43.

138. Cavendish, "A Bisk for Nature's Table," *Poems, and Fancies*, sig. S1[v].

139. "raspis, *n.*," and "rasp, *v.*" (*OED*).

140. Jeffrey Masten, "Material Cavendish: Paper, Performance, 'Sociable Virginity,'" *Modern Language Quarterly* 65, no. 1 (2004): 54.

141. Lisa T. Sarasohn, for instance, argues that the "description of fairyland" in Part V "is actually a recasting of [Cavendish's] initial atomistic theory" in Part I (*The Natural Philosophy of Margaret Cavendish*, 36).

142. Megan J. Fung, for example, argues, "in *Poems and Fancies*, Cavendish fashions a domestic space that serves to authorize and affirm women's writing" ("Art, Authority, and Domesticity," 47).

143. Catharine Gallagher, "Embracing the Absolute: The Politics of the Female Subject in Seventeenth-Century England," *Genders* 1 (1988): 24–39. For literature and science studies work

that productively challenges Gallagher's argument, see Eve Keller, "Producing Petty Gods: Margaret Cavendish's Critique of Experimental Science," *ELH* 64, no. 2 (1997): 447–71; and Mary Baine Campbell, *Wonder and Science: Imagining Worlds in Early Modern Europe* (Ithaca, NY: Cornell University Press, 1999), especially pages 202–18. Other notable work directly responding to Gallagher includes Jonathan Goldberg, "Margaret Cavendish, Scribe," *GLQ: A Journal of Lesbian and Gay Studies* 10, no. 3 (2004): 433–52; Sujata Iyengar, "Royalist, Romancist, Racialist: Rank, Gender, and Race in the Science and Fiction of Margaret Cavendish," *ELH* 69, no. 3 (2002): 649–72; and John Rogers, *The Matter of Revolution: Science, Poetry, and Politics in the Age of Milton* (Ithaca, NY: Cornell University Press, 1996), Chapter 6.

144. Ingram, "First Words and Second Thoughts," 114.

145. Ingram, "First Words and Second Thoughts," 114.

146. Cavendish, "The Circle of the Brain cannot be Squared," *Poems, and Fancies*, sig. G4^{r}.

147. Cavendish, *Poems, and Fancies*, sig. P3^{v}.

148. L. E. Semler, "Nation, Nature, and Poetics: Transitions and Claspes in Denham's 'Cooper's Hill' and Cavendish's *Poems and Fancies*," in *Home and Nation in British Literature from the English to the French Revolutions*, ed. A. D. Cousins and Geoffrey Payne (Cambridge: Cambridge University Press, 2015), 30.

149. Semler, "Nation, Nature, and Poetics," 30.

150. Cavendish, "Of small Creatures, such as we call Fairies," *Poems, and Fancies*, sig. Dd1^{v}; and "The City of the Fairies," *Poems, and Fancies*, sig. Dd2^{r}.

151. Cavendish, "To Poets," *Poems, and Fancies*, sig. R1^{r}.

152. Masten, "Paper, Performance," 54.

153. Cavendish, "Phantasmes Masque," *Poems, and Fancies*, sig. X2^{r}.

154. Cavendish, "A Lady dressed by Youth," *Poems, and Fancies*, sig. X3^{v}; "A Soldier armed by Mars," *Poems, and Fancies*, sig. X3^{r}; and "A Lady dressed by Love," *Poems, and Fancies*, sig. X3^{r}.

155. On this point, see Blake, "Margaret Cavendish's Forms," 39; Hock, "Fanciful Poetics," 79; and Bronwen Price, "Feminine Modes of Knowing and Scientific Enquiry: Margaret Cavendish's Poetry as Case Study," in *Women and Literature in Britain, 1500–1700*, ed. Helen Wilcox (Cambridge: Cambridge University Press, 1996), 133.

156. Blake, "Margaret Cavendish's Forms," 41.

157. Cavendish, "Phantasmes Masque," *Poems, and Fancies*, sig. X2^{r}.

158. For more on Cavendish's materializing of letters, see Jonathan Goldberg, *The Seeds of Things: Theorizing Sexuality and Materiality in Renaissance Representations* (New York: Fordham University Press, 2009), especially Chapter 4.

159. The Royal Society's meeting minutes for May 23, 1667, record: "Lord Berkeley mentioned, that the duchess of Newcastle had expressed a great desire to come to the society, and to see some of their experiments; but that she desired to be invited. This was seconded by the earl of Carlisle and Dr. Charleton, who pressing, that it might be put to the vote accordingly, whether the duchess of Newcastle should at her desire be invited to be present at the meeting on the Thursday following; it was carried in the affirmative. The ceremonies and the subjects for her entertainment were referred to the council" (Thomas Birch, *The History of the Royal Society of London*, vol. 2 [London, 1756], 176). For more on Cavendish's attendance at the Royal Society meeting, see Emma Wilkins, "Margaret Cavendish and the Royal Society," *Notes and Records of the Royal Society of London* 68, no. 3 (2014): 245–60.

160. For more on Cavendish's engagements (in-person and through text) with contemporary scientific and philosophical thinkers, see Anna Battigelli, *Margaret Cavendish and the Exiles of the Mind* (Lexington: University Press of Kentucky, 1998), especially Chapter 5; Justin Begley,

"'The Minde Is Matter Moved': Nehemiah Grew on Margaret Cavendish," *Intellectual History Review* 27, no. 4 (2017): 493–514; Baine Campbell, *Wonder and Science*, especially Chapter 4; Stephen Clucas, "The Atomism of the Cavendish Circle: A Reappraisal," *The Seventeenth Century* 9, no. 2 (1994): 247–73; Susan James, "The Philosophical Innovations of Margaret Cavendish," *British Journal for the History of Philosophy* 7, no. 2 (1999): 219–44; and Tillery, "Discourse and Practice of Natural Philosophy." Also of note here is a curious manuscript of recipes and letters in the University of Nottingham archives (MS Pw V90) that Justin Begley and Benjamin Goldberg have attributed to Margaret and William Cavendish, with significant influence from Théodore de Mayerne and Kenelm Digby. See Begley and Goldberg, eds., *The Medical World of Margaret Cavendish: A Critical Edition* (London: Palgrave Macmillan, 2022).

161. Crupi, "'Mirabili Visioni,'" 49.

CHAPTER 2

1. John Aubrey, *Aubrey's Brief Lives*, ed. Oliver Lawson Dick (London: Secker & Warburg, 1949), 139.

2. Hilary Rose, "Foreword," in *Women, Science and Medicine, 1500–1700: Mothers and Sisters of the Royal Society*, ed. Lynette Hunter and Sarah Hutton (Gloucestershire: Sutton Publishing, 1997), xii. For accounts of Pembroke's scientific work at Wilton House, see Margaret P. Hannay, "'How I These Studies Prize': The Countess of Pembroke and Elizabethan Science," in *Women, Science and Medicine, 1500–1700: Mothers and Sisters of the Royal Society*, ed. Lynette Hunter and Sarah Hutton (Gloucestershire: Sutton Publishing, 1997), 108–21; and Anna Marie Roos, *Web of Nature: Martin Lister (1639–1712), The First Arachnologist* (Leiden: Brill, 2011), 22–26.

3. See Hannay, "'How I These Studies Prize,'" 112; Margaret P. Hannay, *Philip's Phoenix: Mary Sidney, Countess of Pembroke* (New York: Oxford University Press, 1990), 131; and Cristina Malcolmson, "'The Empire of Man over the Inferior Creatures': British Women, Race, and Seventeenth-Century Science," in *The History of British Women's Writing, 1610–1690*, vol. 3, ed. Mihoko Suzuki (New York: Palgrave Macmillan, 2011), 241. It is also worth mentioning a curious eighteenth-century German volume, titled *Der englischen Gräfin Maria von Pembrock seltene Geheimnisse bestehend in einer wunderbaren Punctir-kunst* (*The English Countess Mary of Pembroke's Rare Secrets of the Wonderful Art of Geomancy*), which includes a recipe for disappearing ink, along with instructions for communicating with the spirit world and questions for specific spirits. While this volume is often attributed to Pembroke in library catalogues because of its title, it is, rather, a text about Pembroke as geomancer that represents her as an eccentric occult figure. See Mary Herbert of Pembroke, *Seltene Geheimnisse* (Ulm and Memmingen: Joh. Friederich Gaums, 1752), Phys.m. 434 d, Bavarian State Library, Munich; Countess Mary of Pembroke, *Der englischen Gräfin Maria von Pembrock* (1814), Abraham H. Cassel Collection, Historical Society of Pennsylvania; and Sarah Newhouse, "Stories from the Archives: The Countess of Pembroke and Geomancy," Historical Society of Pennsylvania, October 31, 2012, YouTube video, 2:29, https://youtu.be/aZhohxEKUMU, accessed September 19, 2022.

4. See Thomas Moffet, *Nobilis or a View of the Life and Death of a Sidney and Lessus Lugubris*, ed. Virgil B. Heltzel and Hoyt H. Hudson (San Marino: Huntington Library, 1940).

5. Aubrey, *Brief Lives*, 139. Hannay remarks that "any attempt to recover Mary Sidney's scientific work is frustrated by reliance on John Aubrey," whose biographical notes are "as much gossip as fact" (Hannay, "'How I These Studies Prize,'" 111). For more on early modern women's education, see Margaret W. Ferguson and Mihoko Suzuki, "Women's Literacies and Social

Hierarchy in Early Modern England," *Literature Compass* 12, no. 11 (2015): 575–90; Elizabeth Mazzola, *Learning and Literacy in Female Hands, 1520–1698* (London: Routledge, 2013); and Kathryn M. Moncrief and Kathryn Read McPherson, eds., *Performing Pedagogy in Early Modern England: Gender, Instruction, and Performance* (Burlington, VT: Ashgate, 2011).

6. Mary Sidney Herbert, *The Collected Works of Mary Sidney Herbert, Countess of Pembroke: The Psalms of David*, vol. 2, ed. Margaret P. Hannay, Noel J. Kinnamon, and Michael G. Brennan (Oxford: Clarendon Press, 1998). All references to Pembroke's psalms are from this edition and will be cited parenthetically in the text by psalm and line number. I maintain the stanza structures and line spacing used by the editors in the *Collected Works*.

7. Hannay, "'How I These Studies Prize,'" 116.

8. "Chymistry" was a form of early modern chemistry, more closely related to alchemy than modern-day chemistry. On Paracelsus's work in relation to chymistry and alchemy, see Daniel Garber, "Physics and Foundations," in *The Cambridge History of Science: Early Modern Science*, vol. 3, ed. Katharine Park and Lorraine Daston (Cambridge: Cambridge University Press, 2006), 21–69; and William R. Newman, "From Alchemy to 'Chymistry,'" in *The Cambridge History of Science: Early Modern Science*, vol. 3, ed. Katharine Park and Lorraine Daston (Cambridge: Cambridge University Press, 2006), 497–517.

9. Roos, *Web of Nature*, 24.

10. See Josephine A. Roberts, "The Huntington Manuscript of Lady Mary Wroth's Play, 'Loves Victorie,'" *Huntington Library Quarterly* 46, no. 2 (1983): 167. Anna Maria Roos speculates that Pembroke and Lister's relationship extended to collaborative chemical and medicinal work (Roos, *Web of Nature*, 24).

11. Naomi J. Miller, "Imagining Shakespeare's Sisters: Fictionalizing Mary Sidney Herbert and Mary Sidney Wroth," in *Authorizing Early Modern European Women: From Biography to Biofiction*, ed. James Fitzmaurice, Naomi J. Miller, and Sara Jayne Steen (Amsterdam: Amsterdam University Press, 2021), 131. Miller takes Pembroke as the protagonist for her novel *Imperfect Alchemist*, arguing that it is precisely the lack of biographical evidence on Pembroke's scientific work that makes her a perfect candidate for biofiction. Fictional Pembroke fandom also includes Deborah Harkness's *All Souls* series, now a British fantasy television show produced by Bad Wolf and Sky Studios. In Book 2 of the series, *Shadow of Night*, Pembroke features as an eccentric, brash, well-connected scientist with a fully functioning alchemy lab. See Naomi J. Miller, *Imperfect Alchemist* (London: Allison & Busby, 2020); and Deborah Harkness, *Shadow of Night* (New York: Penguin Books, 2012).

12. Joan W. Scott, "The Evidence of Experience," *Critical Inquiry* 17, no. 4 (1991): 777.

13. Margaret Healy, "Paracelsian Medicine and Female Creativity: Aemilia Lanyer's *Salve Deus Rex Judaeorum*," *Renaissance and Reformation* 36, no. 2 (2013): 85; and Scott, "Evidence of Experience," 790.

14. Scott, "Evidence of Experience," 779.

15. Here, I think with scholars who have productively troubled the imperative of recovery that long dominated critical approaches to feminist literary histories. See, for instance, Margaret J. M. Ezell, "Recovery," in *The Palgrave Encyclopedia of Early Modern Women's Writing*, ed. Patricia Pender and Rosalind Smith (London: Palgrave Macmillan, 2021), https://doi.org/10.1007/978-3-030-01537-4_246-2; Jamie Goodrich and Paula McQuade, "Beyond Canonicity: The Future(s) of Early Modern Women Writers," *Criticism* 63, no. 1–2 (2021): 1–21; Kate Ozment, "From Recovery to Restoration: Aphra Behn and Feminist Bibliography," *Early Modern Women: An Interdisciplinary Journal* 13, no. 1 (2018): 106–17; and Laura J. Rosenthal, "Introduction: Recovering from Recovery," *The Eighteenth Century* 50, no. 1 (2009): 1–11.

16. Scott, "Evidence of Experience," 780.

17. On Pembroke's sustained interest in the topic of death, see Mary Ellen Lamb, "The Countess of Pembroke and the Art of Dying," in *Women in the Middle Ages and the Renaissance: Literary and Historical Perspectives*, ed. Mary Beth Rose (Syracuse: Syracuse University Press, 1986), 207–26; and Gavin Alexander, *Writing After Sidney: The Literary Response to Sir Philip Sidney, 1586–1640* (Oxford: Oxford University Press, 2006), especially Chapter 3.

18. Pembroke's translation of *The Triumph of Death* survives only in a manuscript miscellany, which also contains three of Pembroke's psalm translations (*The Triumph of Death translated out of Italian by the Countess of Pembroke*, Library of the Inner Temple, MS 538.43.14, ff. 286r–289r). For more on Pembroke's *Triumph* and the manuscript, see Gavin Alexander, "The Triumph of Death: A Critical Edition in Modern Spelling of the Countess of Pembroke's Translation of Petrarch's *Trionfo della Morte*," *Sidney Journal* 17, no. 1 (1999): 2–18.

19. Luciana Mascarenhas Fonseca and Ines Testoni, "The Emergence of Thanatology and Current Practice in Death Education," *Omega: Journal of Death and Dying* 64, no. 2 (2012): 157. The field gained traction after World War II, amid what Philippe Ariès describes as an increasingly widespread denial of death in the twentieth century. See Philippe Ariès, *The Hour of Our Death*, trans. Helen Weaver (New York: Alfred A. Knopf, 1981), 559–601.

20. For more on the extant manuscripts, along with their relationship to each other and Pembroke's lost working copies, see *Collected Works*, vol. 2, 308–57; and Gavin Alexander, "A New Manuscript of the Sidney Psalms: A Preliminary Report," *Sidney Journal* 18, no. 1 (2000): 43–56. I also discuss differences between manuscript editions more thoroughly in n136.

21. Early scholarly treatment of the Sidney Psalter frequently diminished or completely elided Pembroke's work on the collection. See, for instance, William Ringler Jr.'s edition of Philip Sidney's writings, which includes Psalms 1–42 and one version of Pembroke's "To the Angell spirit." See Ringler, ed., *The Poems of Sir Philip Sidney* (Oxford: Clarendon Press, 1962). Ringler introduces Pembroke as an adoring younger sibling: "she admired, almost worshipped, her brilliant older brother, and he had a profound affection for her" (xxiv). Ringler also neglects to list Pembroke in his account of Sidney's "Influences and Sources." For a thoughtful critique of Ringler's positioning of Pembroke in relation to her brother's oeuvre, see Danielle Clarke, "Nostalgia, Anachronism, and the Editing of Early Modern Women's Texts," *Text* 15 (2003): 187–209.

22. Michael Neill, *Issues of Death: Mortality and Identity in English Renaissance Tragedy* (Oxford: Clarendon Press, 1997), 87. On the intersecting events that destabilized conceptions of and beliefs surrounding death in early modern England, see Ariès, *Hour of Our Death*, Chapter 6; Neill, *Issues of Death*, especially pages 38–42; Nigel Llewellyn, *Art of Death: Visual Culture in the English Death Ritual c. 1500–c. 1800* (London: Reaktion Books, 2013); and Lucinda McCray Beier, "The Good Death in Seventeenth-Century England," in *Death, Ritual, and Bereavement*, ed. Ralph Houlbrooke (London: Routledge, 1989), 43–61.

23. Hannibal Hamlin, *Psalm Culture and Early Modern English Literature* (Cambridge: Cambridge University Press, 2004), 12.

24. Mary Sidney Herbert, "To the Angell spirit of the most excellent Sir Phillip Sidney," *The Collected Works of Mary Sidney Herbert, Countess of Pembroke: Poems, Translations, and Correspondence*, vol. 1, ed. Margaret P. Hannay, Noel J. Kinnamon, and Michael G. Brennan (Oxford: Clarendon Press, 1998). All references to Pembroke's original poems and ars moriendi translations are from this edition and will be cited parenthetically in the text by page number or (for poems) by page and line number.

25. On Pembroke's biography, see Hannay, *Philip's Phoenix*.

26. On Pembroke's relationship with Queen Elizabeth, see Michael G. Brennan, "The Queen's Proposed Visit to Wilton House in 1599 and the 'Sidney Psalms,'" *Sidney Journal* 20, no. 1 (2002): 27–53. On the political implications of Pembroke's "Even now that care," see Margaret P. Hannay, "'Princes You as Men Must Die': Genevan Advice to Monarchs in the Psalms of Mary Sidney," *English Literary Renaissance* 19, no. 1 (1989): 22–41.

27. Pembroke published an edition of Sidney's *The Countess of Pembroke's Arcadia* in 1593, followed by a composite volume of *Arcadia, Defence of Poesy, Certain Sonnets*, the *Lady of May*, and *Astrophel and Stella* in 1598. For more on Pembroke as the caretaker of Sidney's textual corpus, see Joel B. Davis, *The Countess of Pembroke's Arcadia and the Invention of English Literature* (New York: Palgrave Macmillan, 2011), especially Chapters 4 and 5.

28. Abraham Fraunce, *The Third part of the Countesse of Pembrokes Yuychurche* (London, 1592), Early English Books Online Text Creation Partnership, 2011, https://quod.lib.umich.edu/e/eebo/A01228.0001.001, accessed October 5, 2022; and Alexander, *Writing After Sidney*, 81.

29. Elizabeth Harris Sagaser, "Elegiac Intimacy: Pembroke's 'To the Angell Spirit of the Most Excellent Sir Philip Sidney,'" *Sidney Journal* 23, no. 1–2 (2005): 112; and Marion Wynne-Davies, "'More Women: More Weeping'; The Communal Lamentation of Early Modern Women in the Works of Mary Sidney Herbert and Mary Wroth," *Early Modern Literary Studies* 17, no. 2 (2014): 2.

30. Jonathan Goldberg, *Desiring Women Writing: English Renaissance Examples* (Stanford: Stanford University Press, 1997), 119.

31. Patricia Phillippy, *Women, Death and Literature in Post-Reformation England* (Cambridge: Cambridge University Press, 2010), 4.

32. Kathleen M. Brown, *Foul Bodies: Cleanliness in Early America* (New Haven, CT: Yale University Press), 5; and Mary E. Fissell, "Introduction: Women, Health, and Healing in Early Modern Europe," *Bulletin of the History of Medicine* 82, no. 1 (2008): 10.

33. Fissell, "Introduction," 13. On Pembroke's consistent care of her own body due to chronic illness, see Catherine Medici, "The Extremity of Illness: Mary Sidney, Early Modern Women's Chronic Illness, and Disability Studies," *Early Modern Women* 15, no. 1 (2020): 107–18.

34. Lady Elizabeth Southwell, recorded by Father Robert Persons in "The Relation of the Lad[y] Southwell of late Q. death, po Aprilis 1607," Archives of the British Province of the Society of Jesus, Stonyhurst Ang. iii. 77. All quotations from Southwell's account are cited from the full transcript of her report reproduced in Catherine Loomis, *The Death of Elizabeth I: Remembering and Reconstructing the Virgin Queen* (New York: Palgrave Macmillan, 2010), 88–91.

35. It is this detail that makes historians skeptical of the veracity of Southwell's account (and, perhaps, why Southwell's account is redacted in Person's manuscript). On this controversy, see Catherine Loomis, "Elizabeth Southwell's Manuscript Account of the Death of Queen Elizabeth," *English Literary Renaissance* 26, no. 3 (1996): 482–509; Louis Montrose, *The Subject of Elizabeth: Authority, Gender, and Representation* (Chicago: University of Chicago Press, 2006), 245–47; and J. E. Neale, *Essays in Elizabethan History* (Ann Arbor: University of Michigan Press, 1958), 84–112.

36. Phillippy, *Women, Death and Literature*, 4.

37. Phillippy, *Women, Death and Literature*, 4.

38. Cited in Phillippy, *Women, Death and Literature*, 20.

39. See Katharine Park, *Secrets of Women: Gender, Generation, and the Origins of Human Dissection* (Baltimore: MIT Press, 2006), 14–17. Critical approaches focused on academic and professional settings also limit the racial and class-based contexts for the history of science. Figures who earned the title "anatomist" in the Renaissance period were exclusively white, educated men,

most often from aristocratic families. There is some evidence that cadavers primarily would have been from lower social classes, particularly given the regular practice of using the bodies of executed criminals for dissection and study. Anatomists' accounts suggest that female corpses were especially difficult to procure, a topic I take up in more detail in Chapter 4. For an overview of these historical contexts, see Giovanna Ferrari, "Public Anatomy Lessons and the Carnival: The Anatomy Theatre of Bologna," *Past & Present* 117 (1987): 56–61; Cynthia Klestinac, *Theaters of Anatomy: Students, Teachers, and Traditions of Dissection in Renaissance Venice* (Baltimore: Johns Hopkins University Press, 2011); and Jonathan Sawday, *The Body Emblazoned: Dissection and the Human Body in Renaissance Culture* (London: Routledge, 1995), Chapter 4. For recent work that brings the historical circumstances of Renaissance anatomy into conversation with the histories of race making and colonialism, see Averyl Dietering, "Abject Black Flesh and the Manufacture of White Fear in Early Modern Anatomical Illustrations," *Shakespeare Studies* 49 (2021): 153–60; Dietering, "Black Ink, White Feelings: Early Modern Print Technology and Anti-Black Racism," in *Race and Affect in Early Modern English Literature*, ed. Carol Mejia LaPerle (Phoenix: ACMRS Press, 2022), https://asu.pressbooks.pub/race-and-affect/, accessed January 24, 2023; and Valerie Traub, "Anatomy, Cartography, and the New World Body," in *Geographies of Embodiment in Early Modern England*, ed. Mary Floyd-Wilson and Garrett A. Sullivan (Oxford: Oxford University Press, 2020), 64–112.

40. Park, *Secrets of Women*, 15–16.

41. Cited in Park, *Secrets of Women*, 40–41.

42. Here, I also think with Judith H. Anderson's work on death, light, and figuration. Citing C. P. Snow's famous argument on the "two cultures" of science and the humanities, Anderson posits that "death is one place where the cultural fissure is bridged." See Anderson, *Light and Death: Figuration in Spenser, Kepler, Donne, Milton* (New York: Fordham University Press, 2017), 4; and C. P. Snow, *The Two Cultures and the Scientific Revolution* (New York: Cambridge University Press, 1959).

43. Richelle Munkhoff, "Reckoning Death: Women Searchers and the Bills of Mortality in Early Modern London," in *Rhetorics of Bodily Disease and Health in Medieval and Early Modern England*, ed. Jennifer C. Vaught (Farnham: Ashgate, 2010), 119–34.

44. These questions and my interdisciplinary work in this chapter build on a long critical history of using women's poetry to better understand cultures surrounding loss, grief, and mourning in the early modern period. See Elizabeth Hodgson, *Grief and Women Writers in the English Renaissance* (Cambridge: Cambridge University Press, 2015); Elizabeth Zeman Kolkovich, "In Defense of Indulgence: Hester Pulter's Maternal Elegies," *Journal for Early Modern Cultural Studies* 20, no. 2 (2020): 43–70; Kate Lilley, "True State Within: Women's Elegy 1640–1700," in *Women Writing History, 1640–1740*, ed. Isobel Grundy and Susan Wiseman (London: Batsford, 1992), 72–92; Phillippy, *Women, Death and Literature*; Lauren Shohet, "Women's Elegy: Early Modern," in *The Oxford Handbook of the Elegy*, ed. Karen Weisman (Oxford: Oxford University Press, 2010), 433–41; Susan Wiseman, "No 'Publick funerall'? Lucy Hutchinson's Elegy, Epitaph, Monument," *The Seventeenth-Century* 30, no. 2 (2015): 207–28; and Wynne-Davies, "'More Women: More Weeping.'"

45. Jonathan P. Lamb, *Shakespeare in the Marketplace of Words* (Cambridge: Cambridge University Press, 2017), 68.

46. See Lamb, *Marketplace of Words*, Chapter 3.

47. For more on Pembroke's source texts, see *Collected Works*, vol. 2, 3–32; Ringler, ed., *The Poems of Sir Philip Sidney*, 505–9; and J. C. A. Rathmell, ed., *The Psalms of Sir Philip Sidney and the Countess of Pembroke* (New York: New York University Press, 1963), xv–xix.

48. Ariès, *Hour of Our Death*, 354.

49. Ariès, *Hour of Our Death*, 354.

50. Alexander, *Writing After Sidney*, 99. The opening lines of Psalm 51, which begin "miserere mei Deus" in the Latin vulgate translation, were famously known as the "neck-verse," often recited by condemned criminals praying for God's mercy on the execution block or by the accused claiming benefit of clergy. John Foxe records that Lady Jane Grey recited *Miserere mei Deus* in English as she awaited execution. See John Foxe, *The Acts and Monuments of John Foxe*, ed. George Townsend (New York: AMS Press, 1965), VI. 700; and Lydia Whitehead, "A Poena et Culpa: Penitence, Confidence and the *Miserere* in Foxe's *Actes and Monuments*," *Renaissance Studies* 4, no. 3 (1990): 287–99.

51. See Anne Locke, *The Collected Works of Anne Vaughan Lock*, ed. Susan M. Felch (Tempe: Renaissance English Text Society, 1999). Pembroke's editors suggest that she was likely familiar with Locke's work (*Collected Works*, vol. 2, 7–8).

52. In Philbert Guibert's popular *The charitable physitian* (first published in French in 1625), he calls for eight handfuls of hyssop, along with thyme, sage, lavender, rosemary, wormwood, marjoram, rue, and scordeum in one of his embalming powder recipes. See Guibert, *The charitable physitian With the charitable apothecary*, trans. I.W. (London, 1639), 143–49, Early English Books Online Text Creation Partnership, 2011, https://quod.lib.umich.edu/e/eebo/A02327.0001.001, accessed November 9, 2020.

53. Guibert, *The charitable physitian*, 144.

54. Guibert, *The charitable physitian*, 146.

55. Hodgson, *Grief and Women Writers*, 4. For work that reads "Angell spirit" as a strategic political act, see Margaret P. Hannay, "'Doo What Men May Sing': Mary Sidney and the Tradition of Admonitory Dedication," in *Silent but for the Word: Tudor Women as Patrons, Translators, and Writers of Religious Works*, ed. Margaret P. Hannay (Kent: Kent State University Press, 1985), 149–65; Hodgson, *Grief and Women Writers*; Shannon Miller, "Mary Sidney and Gendered Strategies for the Writing of Poetry," in *Write or Be Written: Early Modern Women Poets and Cultural Constraints*, ed. Barbara Smith and Ursula Appelt (Aldershot: Ashgate, 2001), 155–76; Shohet, "Women's Elegy: Early Modern"; Beth Wynne Fisken, "'To the Angel Spirit . . .': Mary Sidney's Entry into the 'World of Words,'" in *The Renaissance Englishwoman in Print: Counterbalancing the Canon*, ed. Anne M. Haskelkorn and Betty S. Travitsky (Amherst: University of Massachusetts Press, 1990), 263–75. For more on the literary import of "Angell spirit," see Sagaser, "Elegiac Intimacy"; Clare R. Kinney, "'Love Which Hath Never Done': The Countess of Pembroke's Elegies and the Apology for Copia," *Sidney Journal* 21 (2003): 31–40; and Patricia Pender, "The Ghost and the Machine in the Sidney Family Corpus," *Studies in English Literature, 1500–1900* 51, no. 1 (2011): 65–85.

56. Hodgson, *Grief and Women Writers*, 21. This reminder is particularly compelling when we read Pembroke's elegy to Sidney alongside her dedication to Queen Elizabeth, preserved together in the Tixall Manuscript, which William A. Ringler Jr. speculates was prepared by John Davies of Hereford for the queen's planned visit to Wilton House in 1599. Margaret P. Hannay argues that the two poems together make "a powerful political statement" (*Philip's Phoenix*, 90). See Ringler Jr., *The Poems of Sir Philip Sidney*, 546–47; and Michael G. Brennon, "The Queen's Proposed Visit to Wilton House in 1599 and the 'Sidney Pslams,'" *Sidney Journal* 20, no. 1 (2002): 27–53.

57. Celeste M. Schenck, "Feminism and Deconstruction: Re-Constructing the Elegy," *Tulsa Studies in Women's Literature* 5, no. 1 (1986): 14.

58. Schenck, "Feminism and Deconstruction," 14–15.

59. Schenck, "Feminism and Deconstruction," 14–15.

60. Schenck, "Feminism and Deconstruction," 24.

61. Kinney, "'Love Which Hath Never Done,'" 34; and Danielle Clarke, *The Politics of Early Modern Women's Writing* (London: Routledge, 2001), 140, 169.

62. The only extant manuscript copy of "Angell spirit" is included in the preliminary leaves of the Tixall manuscript (Manuscript J), currently the property of Dr. Bent E. Juel-Jensen, DM, FRCP (Headington, Oxford). The version known as the Daniel Variant was printed in *The whole workes of Samuel Daniel Esquire in Poetrie* (London, 1623). For more on the two different versions of "Angell spirit," see *Collected Works*, vol. 1, 74–77, 303–4.

63. Katharine Park, "The Criminal and the Saintly Body: Autopsy and Dissection in Renaissance Italy," *Renaissance Quarterly* 47, no. 1 (1994): 4.

64. Cited in Robert G. Mayer, ed., *Embalming: History, Theory, and Practice*, 5th ed. (New York: McGraw-Hill Education, 2012), 478.

65. Embalming became central to early modern burial practices because of the elaborate funeral rituals increasingly common to the aristocracy and ruling classes in the wake of the Reformation. As Michael Neill demonstrates, such rituals were part of "a wider preoccupation with the importance of remembrance in a culture forced to devise new ways of accommodating itself to the experience of mortality" (*Issues of Death*, 38).

66. Sarah Tarlow, *Ritual, Belief and the Dead in Early Modern Britain and Ireland* (Cambridge: Cambridge University Press, 2010), 98.

67. David Harley, "Political Post-Mortems and Morbid Anatomy in Seventeenth-Century England," *Social History of Medicine* 7, no. 1 (1994): 5. On this point, also see Phillippy, *Women, Death and Literature*, 19; Park, "The Criminal and the Saintly Body," 6; and Tarlow, *Ritual, Belief and the Dead*, 98.

68. J. P. Gilson, ed., *Lives of Lady Anne Clifford Countess of Dorset, Pembroke and Montgomery (1590–1676) and of Her Parents* (Arundel: Roxburghe Club Books, 1916), 2.

69. Gilson, *Lives of Lady Anne Clifford*, 5. For more on the long tradition of domestic autopsies and women's roles in the practice, see Park, *Secrets of Women*, Chapter 3. Another interesting example of the practice of domestic autopsy comes from Lucy Hutchinson's description of her husband's embalming in *Memoirs of the Life of Colonel Hutchinson*. "And when he was embowelled," Hutchinson records, "all his inwards were found exceeding sound, and no taint in any part, only two or three purple spots on his lungs; his gall, the Doctor said, was the largest that ever he saw in any man, and observed it to be a miracle of grace that he had been so patient as he had seen him" (Hutchinson, *Memoirs of the Life of Colonel Hutchinson, with the Fragment of an Autobiography of Mrs Hutchinson*, ed. J. Sutherland [London: Oxford University Press, 1973], 273).

70. Kinney, "'Love Which Hath Never Done,'" 33. This poem is included in an elegiac procession for Sidney compiled and published with Spenser's *Colin Clouts Come Home Again* (London, 1595). On the debate surrounding authorship attribution, see Danielle Clarke, "'In Sort as She It Sung': Spenser's 'Doleful Lay' and the Construction of Female Authorship," *Criticism* 42, no. 4 (2000): 451–68; Hannay, *Philip's Phoenix*, 63–69; and Kinney, "'Love Which Hath Never Done,'" 41n8.

71. Patricia Phillippy suggests that Pembroke would not have been able to participate in her brother's funeral because of England's gendered regulations surrounding mourning rituals (*Women, Death and Literature*, 21).

72. Wendy Wall, *The Imprint of Gender: Authorship and Publication in the English Renaissance* (Ithaca, NY: Cornell University Press, 1993), 316.

73. Schenck, "Feminism and Deconstruction," 24.

74. William Shakespeare, *The Narrative and Other Poems*, ed. Katherine Duncan-Jones and H. R. Woudhuysen, 3rd ed. (London: Bloomsbury Arden Shakespeare, 2007), 107.13–14. Samuel Daniel draws on the exegi monumentum convention in his praise of Pembroke and her psalm translations at the start of his *The Tragedie of Cleopatra*. Her work, he asserts, is a monument not only to Sidney (his "eternall songs") but also to Pembroke herself: "here thou survivest thyself, here thou art found." See Samuel Daniel, "To the Right Honorable, the Lady Marie, Countess of Pembroke," in *Delia and rosamond augmented. Cleopatra* (London, 1594), Early English Books Online Text Creation Partnership, 2011, http://name.umdl.umich.edu/A19816.0001.001, accessed November 5, 2020.

75. See Danielle Clarke on how the Psalter is "indebted both to *acts* of memory (remembering, recall, reworking, rewriting) and to a *process* of memorialization" (Clarke, "Memory and Memorialization in the Psalms of Mary Sidney, Countess of Pembroke," *Memory Studies* 11, no. 1 [2017]: 92).

76. In Dennis Kay's study of the English elegy, he notes that early moderns sometimes incorporated elegies into the funeral process, pinning poems onto the coffin as "a literary counterpart of decorating the hearse with flowers." Elizabeth Cooke Russell was famous for her tombstone elegies, actually carved into the monuments placed at the burial site, such as these lines from the inscription on her second husband's tombstone at Westminster Abbey: "Alike the beauteous face, the comely air, / the tongue persuasive and the actions fair / decay: so learning too in time shall waste. / But faith, chaste lovely faith, shall ever last." See Dennis Kay, *Melodious Tears: The English Funeral Elegy from Spenser to Milton* (Oxford: Clarendon Press, 1990), 61–62; Shohet, "Women's Elegy: Early Modern," 436; and Peter Sherlock, *Monuments and Memory in Early Modern England* (Aldershot: Ashgate, 2008).

77. "arrest, *n*.1" (*OED*).

78. Peter M. Sacks, *The English Elegy: Studies in the Genre from Spenser to Yeats* (Baltimore: Johns Hopkins University Press, 1987), 1. Also see Dennis Kay on how Spenser and Sidney in particular used their elegies to explore the work of the elegiac mode itself (Kay, *Melodious Tears*, 29).

79. Mayer, *Embalming*, 468.

80. "artificially *adv*.2a" (*OED*).

81. In Clifton D. Bryant and Dennis L. Peck's extensive, two-volume *Encyclopedia of Death and the Human Experience* (Thousand Oaks, CA: Sage, 2009), the entry on "embalming" comes directly after the entry on "elegy," so readers will always encounter one while reading about the other—a happy coincidence of alphabetization. For more on the genre of the elegy, see Kay, *Melodious Tears*; Sacks, *The English Elegy*; and Karen Weisman's introduction to *The Oxford Handbook of the Elegy* (New York: Oxford University Press, 2010), 1–9.

82. Ariès, *Hour of Our Death*, 355.

83. Ariès, *Hour of Our Death*, 355.

84. For more on the Daniel Variant as an early version of the Tixall manuscript elegy, see *Collected Works*, vol. 1, 74–77; and Kinney, "'Love Which Hath Never Done.'" In making this argument, I am not suggesting that one version is more complete or authorized than the other. Pembroke's project is *both and*—an ongoing experiment with how poetry might deal in and become part of the matter of death.

85. Deborah Uman, *Women as Translators in Early Modern England* (Newark: University of Delaware Press, 2010), 56.

86. For more on early modern corpse preservation techniques, see Erich Brenner, "Human Body Preservation—Old and New Techniques," *Journal of Anatomy* 224, no. 3 (2014): 316–44; and Mayer, *Embalming*, 467–510.

87. Pembroke's use of "render" adds another postmortem action to these lines. To "render" is "to obtain or extract (fat) by heating" animal remains or carcasses ("render, *v.*20a"). The verb has several other early modern connotations that resonate with Pembroke's project: "to reproduce or express in another language; to translate" ("render, *v.*2a"); "to represent or reproduce, esp. artistically; to depict, portray" ("render, *v.*3b"); and "to emit, give out, discharge; spec. to shed (tears)" ("render, *v.*9a)." See "render, *v.*" (*OED*). I am grateful to Catherine Loomis for pointing out the rich contexts of this term.

88. On Pembroke's writing, Diane Bornstein has lamented, "one can only regret that the countess limited herself to the silent art of translation and did not write her own meditations" ("The Style of the Countess of Pembroke's Translation of Philippe de Mornay's *Discourse de La Vie et de La Mort*," in *Silent but for the Word: Tudor Women as Patrons, Translators, and Writers of Religious Works*, ed. Margaret P. Hannay [Kent: Kent State University Press, 1985], 134). Scholars of women's writing, as Helen Smith notes, still tend to "describ[e] translation as secondary: the lesser half of a binary which privileges original composition" (Smith, *'Grossly Material Things': Women and Book Production in Early Modern England* [Oxford: Oxford University Press, 2012], 32). Smith's point extends to arguments that devalue the Sidney Psalter more broadly because of its status as translation. Coburn Freer lamented in 1972 that Sidney had received "little or poor notice for his translation of the Psalms" (*Music for a King: George Herbert's Style and the Metrical Psalms* [Baltimore: Johns Hopkins University Press, 2020], 72), citing for instance Robert L. Montgomery Jr.'s dismissal of the psalms as "more primitive than the general run of poems in the *Arcadia*" (*Symmetry and Sense: The Poetry of Sir Philip Sidney* [Austin: University of Texas Press, 1961], 23). Freer's aims to recuperate appreciation for the psalms as a rigorous poetic collection, however, came at the expense of Pembroke's work: "unfortunately for students of Sidney, we have to rely on his sister . . . who repeatedly revised not only her own psalms but Sidney's as well" (Freer, *Music for a King*, 74). Although Freer ultimately acknowledges that Pembroke is "historically important" (adding that she was "not a particularly good poet"), his argument relies on "working backward through her revisions" in order to recover "Sidney's versions" of the psalms (*Music for a King*, 90, 74). For work on Pembroke's writings that productively counters such assessments, see Danielle Clarke, "'Lover's Songs Shall Turne to Holy Psalms': Mary Sidney and the Transformation of Petrarch," *Modern Language Review* 92 (1997): 282–94; Goldberg, *Desiring Women Writing*; Debra Rienstra and Noel Kinnamon, "Circulating the Sidney-Pembroke Psalter," in *Women's Writing and the Circulation of Ideas: Manuscript Publication in England, 1550–1800*, ed. George L. Justice and Nathan Tinker (Cambridge: Cambridge University Press, 2003), 50–72; and Theodore L. Steinberg, "The Sidneys and the Psalms," *Studies in Philology* 92, no. 1 (1995): 1–17.

89. Smith, *Grossly Material Things*, 23, my emphasis. Also see Jonathan Goldberg, "The Countess of Pembroke's Literal Translation," in *Subject and Object in Renaissance Culture*, ed. Margreta De Grazia, Maureen Quilligan, and Peter Stallybrass (Cambridge: Cambridge University Press, 1996), 321–36; and Jaime Goodrich, *Faithful Translators: Authorship, Gender, and Religion in Early Modern England* (Evanston: Northwestern University Press, 2013).

90. On this point, see Andrew Cunningham, *The Anatomical Renaissance: The Resurrection of the Anatomical Projects of the Ancients* (Aldershot: Scolar Press, 1997).

91. John Banister, *The Historie of Man, sucked from the sappe of the most approved Anathomistes, in this present age, compiled in most compendious fourme, and now published in English, for the utilitie of all godly Chirurgians, within this Realme* (London: John Daye, 1578), Wellcome Library, EPB/D/657. Banister's text is translated primarily from Realdo Columbo's *De re anatomica* (1559), and includes several illustrations from Andreas Vesalius's *De humani corporis fabrica* (1543).

92. Banister sums up his translation choices with this statement: "I refused to bynde my selfe to any peculiar translation, chusing rather to picke a posie of the chiefest flowers from all their Gardens." In this same preface, he cites Vesalius, Galen, Columbo, Fuchsius [Leonhart Fuchs], and Fernelius [Jean François Fernel] as his source texts (*The Historie of Man*, fol. A.iiij).

93. Helkiah Crooke, "To the worshipfull company of the Barber-Chyrurgeons," in *Mikrokosmographia. A description of the Body of Man. Together with the controversies thereto belonging. Collected and Translated out of the Best Authors of Anatomy, Especially out of Gasper Bauhinus and Andreas Laurentius* (London: William Jaggard, 1615), Wellcome Library, EPB/D/1683. Like Banister's text, *Mikrokosmographia* announces its translation status in its title, and Crooke explicitly references Banister's *The History of Man* in this same preface: "because it is now a long time since your Banister (that good old man) first presented you with a service of this kinde, and no man hath seconded him; I have adventured to commit unto you these first fruites of my untainted fame."

94. On the history of English anatomists' access to cadavers and necessary resources for study, see Sawday, *The Body Emblazoned*, 54–60; Phyllis Allen, "Medical Education in Seventeenth-Century England," *Journal of the History of Medicine and Allied Sciences* 1, no. 1 (1946): 115–43; and W. S. Copeman, "The Evolution of Anatomy and Surgery under the Tudors," *Annals of the Royal College of the Surgeons of England* 32 (1964): 1–21.

95. Ambrose Paré, *The works of Ambrose Parey, chyrurgeon to Henry II. Francis II. Charles IX. and Henry III. Kings of France*, trans. Thomas Johnson (London, 1691), Wellcome Library, 66238/D, fol. A2r.

96. Paré, *The works of Ambrose Parey*, A2v.

97. Paré, *The works of Ambrose Parey*, A2v.

98. Goldberg, "Literal Translations," 325.

99. Holinshed records: "in this her last action and ending of her life . . . she so far surpassed herself in discrete, wise, effectual, sound, and grounded reasons, all tending to zeal and piety, as the same almost amazed and astonished the hearers to hear and conceive such plenty of goodly and pithy matter to come from such a creature" (cited in Lamb, "The Countess of Pembroke and the Art of Dying," 225n13). For more on the connection between Lady Mary Sidney's (née Dudley) death and Pembroke's ars moriendi texts, see *Collected Works*, vol. 1, 139–46.

100. Anatomy was (and is) a science of death as much as it was a science of the body. Caspar Barlaeus, Dutch philosopher and theologian, visited the Leiden anatomy theater in 1639 and described the space as "this house we found for death." "Here you can learn," he reported, "the ways of death and a desire to shun death" (cited in Neill, *Issues of Death*, 119). Using accounts like Barlaeus's, along with comparisons between anatomical illustrations and memento mori imagery, Michael Neill demonstrates how dissection practices "helped to produce an entirely new understanding of the human body and its processes of morbidity" (*Issues of Death*, 102). "For what the science of dissection ultimately seemed to disclose," Neill observes, "was nothing less than the 'original of death'—the death that is always already inside us" (133). Anatomy, that is, helped establish in practice what Pembroke's ars moriendi translations outlined in theory: "if we call the last breath death, so must we all the rest" because "all proceed[s] from one place, and all in one manner" (248).

101. Goldberg, "Literal Translations," 321, 324.

102. These numbers correspond to the penitential psalms in the Hebrew Masoretic Text and most Protestant bibles, but the numbers can vary and are different in the Greek Septuagint and Latin Vulgate. See Clare Costley King'oo, *Miserere Mei: The Penitential Psalms in Late Medieval and Early Modern England* (Notre Dame: Notre Dame Press, 2012), 1–2.

103. King'oo, *Miserere Mei*, 1.

104. Anne Lake Prescott, "The Countess of Pembroke's Ruins of Rome," *Sidney Journal* 23, no. 1–2 (2005): 4; 2.

105. Prescott, "Ruins of Rome," 6.

106. The imagery of the withered corpse returns again in Psalm 109. The poet describes "the withered knots" of her "disjointed knees" (Ps 109.65) and claims to be nothing but "dried flesh," having lost "all juice and moisture" (Ps 109.66) at the hands of her "malicious" enemies (Ps 109.4).

107. "Woe" was a term thick with meaning in early modern parlance. It could refer to a general state of suffering and emotional distress, but also to physical disease and infirmity, or to lamentations specific to mourning and loss (*OED*).

108. Hannay, *Philip's Phoenix*, 99. Theodore L. Steinberg also foregrounds Pembroke's attention to repetition and doubling, connecting it directly to the "system of parallelism" that governs ancient Hebrew verse (Steinberg, "The Sidneys and the Psalms," 3).

109. Roy T. Eriksen points out that Pembroke's verse structure in Psalm 130 deviates from how the poem is structured in her source texts. See Eriksen, "George Gascoigne's and Mary Sidney's Versions of Psalm 130," *Cahiers Élisabéthains: Late Medieval and Renaissance Studies* 36, no. 1 (1989): 3.

110. Kimberly Johnson, *Made Flesh: Sacrament and Poetics in Post-Reformation England* (Philadelphia: University of Pennsylvania Press, 2014), 44. On the connection between Pembroke and Herbert's styles, see Freer, *Music for a King*, 72–108; Louis L. Martz, *The Poetry of Meditation: A Study in English Religious Literature of the 17th Century* (New Haven, CT: Yale University Press, 1954), 273; and Rathmell, *The Psalms of Sir Philip Sidney*, xii.

111. Piero Camporesi, *Bread of Dreams: Food and Fantasy in Early Modern Europe*, trans. David Gentilcore (Chicago: University of Chicago Press, 1996), 47. Also see Margaret Pelling, "Review of *Bread of Dreams: Food and Fantasy in Early Modern Europe* by Piero Camporesi," *Bulletin of the History of Medicine* 66, no. 3 (1992): 477–78.

112. John Donne, "Upon the Translation of the Psalms by Sir Philip Sidney, and the Countess of Pembroke, His Sister," *John Donne: The Complete English Poems*, ed. A. J. Smith (London: Penguin Books, 1996), 332–34. On Pembroke and Sidney's collaboration, see Alexander, *Writing After Sidney*, Chapter 3; Maureen Quilligan, *Incest and Agency in Elizabeth's England* (Philadelphia: University of Pennsylvania Press, 2005), Chapter 4; and Steinberg, "The Sidneys and the Psalms." There is also room here to consider the siblings' collaboration within the "proto-scientific" context Henry S. Turner reads in Philip Sidney's *Defence of Poesy*. In Sidney's famous essay, Turner argues, "we may observe a shift from the neo-scholastic understanding of *scientia* as certain knowledge of metaphysical causes to an empirical meaning that anticipates several modern scientific presuppositions" (Turner, *The English Renaissance Stage: Geometry, Poetics, and the Practical Spatial Arts 1580–1630* [Oxford: Oxford University Press, 2006], 83–84).

113. Donne, "Upon the Translation," ll. 13–15.

114. Donne, "Upon the Translation," ll. 31. Margaret P. Hannay notes that Pembroke's contemporary readers would have read her psalms "simultaneously as the words of the Sidneys and the words of David." Hannay, "Re-revealing the Psalms: The Countess of Pembroke and Her Early Modern Readers," *Sidney Journal* 23, no. 1–2 (2005): 20.

115. Donne, "Upon the Translation," ll. 9–11.

116. Donne, "Upon the Translation," ll. 10; 17.

117. Ryan J. Croft, "Sidney's Wounds: Poetic Physicality, Revision, and Remembrance in the Sidney Circle," *Sidney Journal* 31, no. 2 (2013): 38. "By emphasizing the body of the writer," Croft argues, "[Pembroke] fashions a manual poetic physicality that contrasts with the oral world of her

Psalms translation" (32). Croft ultimately goes on to read the psalms as diverging from the manual focus of "Angell spirit" to instead foreground oral and aural transmission in the form of song.

118. Croft, "Sidney's Wounds," 31. On how the psalm translations serve as "a source of ongoing intimacy" between Pembroke and Sidney, see Sagaser, "Elegiac Intimacy."

119. John Calvin, "John Calvin to the godly Readers sendeth greeting," in *The Psalmes of David and others. With M. John Calvin's Commentaries*, trans. Arthur Golding (London, 1571), sig. 6v, Princeton Theological Seminary Library, 11329. We find very literal interpretations of Calvin's description in some of the psalm translations produced in the sixteenth and seventeenth centuries, such as Anne Locke's *A Meditation of a Penitent Sinner*. In Locke's sonnet sequence, based on Psalm 51 and appended to her translations of Calvin's sermons, the poet's conscience wields a knife and performs a vivisection of her body: "my cruell conscience with sharpned knife, / doth splat my ripped hert." Her conscience then "layes abrode [her] lothesome secretes" and "spredes them forth before the face of God." See Locke, "Againste thee onely have I sinned," *Collected Works*. In tracing these connections, it is also worth noting that Helkiah Crooke references King David ("that Kingly Prophet") and the psalms in the first section of *Mikrokosmographia*, citing Psalm 139, which reads in the King James Bible version, "I will praise thee; for I am fearfully and wonderfully made" (14).

120. Crooke, *Mikrokosmographia*, 12.

121. Crooke, *Mikrokosmographia*, 2.

122. Crooke, *Mikrokosmographia*, 13.

123. Katherine Rowe, *Dead Hands: Fictions of Agency, Renaissance to Modern* (Stanford: Stanford University Press, 1999), 24.

124. On "handwork," see Pamela H. Smith, *The Body of the Artisan: Art and Experience in the Scientific Revolution* (Chicago: University of Chicago Press, 2004).

125. See Hannay, "'How I These Studies Prize'"; *Collected Works*, vol. 1, 73; Margaret Healy, "Paracelsian Medicine and Female Creativity: Aemilia Lanyer's *Salve Deus Rex Judaeorum*," *Renaissance and Reformation / Renaissance et Réforme* 36, no. 2 (2013): 85; and Katherine R. Larson, "From Inward Conversation to Public Praise: Mary Sidney Herbert's *Psalmes*," *Sidney Journal* 24, no. 1 (2006): 21–43.

126. Crooke, *Mikrokosmographia*, 12.

127. Beth Wynne Fisken, "Mary Sidney's *Psalmes*: Education and Wisdom," in *Silent but for the Word: Tudor Women as Patrons, Translators, and Writers of Religious Works*, ed. Margaret P. Hannay (Kent: Kent State University Press, 1985), 177.

128. Alexander Read, *The Manuall of the Anatomy or Dissection of the Body of Man, Which Usually Are Shewed in the Publike Anatomicall Exercises. Methodically Digested into 6 Books* (London: J.H., 1638), 1–2, Wellcome Library, EPB/A/5355.

129. For more language of "forming" and "framing," see Pembroke's Psalm 148: "All these I say advaunce that name, / that doth eternall beeing show: / who bidding, into forme and frame, / not beeing yet, they all did grow. / all formed, framed, founded so, / tyll ages uttmost date, / they place retaine they order know, / they keepe their first estate" (Ps 148.17–24).

130. On this point, see Michelle Osherow's argument on Psalm 139 as an example of Pembroke's "linguistic embroidery" ("Mary Sidney's Embroidered Psalms," *Renaissance Studies* 29, no. 4 [2015]: 652).

131. See *Collected Works*, vol. 1, 55–77; vol. 2, 469–83.

132. *Collected Works*, vol. 1, 58. Hallett Smith describes the Sidney Psalter as a "school of English Versification" (Smith, "English Metrical Psalms in the Sixteenth Century and their Literary Significance," *Huntington Library Quarterly* 9, no. 1 [1946]: 269).

133. Hannibal Hamlin, "The Influence of the Sidney Psalter," in *The Ashgate Research Companion to The Sidneys, 1500–1700*, vol. 2, ed. Mary Ellen Lamb and Margaret P. Hannay (London: Routledge, 2015), 135. Also see Ramie Targoff, *Common Prayer: The Language of Public Devotion in Early Modern England* (Chicago: University of Chicago Press, 2001), chapter 3.

134. Jenny C. Mann and Debapriya Sarkar, "Introduction: 'Capturing Proteus,'" *Philological Quarterly* 98, no. 1–2 (2019): 3. On this point, also see Henry S. Turner, "Lessons from Literature for the Historian of Science (and Vice Versa): Reflections on 'Form,'" *Isis* 101, no. 3 (2010): 578–89.

135. On revision and in-process poetic practice in early modern England, see Adhaar Noor Desai, *Blotted Lines: Early Modern English Literature and the Poetics of Discomposition* (Ithaca, NY: Cornell University Press, 2023).

136. There are, for instance, some key variations across the manuscripts that complicate some of the language we tracked in the penitential psalms. Manuscripts *A*, *M*, and *C* contain variations in word choice and syntax in Psalm 102, where Pembroke is describing grief's effects on the poet's body. The descriptive adjectives are flipped in Manuscript *M*, so "my hart is withered like the wounded grass" reads, "my hart is wounded like withered grass" (11). Also in Manuscript *M* "flesh" is erased and "skynne" is inserted in line 14: "to my bones my ~~flesh~~ skynne do cleave." Manuscripts *M* and *C*, Pembroke's editors argue, were likely copied from a lost, heavily marked working copy (Manuscript *X*). See *Collected Works*, vol. 2, 348–51. Comparing all eighteen extant manuscripts, editors usually take Manuscript *A* (also known as the Penshurst MS) as the most authoritative text (it is the base text for the *Collected Works*). But none of these extant manuscripts are written in Pembroke's hand, so what does "authoritative" mean in this case? "Comparison of the Penshurst manuscript (MS *A*) with the other manuscripts," Hannay, Kinnamon, and Brennan observe, "reveals a gradual deterioration of the text, sometimes because of careless scribes, but also, particularly in *G* and *M*, because of apparent 'tinkering' by someone other than the author" (*Collected Works*, vol. 2, 344). As Danielle Clarke rightly points out, however, such an approach (following William A. Ringler Jr.'s work on Philip Sidney's writings) privileges completion, certainty, and authorization over incompletion, uncertainty, and collaboration (Clarke, "Nostalgia, Anachronism," 194–97).

137. Mann and Sarkar, "Introduction," 6.

138. For more on how Pembroke's writing circulated during her lifetime and in the decades following her death, see *Collected Works*, vol. 1, 47–53; Hannay, "Re-revealing the Psalms"; and Rienstra and Kinnamon, "Circulating the Sidney-Pembroke Psalter."

139. *Collected Works*, vol. 1, 51.

140. *Collected Works*, vol. 1, 211. An early use of the phrase "scientia mortis" can be found in Parisian Chancellor Jean Gerson's treatise of that same title (1408), which he translated himself into French as *La Médicine de l'Âme*. Gerson's text circulated widely in Latin, French, Dutch, English, Spanish, Catalan, and Italian translations, and was an important foundation for the ars moriendi tradition. Gerson not only argues for the importance of finding comfort in the face of death, but also insists on the value of a lifelong meditation on death, a meditation framed scientifically in both his Latin and French titles—*Scientia Mortis* (*Knowledge of Death*) and *La Médicine de l'Âme* (*The Medicine of the Soul*). See Scott L. Taylor, "'L'aage plus Fort Ennaye': *Scientia Mortis, Ars Moriendi* and Jean Gerson's Advice to an Old Man," in *Old Age in the Middle Ages and the Renaissance: Interdisciplinary Approaches to a Neglected Topic*, ed. Albrecht Classen (Berlin: De Gruyter, 2007), 407–19.

141. Park, *Secrets of Women*, 131.

CHAPTER 3

1. Emily J. Peters, *The Brilliant Line: Following the Early Modern Engraver 1480–1650*, Museum of Art, Rhode Island School of Design, 2009, https://risdmuseum.org/exhibitions-events/exhibitions/brilliant-line, accessed May 15, 2024.

2. Cited in Domenico Laurenza, "Art and Anatomy in Renaissance Italy: Images from a Scientific Revolution," *The Metropolitan Museum of Art Bulletin* 69, no. 3 (2012): 27.

3. Paul G. McMenamin, "Art and Anatomy in the Renaissance: Are the Lessons Still Relevant Today?," *ANZ Journal of Surgery* 92, no. 1–2 (2021): 44. Also see Caroline Fowler, "*Res Papirea*: Mantegna's Paper Things," *The Art Bulletin* 99, no. 1 (2017): 8–35; and Laurenza, "Art and Anatomy."

4. Also extant in the archive of Wroth's writing are several autograph letters and her closet drama *Love's Victory*, which exists in two different manuscripts, one held at the Huntington Library (HM 600) and one currently owned by the Viscount De L'Isle and held at Penshurst Place. Wroth's only published text was the 1621 *Urania*, with appended *Pamphilia to Amphilanthus* sonnet sequence. The printed *Urania* contains a title page with a copperplate engraving by Dutch artist Simon van de Passe (famous, at that point, for his 1616 engraving of Pocahontas), but no additional prefatory material. Shortly after the book's publication, Edward Denny, Baron of Waltham, charged Wroth with slander, publicly attacking her for defaming him in the story. For more on Denny's critique and Wroth's publication history, see Ilona Bell, "Introduction," in *Lady Mary Wroth: Pamphilia to Amphilanthus in Manuscript and Print*, ed. Ilona Bell and Steven W. May (Toronto: Iter Press, 2017), 58–64.

5. Sarah Rodgers describes the first part of *Urania* as a text "positioned at the intersection of print and manuscript cultures," a status reinforced by evidence of readers adding handwritten endings to printed copies. The printed *Urania* ends with a dangling "And" and, as Rodgers notes, "a large enough blank space to invite writing." Of the twenty-two extant print copies, two include readers' handwritten endings (one prose, one poetry) and three have the final "And" crossed out, replaced by a period. See Rodgers, "Embedded Poetry and Coterie Readers in Mary Wroth's *Urania*," *Studies in Philology* 111, no. 3 (2014): 470; 483–84. For work that takes up the "endlessness" of Wroth's prose construction in relation to the unfinished first and second parts of *Urania*, see Colleen Ruth Rosenfeld, "Wroth's Clause," *ELH* 76, no. 4 (2009): 1049–71.

6. On Wroth's biography, see Bell, "Introduction," 19–38; Josephine A. Roberts, "Introduction," *The Poems of Lady Mary Wroth*, ed. Josephine A. Roberts (Baton Rouge: Louisiana State University Press, 1983), 3–40; and Margaret P. Hannay, *Mary Sidney, Lady Wroth* (Farnham: Ashgate, 2010). For more on Sidney and Herbert family influences on Wroth's writing, see Gary Waller, *The Sidney Family Romance: Mary Wroth, William Herbert, and the Early Modern Construction of Gender* (Detroit: Wayne State University Press, 1993), Chapter 4; Maureen Quilligan, "The Constant Subject: Instability and Female Authority in Wroth's *Urania* Poems," in *Soliciting Interpretation: Literary Theory and Seventeenth-Century English Poetry*, ed. Elizabeth D. Harvey and Katharine Eisaman Maus (Chicago: University of Chicago Press, 1990), 307–35; and Margaret P. Hannay, "'Your Virtuous and Learned Aunt': The Countess of Pembroke as Mentor to Mary Wroth," in *Reading Mary Wroth: Representing Alternatives in Early Modern England*, ed. Naomi J. Miller and Gary Waller (Knoxville: University of Tennessee Press, 1991), 15–34.

7. Paul J. Hecht, "Distortion, Aggression, and Sex in Mary Wroth's Sonnets," *Studies in English Literature, 1500–1900* 53, no. 1 (2013): 91.

8. Mary Moore, "The Labyrinth as Style in *Pamphilia to Amphilanthus*," *Studies in English Literature, 1500–1900* 38, no. 1 (1998): 115; Paul Salzman, "Not Understanding Mary

Wroth's Poetry," *Parergon* 29, no. 2 (2012): 148; and Roger Kuin, "More I Still Undoe: Louise Labe, Mary Wroth, and the Petrarchan Discourse," *Comparative Literature Studies* 36, no. 2 (1999): 156.

9. Heather Dubrow, *Echoes of Desire: English Petrarchism and Its Counterdiscourses* (Ithaca, NY: Cornell University Press, 1995), 161. For more on Shakespeare as a Petrarchan poet, see Gordon Braden, "Shakespeare's Petrarchism," in *Shakespeare's Sonnets: Critical Essays*, ed. James Schiffer (New York and London: Routledge, 1999), 163–83; and Richard Strier, *The Unrepentant Renaissance: From Petrarch to Shakespeare to Milton* (Chicago: University of Chicago Press, 2011), especially Chapter 2.

10. Laurie Maguire, *The Rhetoric of the Page* (Oxford: Oxford University Press, 2021).

11. Barbara Kiefer Lewalski remarks, "absent ... are the usual motifs for aggrandizing the poet-lover at the beloved's expense: no blazons for scattering his parts," and Paul J. Hecht emphasizes this point parenthetically, noting "the almost total lack of the physical presence of the beloved (no blazons here)." See Lewalski, "Writing Women and Reading the Renaissance," *Renaissance Quarterly* 44, no. 4 (1991): 811; and Hecht, "Distortion, Aggression, and Sex," 101. For additional work that comments on the lack of a blazon in Wroth's poetry, see Clare R. Kinney, "Mary Wroth's Guilty 'Secrett Art': The Poetics of Jealousy in *Pamphilia to Amphilanthus*," in *Write or Be Written: Early Modern Women Poets and Cultural Constraints*, ed. Barbara Smith and Ursula Appelt (Aldershot: Ashgate, 2001), 69; Natasha Distiller, *Desire and Gender in the Sonnet Tradition* (London: Palgrave Macmillan, 2008), 88; Moore, "Labyrinth as Style," 110; and Wendy Wall, *The Imprint of Gender: Authorship and Publication in the English Renaissance* (Ithaca, NY: Cornell University Press, 1993), 330.

12. Bell, "Introduction," 12, original emphasis. Margaret P. Hannay describes one of Wroth's poems as "the closest that [she] comes to a blazon." The poem in question is "Oft did I wounder why the sweets of love" (MS 113), in which Wroth references "two sparkling eyes" and "love-begetting lips." See Margaret P. Hannay, "The 'Ending End' of Lady Mary Wroth's Manuscript of Poems," *Sidney Journal* 31, no. 1 (2013): 17.

13. Lady Mary Wroth, *Lady Mary Wroth: Pamphilia to Amphilanthus in Manuscript and Print*, ed. Ilona Bell and Steven W. May (Toronto: Iter Press, 2017). All references to poetry from Folger MS V.a.104 will be cited from this edition and referenced in the body of the essay by poem and line number.

14. Fowler, "Mantegna's Paper Things," 22.

15. Domenico Bertolini Meli, *Visualizing Disease: The Art and History of Pathological Illustrations* (Chicago: University of Chicago Press, 2017), 57.

16. Lady Mary Wroth, *The First Part of the Countess of Montgomery's Urania*, ed. Josephine A. Roberts (Tempe: Arizona Center for Medieval and Renaissance Studies, 1995), 78. All references to the first part of *Urania* will be cited from this edition, marked parenthetically by volume (I) and page number.

17. Ollorandus's description, full of qualifiers and contradictory conjunctions, is worth reading here: "me thought I saw a Creature, for shape a woman, but for excellencie, such as all the rarenes in that sexe, curiously, and skilfully mixed, could but frame such an one; and yet but such a one in shew, like a Picture well drawne, but the subject more perfect" (I.78).

18. For more on the etymological roots of "tact" and its many early modern connotations, see Carla Mazzio, "Acting with Tact: Touch and Theater in the Renaissance," *Sensible Flesh: On Touch in Early Modern Culture*, ed. Elizabeth D. Harvey (Philadelphia: University of Pennsylvania Press, 2003), 159–86.

19. Thomas Wyatt, "Whoso List to Hunt," *Poetry Foundation*, accessed January 18, 2023.

20. Carla Freccero, "Tangents (of Desire)," *Journal for Early Modern Cultural Studies* 16, no. 2 (Spring 2016): 97. On the biblical and "spiritual anatomy" origins of *noli me tangere*, see Christian K. Kleinbub, "To Sow the Heart: Touch, Spiritual Anatomy, and Image Theory in Michelangelo's *Noli Me Tangere*," *Renaissance Quarterly* 66, no. 1 (2013): 81–129.

21. Francesco Petrarca, *Petrarch's Lyric Poems: The* Rime sparse *and Other Lyrics*, trans. and ed. Robert M. Durling (Cambridge, MA: Harvard University Press, 1976), 190.9–10.

22. On beauty, whiteness, and the blazon in *Urania*, see Kim F. Hall, "'I Rather Would Wish to Be a Black-Moor': Beauty, Race, and Rank in Lady Mary Wroth's *Urania*," in *Women, "Race," and Writing in the Early Modern Period*, ed. Margo Hendricks and Patricia Parker (London: Routledge, 1994), 178–94; and *Things of Darkness: Economies of Race and Gender in Early Modern England* (Ithaca, NY: Cornell University Press, 1995), 187–210. Hall expertly tracks how Wroth uses "darkness to reify whiteness," particularly in her descriptions of the women characters, as well as how beauty is dependent on rank throughout the romance (*Things of Darkness*, 197).

23. For an excellent reading of the early modern blazon within a broader culture of thigh sex, see Will Fisher, "'Wantoning with the Thighs': The Socialization of Thigh Sex in England, 1590–1730," *Journal of the History of Sexuality* 24, no. 1 (2015): 1–24.

24. Jonathan Sawday, *The Body Emblazoned: Dissection and the Human Body in Renaissance Culture* (London: Routledge, 1995), 2.

25. Nancy J. Vickers, "Members Only: Marot's Anatomical Blazons," in *The Body in Parts: Fantasies of Corporeality in Early Modern Europe*, ed. David Hillman and Carla Mazzio (New York: Routledge, 1997), 7. Also see Vickers, "'The Blazon of Sweet Beauty's Best': Shakespeare's *Lucrece*," in *Shakespeare and the Question of Theory*, ed. Patricia Parker and Geoffrey Hartman (New York: Methuen, 1985), 95–115; and "Diana Described: Scattered Woman and Scattered Rhyme," *Critical Inquiry* 8, no. 2 (1981): 265–79.

26. Towards the end of *Urania*, Part I, Amphilanthus is ambushed in the woods, and in a wounded stupor he envisions Pamphilia "slaine . . . by his hand," her body then dragged "into a stone" that lies "in the middest" of a "Crowne of stones" (I.654–55). Amphilanthus "open[s] the stone" (I.655) to find Pamphilia lying dead and his name carved on her exposed heart: "lying within an arch, her breast open and in it his name made, in little flames burning like pretty lamps which made the letters, as if set round with diamonds, and so cleare it was, as hee distinctly saw the letters ingraven at the bottome in Characters of bloud" (I.655).

27. Nancy J. Vickers, "'This Heraldry in Lucrece' Face,'" *Poetics Today* 6, no. 1/2 (1985): 181. By referring to Steriamus's offence as a "glasse discovery," Wroth signals her recognition that the blazon functioned at the intersection of geographic, economic, and scientific aims. Importantly, Pamphilia is also a *place* in Wroth's romance. The King of Pamphilia, uncle to Pamphilia, comes to visit "his Neece, who by his gift was to enjoy that kingdome after his decease, and therefore bore that name likewise given by him" (I.100). Sawday notes that the English blazon was so deeply rooted in the rhetoric of discovery that "poetic language seems naturally embedded within a colonizing process of appropriation and exploitation" (Sawday, *The Body Emblazoned*, 198).

28. Josephine A. Roberts reminds us that "Pamphila" was the name of a Roman poet, writing during the reign of Nero, and suggests that Wroth may be evoking this figure ("Introduction," 42).

29. Dubrow, *Echoes of Desire*, 135.

30. Maguire, *Rhetoric of the Page*, 81.

31. Mary Sidney Herbert, "To the Angell spirit of the most excellent Sir Phillip Sidney," *The Collected Works of Mary Sidney Herbert, Countess of Pembroke: Poems, Translations, and Corre-*

spondence, vol. 1, ed. Margaret P. Hannay, Noel J. Kinnamon, and Michael G. Brennan (Oxford: Clarendon Press, 1998).

32. Maguire, *Rhetoric of the Page*, 103; and Damian Judge Rollison, "The Poem on the Page: Graphical Prosody in Postmodern American Poetry," *Text* 15 (2003): 302–3.

33. Wall, *Imprint of Gender*, 335.

34. Wall, *Imprint of Gender*, 336.

35. On Pamphilia's cabinet, see Bernadette Andrea, "Pamphilia's Cabinet: Gendered Authorship and Empire in Lady Mary Wroth's *Urania*," *ELH* 68, no. 2 (2001): 335–58; and Caro Pirri, "Cabinet Work: Mary Wroth and the World," *Exemplaria* 32, no. 1 (2020): 51–71.

36. Juliet Fleming, *Graffiti and the Writing Arts of Early Modern England* (Chicago: University of Chicago Press, 2001), 43.

37. Maguire, *Rhetoric of the Page*, 5.

38. Laurenza, "Art and Anatomy," 14.

39. Laurenza, "Art and Anatomy," 9.

40. Laurenza, "Art and Anatomy," 26.

41. James Elkins, *Pictures of the Body: Pain and Metamorphosis* (Stanford: Stanford University Press, 1999), 139.

42. Elkins, *Pictures of the Body*, 149.

43. William Faithorne, *The art of graveing and etching wherein is exprest the true way of Graveing in Copper* (London: William Faithorne, 1662), Early English Books Online Text Creation Partnership, 2011, https://quod.lib.umich.edu/e/eebo/A40780.0001.001?view=toc, accessed January 9, 2023. For an overview of the engraving process, see Eric Paul Meier, "From Paper to Copper: The Engraver's Process (2009)," Rhode Island School of Design, September 28, 2009, YouTube video, 7:15, https://youtu.be/fQvghHs15hA, accessed October 17, 2022.

44. Faithorne, *The art of graveing* (1662), 41.

45. John Evelyn, *Sculptura: or, the history, and art of chalcography and Engraving in Copper* (London: J. C., 1662), 21, Early English Books Online Text Creation Partnership, 2011, https://quod.lib.umich.edu/e/eebo/A69993.0001.001, accessed May 14, 2023.

46. Peters, *The Brilliant Line*, 9; and Laurenza, "Art and Anatomy," 9. Peters provides an exceptional study of engraving as an artisanal, artistic, and scientific medium and the entire exhibit catalog is well worth a look. Caroline Fowler also makes the important point that the technique of engraving uses "the repetitive application of line to build dimensionality" (Fowler, "Mantegna's Paper Things," 17).

47. "substrate, *n.*" and "substrate, *adj.*" (*OED*).

48. William Faithorne, *The Art of graveing and etching. Wherein is express'd the true way of Graveing in copper* (London: William Faithorne, 1702), 66, https://archive.org/details/artgraveingandeoofaitgoog/page/n106/mode/2up, accessed January 9, 2023.

49. Faithorne, *The Art of graveing* (1702), 66.

50. Faithorne, *The Art of graveing* (1702), 67.

51. Andrea, "Pamphilia's Cabinet," 335.

52. In Vin Nardizzi and Miriam Jacobson's reading of this poem through an ecofeminist framework, the detail of the "knife" is central to the "material relation that links the arts of the grafter and the tree poet" (Vin Nardizzi and Miriam Jacobson, "The Secrets of Grafting in Wroth's *Urania*," in *Ecofeminist Approaches to Early Modernity*, ed. Jennifer Munroe and Rebecca Laroche [New York: Palgrave Macmillan, 2011], 54). For more on grafting, particularly in relation to writing as a "technology of repetition," see Jessica Rosenberg, *Botanical Poetics: Early Modern Plant*

Books and the Husbandry of Print (Philadelphia: University of Pennsylvania Press, 2022), especially pages 247–48.

53. Nardizzi and Jacobson, "The Secrets of Grafting," 182.

54. The act of tree carving intersects with a slightly different kind of violence in Shakespeare's *Titus Andronicus* when Aaron describes his treatment of dead enemies: "Oft have I digged up dead men from their graves . . . and on their skins, as on the bark of trees, / have with my knife carved in Roman letters / 'Let not your sorrow die, though I am dead'" (5.1.140–42). William Shakespeare, *Titus Andronicus*, ed. Barbara Mowat, Paul Werstine, Michael Poston, and Rebecca Niles (Washington: Folger Shakespeare Library, n.d.).

55. Elaine Scarry, *The Body in Pain: The Making and Unmaking of the World* (New York: Oxford University Press, 1985), 4.

56. Evelyn, *The history and art of chalcography*, 29.

57. Evelyn, *The history and art of chalcography*, 29.

58. Evelyn, *The history and art of chalcography*, 29.

59. Here, I think with Anne Anlin Cheng's brilliant work on tattoos. Cheng poses the question, "what can surface (or skin) be or do if it is *not* just a cover?" (Cheng, "Skins, Tattoos, and Susceptibility," *Representations* 108, no. 1 [2009]: 98).

60. For more on the textual features of Folger MS V.a.104, see Bell, "Introduction," 44–52; and Roberts, "Introduction," 61–65.

61. Many of the manuscript poems are included in the printed 1621 *Pamphilia to Amphilanthus* published with the first part of *Urania*, but the differences between the manuscript and printed poetry collections are significant. The manuscript contains additional poems, a different arrangement of poems, and line-level differences in meter, syntax, and diction. Given the complicated publication history of Wroth's poems, scholars have long debated the status of Folger MS V.a.104. Margaret P. Hannay, for instance, reads it as a "presentation copy," perhaps intended for William Herbert ("The 'Ending End,'" 5), while Heather Dubrow categorizes it as a "work-in-progress" drastically reconceived for the printed *Pamphilia* sequence ("'And Thus Leave Off': Reevaluating Mary Wroth's Folger Manuscript, V.a.104," *Tulsa Studies in Women's Literature* 22, no. 2 [2003]: 285). In the most recent critical edition of Wroth's poetry, edited by Ilona Bell, Bell attends quite carefully to the differences between the two collections of poetry, but ultimately (and, I think, problematically) insists that "Wroth's poems originally existed as individual lyrics and groups of lyrics that she collected and transcribed onto the paper that became MS V.a.104, which she then revised and culled for print" (Bell, "Introduction," 44).

62. See Gavin Alexander, "Constant Works: A Framework for Reading Mary Wroth," *Sidney Newsletter & Journal* 14, no. 2 (1996): 5–32; Ilona Bell, "'Joy's Sports': The Unexpurgated Text of Mary Wroth's *Pamphilia to Amphilanthus*," *Modern Philology* 111, no. 2 (2013): 231–52; Dubrow, "'And Thus Leave Off'"; and Rahel Orgis, "An Author Re-reading: Lady Mary Wroth's Corrections to the Printed *Urania* and the Manuscript Continuation," *English Literary Renaissance* 47, no. 3 (2017): 412–48. Also see Margaret Simon on the "material" and "multi-modal" work of the first song in MS V.a.104 (Simon, "Mary Wroth's Ephemeral Epitaph," *Studies in English Literature 1500–1900* 56, no. 1 [2016]: 55).

63. Marta Straznicky, "Lady Mary Wroth's Patchwork Play: The Huntington Manuscript of *Love's Victory*," *Sidney Journal* 34, no. 1 (2016): 82; 84.

64. Straznicky, "Patchwork Play," 86; 84. Notably for my discussion of Wroth's poetry here, *Love's Victory* exists in two extant versions (the Huntington and Penshurst manuscripts), and the Penshurst text, as the longer and more orderly version (complete with cover and binding), is often

assumed to be the more "authoritative" version of the play, as has been the tendency with the printed, published *Pamphilia* sequence.

65. Hall, *Things of Darkness*, 69.

66. For more on how climate-based theories regarding blackness existed in constant dialogue with other early modern race-making myths, see Sujata Iyengar, *Shades of Difference: Mythologies of Skin Color in Early Modern England* (Philadelphia: University of Pennsylvania Press, 2005). On Wroth's specific use of heliotropism, see Hall, *Things of Darkness*, 105–7; Iyengar, *Shades of Difference*, Chapter 3; and Lynda E. Boose, "'The Getting of a Lawful Race': Racial Discourse in Early Modern England and the Unrepresentable Black Woman," in *Women, "Race," and Writing in the Early Modern Period*, ed. Margo Hendricks and Patricia Parker (London: Routledge, 1994), 35–54.

67. Hall, *Things of Darkness*, 106. For more on premodern racial formation and English colonial nation-building in Wroth's writing, see Andrea, "Pamphilia's Cabinet"; Boose, "'The Getting of a Lawful Race'"; and Pirri, "Cabinet Work." On racecraft, see Noémie Ndiaye, *Scripts of Blackness: Early Modern Performance Culture and the Making of Race* (Philadelphia: University of Pennsylvania Press, 2022); and Karen E. Fields and Barbara J. Fields, *Racecraft: The Soul of Inequality in American Life* (New York: Verso Books, 2012).

68. Miles P. Grier, "Inkface: The Slave Stigma in England's Early Imperial Imagination," in *Scripturalizing the Human: The Written as the Political*, ed. Vincent L. Wimbush (London: Routledge, 2017), 195. Also see Grier, *Inkface: Othello and White Authority in the Era of Atlantic Slavery* (Charlottesville: University of Virginia Press, 2023).

69. Grier, "Inkface," 199–200.

70. Grier, "Inkface," 201.

71. Grier, "Inkface," 198.

72. Hall, *Things of Darkness*, 70–71.

73. On this point, see Bernadette Andrea, "Persia, Tartaria, and Pamphilia," in *The English Renaissance, Orientalism, and the Idea of Asia*, ed. Debra Johanyak and Walter S. H. Lim (New York: Palgrave Macmillan, 2009), 23–50; Andrea, "Pamphilia's Cabinet"; Pirri, "Cabinet Work"; and Hall, *Things of Darkness*, Chapters 4 and 5. Also of note here is Dennis Britton's work on English engagements with the infidel-conversion motif of the early modern romance. In their use of this motif, Britton argues, English writers "responded to controversies about the means by which individuals acquire both racial and religious identities" (Britton, *Becoming Christian: Race, Reformation, and Early Modern English Romance* [New York: Fordham University Press, 2014], 5; and especially Chapter 3).

74. Andrea, "Pamphilia's Cabinet," 338, original emphasis.

75. Faithorne, *The Art of graveing* (1702), 67.

76. Elkins, *Pictures of the Body*, 139.

77. I treat *Anatomia* as a collaborative project between Bidloo and Lairesse, but there is much debate among historians of science about how much control Bidloo exercised over the book's illustrations. See Susan Donahue Kuretsky, "Lairesse Meets Bidloo, or the Case of the Absent Anatomist," *Midwestern Arcadia: Essays in Honor of Alison Kettering* (2015), https://apps.carleton.edu/kettering/, accessed September 12, 2019; and K. B. Roberts and J. D. W. Tomlinson, *The Fabric of the Body: European Traditions of Anatomical Illustration* (Oxford: Clarendon Press, 1992), 309–19.

78. As K. B. Roberts and J. D. W. Tomlinson document, Bidloo and Lairesse's text was also the first important shift in the style of anatomical illustration, an attempt "to portray human

anatomy in a quite different way" than the illustrations influenced by Vesalius's *Fabrica* (Roberts and Tomlinson, *Fabric of the Body*, 312).

79. On this point, see Donahue Kuretsky, "Lairesse Meets Bidloo"; Elkins, *Pictures of the Body*, 134–41; and Mechthild Fend, *Fleshing Out Surfaces: Skin in French Art and Medicine, 1650–1850* (Manchester: Manchester University Press, 2017), 45–53.

80. Govard Bidloo, *Anatomia humani corporis* (Amsterdam, 1685), Wellcome Collection, EPB/F/2051; and William Cowper, *The anatomy of humane bodies, with figures drawn after the life By some of the best masters in Europe* (Oxford, 1698), sig. C2ʳ, Wellcome Collection, EPB/F/328/1. A decade after Bidloo's atlas was first printed in Amsterdam, William Cowper published an extended English version without crediting Bidloo or Lairesse. The cited Wellcome Library edition of Cowper's text features a small cutout with Cowper's name simply pasted over Bidloo's attribution on the title page. The publication sparked a heated controversy regarding intellectual property (see Donahue Kuretsky, "Lairesse Meets Bidloo").

81. Dániel Margócsy, "A Museum of Wonders or a Cemetery of Corpses? The Commercial Exchange of Anatomical Collections in Early Modern Netherlands," in *Silent Messengers: The Circulation of Material Objects of Knowledge in the Early Modern Low Countries*, ed. Sven Dupré and Christoph Lüthy (Münster: LIT Verlag, 2011), 192.

82. Margócsy, "A Museum of Wonders," 196.

83. Sarah Kay, "Legible Skins: Animals and the Ethics of Medieval Reading," *Postmedieval: A Journal of Medieval Cultural Studies* 2 (2011): 14. Also see Bruce Holsinger, *On Parchment: Animals, Archives, and the Making of Culture from Herodotus to the Digital Age* (New Haven, CT: Yale University Press, 2023).

84. Elizabeth Hodgson, *Grief and Woman Writers in the English Renaissance* (Cambridge: Cambridge University Press, 2014), 77.

85. David Hillman and Carla Mazzio, "Introduction," in *The Body in Parts: Fantasies of Corporeality in Early Modern Europe*, ed. David Hillman and Carla Mazzio (London: Routledge, 1997), xvii–xviii.

86. Jeffrey Masten, "'Shall I Turne Blabb?': Circulation, Gender, and Subjectivity in Wroth's Sonnets," in *Reading Mary Wroth: Representing Alternatives in Early Modern England*, ed. Naomi J. Miller and Gary Waller (Knoxville: University of Tennessee Press, 1991), 84.

87. Marshall, *The Shattering of the Self*, 76.

88. Vickers, "Diana Described," 265.

89. Megan Heffernan, *Making the Miscellany: Poetry, Print, and the History of the Book in Early Modern England* (Philadelphia: University of Pennsylvania Press, 2021), 132.

90. Edward Philips, "The Triall of wits, a new invented Alphabet of Epithets," in *The mysteries of love & eloquence, or, The arts of wooing and complementing as they are manag'd in the Spring Garden, Hide Park, the New Exchange, and other eminent places* (London: James Rawlins, 1685), 13, Early English Books Online Text Creation Partnership, 2011, https://quod.lib.umich.edu/e/eebo/A54745.0001.001/1:15.7?rgn=div2;view=fulltext, accessed January 17, 2023.

91. Thomas Watson, *The hekatompathia or Passionate centurie of love divided into two parts* (London: John Wolfe, 1582), Poem VII, Early English Books Online Text Creation Partnership, 2011, https://quod.lib.umich.edu/e/eebo/A14822.0001.001/1:12.7?rgn=div2;view=toc, accessed January 24, 2023.

92. William Shakespeare, *The Narrative and Other Poems*, 3rd ed., ed. Katherine Duncan-Jones and H. R. Woudhuysen (London: Bloomsbury Arden Shakespeare, 2007), 130.3–4.

93. Shakespeare, *The Narrative and Other Poems*, 130.11–12.

94. Notably, Wroth revises this line from the original, "I, am that body lives deprived of hart," adding more explicitly anatomical language ("trunk") and a part-root word ("depart") to complete the end rhyme.

95. Margócsy, "A Museum of Wonders," 192.

96. On Wroth's use of the fermesse, see Alexander, "Constant Works"; Dubrow, "'And Thus Leave Off'"; and Megan Heffernan, "Miscellany/Sequence," in *Shakespeare/Text: Contemporary Readings in Textual Studies, Editing and Performance*, Arden Shakespeare Intersections, ed. Claire M. L. Bourne (London: Bloomsbury Publishing, 2021), 143–62.

97. Heffernan, "Miscellany/Sequence," 157.

98. Heffernan, "Miscellany/Sequence," 158.

99. Josephine Roberts points out the significant difference in end-line punctuation between the Folger manuscript and the 1621 printed *Pamphilia* sequence, noting that the printed sequence includes more strong end punctuation marks compared to "Wroth's very light pointing" in her manuscripts (Roberts, "Introduction," 73). The addition of end-line punctuation interrupts the effects of enjambment, resulting in a poetic sequence that formally enacts a sense of completion and closure. The manuscript, as we have been seeing, resists that sense of fixity.

100. Samuel Daniel, *Selected Poetry and a Defense of Rhyme*, ed. Geoffrey G. Hiller and Peter L. Groves (New York: Pegasus Press, 1998), 223.

101. Jerome J. McGann, "Rosetti's Iconic Page," in *The Iconic Page in Manuscript, Print, and Digital Culture*, ed. George Bornstein and Theresa Tinkle (Ann Arbor: University of Michigan Press, 1998), 123; 127.

102. Rosalind Smith, "Lady Mary Wroth's 'Pamphilia to Amphilanthus': The Politics of Withdrawal," *English Literary Renaissance* 30, no. 3 (2000): 417.

103. Smith, "Politics of Withdrawal," 418. For a reading of this sonnet as "a celebration of eroticism," see Susan Lauffer O'Hara, *The Theatricality of Mary Wroth's Pamphilia to Amphilanthus: Unmasking Conventions in Context* (Cranbury: Associated University Presses, 2011), 49–57.

104. Smith, "Politics of Withdrawal"; Madeline Bassnett, "The Politics of Election in Lady Mary Wroth's *Pamphilia to Amphilanthus*," *SEL Studies in English Literature 1500–1900* 51, no. 1 (2011): 111–34; and Dubrow, "'And Thus Leave Off,'" 285.

105. Smith, "Politics of Withdrawal," 418.

106. Daniel, *Selected Poetry*, 223.

107. "impart, *v.*" (*OED*).

108. Hannay, "The 'Ending End,'" 5; Straznicky, "Patchwork Play."

109. Alexander, "Constant Works," 10. Also see Dubrow, "'And Thus Leave Off.'"

110. Vickers, "Members Only," 19.

111. Andrew Fleck, "Anatomizing the Body Politic: The Nation and the Renaissance Body in Thomas Nashe's *The Unfortunate Traveller*," *Modern Philology* 104, no. 3 (2007): 307.

112. Richard Rambuss, *Closet Devotions* (Durham, NC: Duke University Press, 1998), 54.

113. Moore, "Labyrinth as Style," 112; 109.

114. Kinney, "Poetics of Jealousy," 77.

115. Masten, "'Shall I Turne Blabb?,'" 69, 73. Masten's influential argument that Wroth's poetry stages "a withdrawal into an interiorized corporeal space" has prompted scholars to understand Wroth's writing "not only as private, but as privative" (Masten, "'Shall I Turne Blabb?,'" 70; and Wall, *Imprint of Gender*, 331). For more readings of Wroth's poetry through this framework of interiority, see Andrea, "Pamphilia's Cabinet"; Angela Bullard, "Love Melancholy and Creative Inspiration in Mary Wroth's *Pamphilia to Amphilanthus*," *Sidney Journal* 33, no. 2 (2015): 81–102;

Clare R. Kinney, "Turn and Counterturn: Reappraising Mary Wroth's Poetic Labyrinths," in *Re-Reading Mary Wroth*, ed. Katherine R. Larson, Naomi J. Miller, and Andrew Strycharski (New York: Palgrave Macmillan, 2015), 85–102; and Nona Fienberg, "Mary Wroth and the Invention of Female Poetic Subjectivity," in *Reading Mary Wroth: Representing Alternatives in Early Modern England*, ed. Naomi J. Miller and Gary Waller (Knoxville: University of Tennessee Press, 1991), 175–90. For a useful counterargument to such readings, see Daniel Juan Gil, "The Currency of the Beloved and the Authority of Lady Mary Wroth," *Modern Language Studies* 29, no. 2 (1999): 73–92.

116. Moore, "Labyrinth as Style," 112. For Moore, the corona's exploration of these themes suggests Wroth's interest in the links between "Protestant attitudes to inwardness" and "the Petrarchan drama of self-knowledge" (112; 122).

117. Elizabeth D. Harvey, "The Touching Organ: Allegory, Anatomy, and the Renaissance Skin Envelope," in *Sensible Flesh: On Touch in Early Modern Culture*, ed. Elizabeth D. Harvey (Philadelphia: University of Pennsylvania Press, 2003), 86.

118. Gil, "Currency of the Beloved," 82.

119. Gil, "Currency of the Beloved," 82.

120. Gil, "Currency of the Beloved," 87; 84.

121. Cowper's translation of Bidloo's caption is relevant here in how it draws attention to the act of exposure simulated in these plates: "the spinal processes of the vertebrae made bare" (Cowper, *The anatomy of humane bodies*, Tab. 30).

122. Faithorne, *The Art of graveing* (1702), 67.

123. Andrea Gazzaniga, "'This Close Room': Elizabeth Barrett's Proximal Poetics in Sonnets from the Portuguese," *Victorian Poetry* 54, no. 1 (2016): 80.

124. On the corona's unusual page layout, see Dubrow, "'And Thus Leave Off,'" 280; Kinney, "Turn and Counterturn"; and Rebecca L. Fall, "Pamphilia Unbound: Digital Re-Visions of Mary Wroth's Folger Manuscript, V.a.104," in *Re-Reading Mary Wroth*, ed. Katherine R. Larson, Naomi J. Miller, and Andrew Strycharski (London: Palgrave Macmillan, 2015), 193–207.

125. For some foundational arguments on this point, see Ann Rosalind Jones, "The Commodities Dance: Exchange and Escape in Irigaray's 'Quand nos Lèvres Se Parlent' and Catherine Des Roches's 'Dialogue D'Iris Et Pasithée,'" in *Luce Irigaray and Premodern Culture: Thresholds of History*, ed. Elizabeth D. Harvey and Theresa Krier (New York: Routledge, 2004), 147–61; Cynthia Marshall, *The Shattering of the Self: Violence, Subjectivity, and Early Modern Texts* (Baltimore: Johns Hopkins University, 2002); Patricia Parker, *Literary Fat Ladies: Rhetoric, Gender, Property* (London: Methuen, 1987); Vickers, "Members Only"; Vickers, "'The Blazon of Sweet Beauty's Best'"; and Vickers, "Diana Described."

CHAPTER 4

1. Katharine Park, *Secrets of Women: Gender, Generation, and the Origins of Human Dissection* (Baltimore: MIT Press, 2006), 103.

2. Andreas Vesalius, *The Fabric of the Human Body: An Annotated Translation of the 1543 and 1555 Editions of* De Humani Corporis Fabrica Libri Septem, ed. and trans. Daniel H. Garrison and Malcolm H. Hast, 2 vols. (Basel: Karger, 2014), 1081; and Vesalius, *De humani corporis fabrica libri septem* (Basel: Johannes Oporinus, 1543), 539. All English translations of Vesalius's *Fabrica* are from Malcolm and Hast's edition and will be cited in the footnotes by page number along with corresponding pages in the original 1543 Latin text. On "pleading the belly," see James C.

Oldham, "On Pleading the Belly: A History of the Jury of Matrons," *Criminal Justice History* 6 (1985): 1–64; and Sara M. Butler, "More than Mothers: Juries of Matrons and Pleas of the Belly in Medieval England," *Law and History Review* 37, no. 2 (2019): 353–96. Oldham's history of matron juries is quite useful, but it should be noted that he is often overly dismissive of early modern midwifery knowledge (he refers several times to the "primitive nature of the understanding of pregnancy" in the period, a nature remedied by the "sophistication" of medical practice in the nineteenth century [24]).

3. For Wendy Wall, Vesalius's frontispiece is one example of the pervasive tendency in early modern print to use the book's "preliminary apparatus" to "construct[ ] a language of intrusion that designated reading as a prurient activity" (Wall, *The Imprint of Gender: Authorship and Publication in the English Renaissance* [Ithaca, NY: Cornell University Press, 1993], 172). For readings of the *Fabrica*'s title page that have informed my analysis throughout this chapter, see Andrea Carlino, *Books of the Body: Anatomical Ritual and Renaissance Learning*, trans. John Tedeschi and Anne C. Tedeschi (Chicago: University of Chicago Press, 1999), especially pages 39–53; Jennifer F. Kosmin, "Midwifery Anatomized: Vesalius, Dissection, and Reproductive Authority in Early Modern Italy," *Journal of Medieval and Early Modern Studies* 48, no. 1 (2018): especially pages 79–87; Park, *Secrets of Women*, Chapter 5; Jonathan Sawday, *The Body Emblazoned: Dissection and the Human Body in Renaissance Culture* (London: Routledge, 1995), especially pages 66–78; and Luke Wilson, "William Harvey's *Prelectiones*: The Performance of the Body in the Renaissance Theater of Anatomy," *Representations* 17 (1987): especially pages 69–74.

4. Vesalius, *The Fabric of the Human Body*, 1081; *De Humani Corporis Fabrica*, 539.

5. Kosmin, "Midwifery Anatomized," 99.

6. Vivian Nutton, "Historical Introduction," in *The Fabric of the Human Body: An Annotated Translation of the 1543 and 1555 Editions of* De Humani Corporis Fabrica Libri Septem, ed. and trans. Daniel H. Garrison and Malcolm H. Hast (Basil: Karger, 2014), C. As Nutton outlines in his introduction to the Garrison and Hast translation of the *Fabrica*, the copies available in Leipzig from the 1543 printing sold out within the year, and both the *Fabrica* and Vesalius's shorter companion text, *De humani corporis fabrica librorum epitome* (better known as the *Epitome*), were immediately copied, plagiarized, and translated. As I discussed in Chapter 3, as early as 1545 Thomas Geminus printed his Latin *Compendiosa totius anatomiae delineation* using the *Fabrica* plates and some of the *Epitome* text. John Banister's *The Historie of Man sucked from the sappe of the most approved Anathomistes* (London, 1578) also included the Vesalian plates, supplemented with anatomical descriptions translated and compiled from Realdo Columbo's *De re anatomica* (1559) and several other contemporary treatises. In 1555, Vesalius and Oporinus published an extended second edition of the *Fabrica*.

7. Nutton, "Historical Introduction," LXXXV.

8. Nutton, "Historical Introduction," LXXXV.

9. Nutton, "Historical Introduction," LXXXV.

10. Vesalius, *The Fabric of the Human Body*, 7; *De humani corporis fabrica*, IV.

11. "incarnate, *v.*1" (*OED*).

12. Barbara Bowen, "Aemilia Lanyer and the Invention of White Womanhood," in *Maids and Mistresses, Cousins and Queens*, ed. Susan Frye and Karen Robertson (New York: Oxford University Press, 1999), 275.

13. Aemilia Lanyer, *The Poems of Aemilia Lanyer: Salve Deus Rex Judaeorum*, ed. Susanne Woods (New York: Oxford University Press, 1993). All citations to Lanyer's *Salve Deus* reference this edition. Citations from the Passion poem will be cited by line number in the text. Citations from the book's paratextual material will be cited by corresponding title and page number.

14. Jacqueline Pearson, "Women Writers and Women Readers: The Case of Aemilia Lanyer," in *Voicing Women: Gender and Sexuality in Early Modern Writing*, ed. Kate Chedgzoy, Melanie Hansen, and Suzanne Trill (Keele: Keele University Press, 1996), 47.

15. Lanyer has been distinctly absent from critical work in literature and science studies, with a few notable exceptions. See Margaret Healy, "Paracelsian Medicine and Female Creativity: Aemilia Lanyer's *Salve Deus Rex Judaeorum*," *Renaissance and Reformation / Renaissance et Réforme* 36, no. 2 (2013): 75–103; and Yaakov Mascetti, "'With the Eie of Faith': Aemilia Lanyer's Religious and Feminine Sight in Context," *Partial Answers: Journal of Literature and the History of Ideas* 16, no. 1 (2018): 1–25.

16. Helkiah Crooke notes that the Greeks referred to the womb as "*mater*, or the mother, because it is the mother of the infant" (the Latin *matrix* derives from *mater*), and also that Pliny referred to it as the "*locus* or rather *loci*, the place of a woman." See Crooke, *Mikrokosmographia. A description of the Body of Man. Together with the controversies thereto belonging. Collected and Translated out of the Best Authors of Anatomy, Especially out of Gasper Bauhinus and Andreas Laurentius* (London: William Jaggard, 1615), 222, Wellcome Library, EPB/D/1683.

17. Pearson, "Women Writers and Women Readers," 47.

18. The phrase—"the privitays of women"—is the English title given to a translation of the Latin *Trotula* texts, a series of gynecological literature from the early Middle Ages. The *Trotula* texts appeared under a variety of titles and were eventually translated into many different languages. On this history, see Monica H. Green, ed., *The Trotula: An English Translation of the Medieval Compendium of Women's Medicine* (Philadelphia: University of Pennsylvania Press, 2002); and Green, *Women's Healthcare in the Medieval West: Texts and Contexts* (London: Routledge, 2011).

19. For a thorough overview of Lanyer's prefatory material, see Erin A. McCarthy, "Speculations and Multiple Dedications in *Salve Deus Rex Judaeorum*," *Studies in English Literature* 55, no. 1 (2015): 45–72.

20. Danielle Clarke, "Aemilia Lanyer," in *Routledge Companion to Renaissance Literature*, ed. Catherine Bates (forthcoming).

21. Karma Lochrie, *Margery Kempe and Translations of the Flesh* (Philadelphia: University of Pennsylvania Press, 2012), 180. For more on medieval depictions of Mary, see Laura Gallagher, "Stabat Mater Dolorosa: Imagining Mary's Grief at the Cross," *Biblical Women in Early Modern Literary Culture 1550–1700*, ed. Victoria Brownlee and Laura Gallagher (Manchester: Manchester University Press, 2015), 180–96; Julia Kristeva, "Stabat Mater," trans. Arthur Goldhammer, *Poetics Today* 6, no. 1/2 (1985): 133–52; Gary Waller, *The Virgin Mary in Late Medieval and Early Modern English Literature and Popular Culture* (Cambridge: Cambridge University Press, 2011); and Marina Warner, *Alone of All Her Sex: The Myth and the Cult of the Virgin Mary* (New York: Alfred A. Knopf, 1976). On medieval women's affective piety more broadly, see Sara Beckwith, *Christ's Body: Identity, Culture and Society in Late Medieval Writings* (London: Routledge, 1993); and Caroline Walker Bynum, *Holy Feast and Holy Fast: The Religious Significance of Food to Medieval Women* (Berkeley: University of California Press, 1988). Katherine R. Larson argues that the "tradition of medieval affective piety"—an insistence on the physical as a means of accessing the spiritual—was a more significant precursor to Lanyer's text than the contemporary religious poetry of Richard Crashaw, John Donne, and George Herbert (Larson, "Reading the Space of the Closet in Aemilia Lanyer's *Salve Deus Rex Judaeorum*," *Early Modern Women: An Interdisciplinary Journal* 2 [2007]: 73–93).

22. Donna Spivey Ellington, "Impassioned Mother or Passive Icon: The Virgin's Role in Late Medieval and Early Modern Passion Sermons," *Renaissance Quarterly* 48, no. 2 (1995): 254.

23. The ruling emerged in response to petitions for a feast day in honor of *La Madonna della Spasimo* (Our Lady of the Swoon), largely a result of a number of artistic and literary representations of Mary swooning at the foot of the cross. See Harvey E. Hamburgh, "The Problem of *Lo Spasimo* of the Virgin in Cinquecento Paintings of the Descent from the Cross," *The Sixteenth Century Journal* 12, no. 4 (1981): 45–75; Gary Kuchar, "Aemilia Lanyer and the Virgin's Swoon: Theology and Iconography in *Salve Deus Rex Judaeorum*," *English Literary Renaissance* 37, no. 1 (2007): 47–73; Amy Neff, "The Pain of *Compassio*: Mary's Labor at the Foot of the Cross," *The Art Bulletin* 80, no. 2 (1998): 254–73; Victoria Brownlee, "Literal and Spiritual Births: Mary as Mother in Seventeenth-Century Women's Writing," *Renaissance Quarterly* 68 (2015): 1297–1326; and Warner, *Alone of All Her Sex*, especially Chapter 14.

24. Cited in Hamburgh, "The Problem of *Lo Spasimo*," 46. Hamburgh cites excerpts from Cardinal Cajetan's epistolary ruling using a translation provided directly to the author by Father Benedict M. Ashley, O.P.

25. Waller, *The Virgin Mary*, 34; and Mary E. Fissell, *Vernacular Bodies: The Politics of Reproduction in Early Modern England* (Oxford: Oxford University Press, 2004), 39. For more on the histories of theological speculation surrounding Mary's sexuality, particularly the tricky circumstances of her all-too-human flesh, see Rachel Fulton, *From Judgement to Passion: Devotion to Christ and the Virgin Mary, 800–1200* (New York: Columbia University Press, 2002), especially pages 383–97.

26. Waller, *The Virgin Mary*, 38–39. Speculation about Mary's anatomical body intersects in interesting ways with early modern tensions between skepticism and belief. Andrea Carlino attributes the "disintegration of the habit of belief," in part, to anatomy's methodological shift from knowledge gained from texts to knowledge gained through dissection. Such disintegration, Carlino asserts, is "inconceivable without the decisive contribution made by dissection" (Carlino, *Books of the Body*, 213). In David Hillman's influential work on knowledge and belief in early modern England, he outlines the difference in anatomical terms: "belief is precisely *not* the need to enter the body . . . it is rather the offer of access, or the sense of being granted access, to the interior . . . that constitutes faith" (32). Like Carlino, Hillman connects the period's anatomical developments (the "will-to-penetration") to growing skepticism in relation to faith, arguing that both signal "a loss of certainty about access to the divine interior, which could no longer be taken for granted" (36). See Hillman, *Shakespeare's Entrails: Belief, Scepticism and the Interior of the Body* (London: Palgrave Macmillan, 2007).

27. Brownlee, "Literal and Spiritual Births," 1299–1300.

28. John Pettus, *Volatiles from the history of Adam and Eve: Containing Many unquestioned Truths, and allowable Notions of Several Natures* (London, 1674), 135–36, Early English Books Online Text Creation Partnership, 2011, https://quod.lib.umich.edu/cgi/t/text/text-idx?c=eebo2;idno=A54603.0001.001, accessed November 12, 2022.

29. Clarke, "Aemilia Lanyer." Also see Boyd Berry, "'Pardon . . . Though I Have Digrest': Digression as Style in *Salve Deus Rex Judaeorum*," in *Aemilia Lanyer: Gender, Genre, and the Canon*, ed. Marshall Grossman (Lexington: University Press of Kentucky, 1998), 212–33.

30. Brownlee, "Literal and Spiritual Births," 1309.

31. Given the importance of the female community Lanyer generates in her text (particularly in the prefaces, which I will discuss, and the "Description of Cooke-ham"), it is worth noting the subtle authorial interjection Lanyer uses to frame Mary's thoughts here: "Thy virgin thoughts did think, non could impart / this great good hap, and blessing unto thee; / far from desire of any man thou art, / knowing not one, thou art from all men free" (1075–78). Could we read in these lines a queer utopian desire that one could be "from all men free" and still have the

"great good hap" of conception and reproduction? This question is one of the generative, active inquiries driving critical work on the queer resonances of *Salve Deus*, including Bowen, "Inventing White Womanhood," 285; Jonathan Goldberg, *Desiring Women Writing: English Renaissance Examples* (Stanford: Stanford University Press, 1997), 16–41; Amy Greenstadt, "Aemilia Lanyer's Pathetic Phallacy," *Journal for Early Modern Cultural Studies* 8, no. 1 (2008): 67–97; Penelope Anderson and Whitney Sperrazza, "Feminist Queer Temporalities in Aemilia Lanyer and Lucy Hutchinson," in *Gendered Temporalities in the Early Modern World*, ed. Merry E. Wiesner-Hanks (Amsterdam: Amsterdam University Press, 2018), 159–84; and Wall, *The Imprint of Gender*, especially pages 279–340.

32. Boyd Berry distinguishes his use of the word "digression" to describe Lanyer's style from Patricia Parker's term "dilation," but I prefer Parker's term, drawing on her sense that dilation is both "dispersal in space" and "postponement in time" (Berry, "Digression as Style"; and Patricia Parker, *Literary Fat Ladies: Rhetoric, Gender, Property* [New York: Methuen, 1987], 9). Catherine Keohane poses the question "digressions or expansions?" in her reading of Lanyer's Passion narrative (Keohane, "'That Blindest Weakness Be Not Over-Bold': Aemilia Lanyer's Radical Unfolding of the Passion," *ELH* 64, no. 2 [1997]: 362), but dilation seems an especially useful word when we consider Lanyer's book alongside anatomical treatment of the "capacious" womb. In Parker's influential argument, the rich lexicon of Renaissance dilation centers on the relationship between texts and female bodies. Dilation is linked to pregnant and fat bodies, influenced by biblical subtext (the Bible, Parker notes, is "filled with figures for the space and time of such extensions"), and anchored in the "rhetorical tradition of the dilation of discourse" (*Literary Fat Ladies*, 11–13).

33. Pearson, "Women Writers and Women Readers," 47.

34. Pearson, "Women Writers and Women Readers," 45.

35. For foundational proto-feminist readings of Lanyer's *Salve Deus*, see Elaine V. Beilin, *Redeeming Eve: Women Writers of the English Renaissance* (Princeton, NJ: Princeton University Press, 1987), especially Chapter 7; Barbara Lewalski, *Writing Women in Jacobean England* (Cambridge, MA: Harvard University Press, 1993), especially Chapter 8; Lewalski, "Re-writing Patriarchy and Patronage: Margaret Clifford, Anne Clifford, and Aemilia Lanyer," *The Yearbook of English Studies* 21 (1991): 87–106; Lynette McGrath, "'Let Us Have Our Libertie Againe': Amelia Lanier's 17th-Century Feminist Voice," *Women's Studies* 20 (1992): 331–48; McGrath, "Metaphoric Subversions: Feasts and Mirrors in Amelia Lanier's *Salve Deus Rex Judaeorum*," *LIT* 3 (1991): 101–13; Naomi J. Miller, "(M)other Tongues: Maternity and Subjectivity," in *Aemilia Lanyer: Gender, Genre, and the Canon*, ed. Marshall Grossman (Lexington: University Press of Kentucky, 1998), 143–66; and Janel Mueller, "The Feminist Poetics of Aemilia Lanyer's *Salve Deus Rex Judaeorum*," in *Feminist Measures: Soundings in Poetry and Theory*, ed. Lynn Keller and Christianne Miller (Ann Arbor: University of Michigan Press, 1993), 208–36. For more on the women readers and dedicatees of *Salve Deus*, see McCarthy, "Speculations and Multiple Dedications"; Leeds Barroll, "Looking for Patrons," in *Aemilia Lanyer: Gender, Genre, and the Canon*, ed. Marshall Grossman (Lexington: University Press of Kentucky, 1998), 29–48; and John Rogers, "The Passion of a Female Literary Tradition: Aemilia Lanyer's *Salve Deus Rex Judaeorum*," *Huntington Library Quarterly* 63, no. 4 (2000): 435–46. Other scholars have productively destabilized the homogeneous, universalizing "female community" often read into Lanyer's text, particularly along the lines of race and class. See Bowen, "Invention of White Womanhood"; Ann Baynes Coiro, "Writing in Service: Sexual Politics and Class Position in the Poetry of Aemilia Lanyer," *Criticism* 35, no. 3 (1993): 357–76; Kimberly Anne Coles, *Religion, Reform, and Women's Writing in Early Modern England* (New York: Cambridge University Press, 2008), especially pages 149–78; Mary Ellen Lamb, "Patronage and Class in Aemilia Lanyer's *Salve Deus Rex Judaeorum*," in *Women, Writ-*

ing, and the Reproduction of Culture in Tudor and Stuart Britain, ed. Mary E. Burke et al. (Syracuse: Syracuse University Press, 1998), 38–57; Melissa E. Sanchez, "Ain't I a Ladie?: Race, Sexuality, and Early Modern Women Writers," in *Routledge Companion to Women, Sex, and Gender in the Early British Colonial World*, ed. Kimberly Anne Coles and Eve Keller (London: Routledge, 2016), 15–32; Lisa Schnell, "Breaking 'the Rule of *Cortezia*': Aemilia Lanyer's Dedications to *Salve Deus Rex Judaeorum*," *Journal of Medieval and Early Modern Studies* 27 (1997): 77–101; and Schnell, "'So Great a Diffrence Is There in Degree': Aemilia Lanyer and the Aims of Feminist Criticism," *MLQ: Modern Language Quarterly* 57, no. 1 (1996): 23–35.

36. Katharine Eisaman Maus, *Inwardness and Theater in the English Renaissance* (Chicago: University of Chicago Press, 1995), 192.

37. As Lanyer scholars frequently note, autobiographical details about Lanyer beyond the speculative information within the *Salve Deus* text itself come from a problematic source, the diaries of Simon Forman, whom Lanyer consulted as an astrologer on several occasions. For more on Lanyer's life and the skepticism with which we must take Forman's accounts, see Barroll, "Looking for Patrons"; David Bevington, "A.L. Rowse's Dark Lady," in *Aemilia Lanyer: Gender, Genre, and the Canon*, ed. Marshall Grossman (Lexington: University Press of Kentucky, 1998), 10–28; and Barbara Lewalski, "Of God and Good Women: The Poems of Aemilia Lanier," in *Silent but for the Word: Tudor Women as Patrons, Translators, and Writers of Religious Works*, ed. Margaret P. Hannay (Kent: Kent State University Press, 1985), 203–24.

38. On early modern pregnancy, childbirth, and lying-in, see Kathryn R. McPherson and Kathryn M. Moncrief, eds., *Performing Maternity in Early Modern England* (London: Routledge, 2007); Adrian Wilson, *Ritual and Conflict: The Social Relations of Childbirth in Early Modern England* (London: Routledge, 2013); and Amanda Zoch, "Rewriting the Lying-In: Hester Pulter, Katherine Philips, and the Felt Mortality of Pregnancy," *Early Modern Women: An Interdisciplinary Journal* 15, no. 1 (2020): 3–25.

39. In the biblical book of Isaiah, a prophecy about the birth of Jesus Christ establishes his ancestry with Jess, father of King David: "There shall come forth a Rod from the stem of Jesse, And a Branch shall grow out of his roots" (Isaiah 11:1, *King James Bible Online*, 2022, https://www.kingjamesbibleonline.org/, accessed December 20, 2022).

40. On this point, see Julia Kristeva's account of how Mary "passes through all three women's stages in the most restricted of all possible kinship systems" ("Stabat Mater," 139).

41. In medieval accounts, Mary's extreme suffering at the foot of the cross is the pain of child labor. As Amy Neff demonstrates, in art and iconography from the twelfth through fourteenth centuries, Mary "becomes a mother again on Calvary." Her swoon is "not only a swoon of death, it is also the pain of childbirth. Mary is in labor. Her companions are the midwife's attendants, who support her in her travail" (Neff, "The Pain of *Compassio*," 255). On this point, also see Hamburgh, "The Problem of *Lo Spasimo*," 51–53.

42. Caroline Walker Bynum's work on late medieval women mystics demonstrates how their writings represent Mary as "the source and container of Christ's physicality," while also suggesting that "the flesh Christ put on was in some sense female, because it was his mother's." Catherine of Siena, for instance, describes Mary's grief at the crucifixion as an extension of her son's pain: "the son was broken in body, and the mother similarly, for his flesh was from her" (Bynum, *Holy Feast and Holy Fast*, 265).

43. On this history, see Elina Gertsman, "Image as Word: Visual Openings, Verbal Imaginings," *Studies in Iconography* 32 (2011): 51–80; Monica H. Green, "From 'Diseases of Women' to 'Secrets of Women': The Transformation of Gynecological Literature in the Later Middle Ages," *Journal of Medieval and Early Modern Studies* 30, no. 1 (2000): 5–39; and Park, *Secrets of Women*.

44. Gertsman, "Image as Word," 71.

45. Park, *Secrets of Women*, 81.

46. Park, *Secrets of Women*, 91.

47. John Banister, *The Historie of Man sucked from the sappe of the most approved Anathomistes, in this present age, compiled in most compendious fourme, and now published in English, for the utilitie of all godly Chirurgians, within this Realme* (London: John Daye, 1578), fol. 85r, Wellcome Library, EPB/D/657. Banister's section on reproduction spans just over seven pages compared to, for instance, the section on the bones, which stretches to seventy-six pages.

48. Banister, *The Historie of Man*, fol. 88v.

49. There is an important difference in tone between these early published anatomical treatises on uterine anatomy and woman-authored accounts of dissection, specifically reports of autopsies practiced within women's communities (convents, for instance) and later published midwifery manuals, such as Jane Sharp's *The midwives book, or, The whole art of midwifry discovered* (London, 1671). On this point, see Kosmin, "Midwifery Anatomized"; and Park, *Secrets of Women*.

50. I focus on Vesalius's title page here because it is such a canonical fixture in Renaissance histories of science, but there are other examples worth noting. The corresponding male and female anatomy figures on the title page of Helkiah Crooke's *Mikrokosmographia*, for instance, demonstrate the significantly different uses to which male and female cadavers were put in Renaissance anatomy. The flayed male figure lacks genitals and displays the body's muscular system, while the female figure displays only the womb. For more on Crooke's title page, see Karen Newman, *Fashioning Femininity and English Renaissance Drama* (Chicago: University of Chicago Press, 1991), 1–12. We also find a title page tableau directly influenced by Vesalius's *Fabrica* in Nicolai Fontani's translation and English printing of Vesalius's *Epitome* (1642), which Fontani followed with a 1645 publication titled *The womans doctour, or, An exact and distinct explanation of all such diseases as are peculiar to that sex with choise and experimentall remedies against the same*, first printed in Amsterdam and translated into English by 1652. The title page to Fontani's edition of the *Epitome* shows five male anatomists standing in a semicircle around a female cadaver who has been dissected to expose her lower torso and reproductive organs. The anatomists look out toward the reader and their hands actively pull back the cloth covering her body and gesture toward her opened torso, serving to draw the reader's gaze into the dissection site. Their active hands are in sharp contrast to the woman's lifeless hand, exposed by the pulling back of the cloth. The dissected uterus and lower torso reveal much more organs than we see in Vesalius's frontispiece, and the flayed skin around the dissection site is minimal to draw our attention instead to the open cavity. Fontani added three illustrations to his version of Vesalius's *Epitome*: one detailed plate on the uterus, a plate of a dissected infant, and a full-page plate of the viscera. See Fontani, trans. and ed., *Librorum andreae vesalii bruxellensis de humani corporis fabrica epitome* (Amsterdam, 1642), Wellcome Collection, EPB/D/56273.

51. As Dániel Margócsy, Mark Somos, and Stephen N. Joffe outline in their detailed worldwide census of *Fabrica* editions, in the 1555 edition "the chapters on generation underwent the most radical changes, partly because Vesalius had the opportunity to dissect more female cadavers, and partly because . . . there was an active demand for information on especially this part of human anatomy" (Margócsy, Somos, and Joffe, *The Fabrica of Andreas Vesalius: A Worldwide Descriptive Census, Ownership, and Annotations of the 1543 and 1555 Editions* [Leiden: Brill, 2018], 8). In addition to relying heavily on Galen in the 1543 edition, Vesalius also kept his descriptions of the womb imprecise and brief. The section titled "Nature of the Uterus of Pregnant Women," for example, is only four sentences long, summed up with the closing statement, "finally (to put it briefly) every-

thing is more or less changed in pregnant women from those that are not pregnant" (Vesalius, *The Fabric of the Human Body*, 1074; *De Humani Corporis Fabrica*, 534–35). The 1543 edition also relies on long-standing depictions of the womb as a monstrous, agential organ. In describing the "mouth of the uterus," Vesalius writes, "when it desires male semen, it attracts it by a natural sucking motion, and when the uterus is not about to conceive, it regurgitates the male semen together with the female semen" (Vesalius, *The Fabric of the Human Body*, 1070; *De humani corporis fabrica*, 523).

52. One early reader of Vesalius's 1555 edition was especially interested in these five cadavers, numbering each cadaver in the margins next to Vesalius's discussion of each body. The pages on the female sex organs in this volume also include underlining and marginal manicules for indexical reference. See Andreas Vesalius, *De humani corporis fabrica libri septum* (Basileae: Joannem Oporinum, 1555), University College London, FOL. 1555 V 2/1.

53. Vesalius, *The Fabric of the Human Body*, 1081; *De humani corporis fabrica*, 539.

54. Vesalius, *The Fabric of the Human Body*, 1081; *De humani corporis fabrica*, 538.

55. Vesalius, *The Fabric of the Human Body*, 1081; *De humani corporis fabrica*, 538.

56. J. B. de C. M. Saunders and Charles D. O'Malley, *The Illustrations from the Works of Andreas Vesalius of Brussels* (Mineola: Dover Publications, 1950), 170. This uterine excision is, perhaps, somewhat of a hypothesis put forward by Saunders and O'Malley based on the fragmentary details Vesalius offers (the students, he notes, "removed all the skin from the cadaver" [Vesalius, *The Fabric of the Human Body*, 1081; *De humani corporis fabrica*, 538]). The *Fabrica* does not describe this process in full, and Saunders and O'Malley do not include any additional citation information on this point.

57. Sawday, *The Body Emblazoned*, 222.

58. Sawday, *The Body Emblazoned*, 69; 67.

59. Sawday, *The Body Emblazoned*, 70.

60. Park, *Secrets of Women*, 255; and also see Kosmin, "Midwifery Anatomized."

61. Bradin Cormack and Carla Mazzio, *Book Use, Book Theory: 1500–1700* (Chicago: University of Chicago Library, 2005), 4–5. There is debate about how much Vesalius's *Fabrica* would have been used in practice. The folio version was elaborate and would have been expensive to print and clumsy to reference at the dissection table. But certain features of the text—the ones I discuss here, as well as the cut-and-paste anatomical figures at the end of the volume (which I discussed in Chapter 1)—do suggest Vesalius's intention to turn this text into a kind of instrument. There is also good evidence that readers interacted thoroughly with this book. In addition to the reader interested in Vesalius's female cadavers that I noted above (see n52), many extant copies of both the 1543 and 1555 *Fabrica* include extensive notes and other forms of reader markings. The same University College London copy noted above contains reader notes on several of the book's diagrams. One reader used red ink to label one of the full-page skeleton plates, cross-referencing the names from Vesalius's index to the diagram, and another hand added shortened Latin names for the muscles next to the longer descriptions Vesalius offered in the diagram indexes (FOL. 1555 V 2/1). Another copy of the *Fabrica*, held at King's College London, includes heavy marginal notes throughout the manuscript and stars on many entries in the book's index (FOL. QM21 VES [2]).

62. Vesalius, *The Fabric of the Human Body*, 11–12; *De humani corporis fabrica*, VII. On woodcut inking, Vesalius advises Oporinus to carefully render "the thickness of the lines in certain parts, which is the most artful feature of these illustrations and thoroughly delightful for me to view."

63. Whitney Trettien, *Cut/Copy/Paste: Fragments from the History of Bookwork* (Minneapolis: University of Minnesota Press, 2021), 20. Here, I think with Trettien's concept of bookwork and her conceptualization of different kinds of readerly interactions with books as technologies.

64. Vesalius, *The Fabric of the Human Body*, 1065; *De humani corporis fabrica*, 529.

65. Vesalius, *The Fabric of the Human Body*, 4; *De humani corporis fabrica*, III.

66. Vesalius, *The Fabric of the Human Body*, 7; *De humani corporis fabrica*, IV.

67. Vesalius, *The Fabric of the Human Body*, 7; *De humani corporis fabrica*, IV.

68. Park, *Secrets of Women*, 248.

69. Crooke, *Mikrokosmographia*, 199.

70. In Crooke's preface to the "Worshipfull Company of Barber-Chyrurgeons," he references Banister and remarks that the *Mikrokosmographia* does the important work of updating English anatomy: "because it is now a long time since your Banister (that good old man) first presented you with a service of this kind, and no man hath seconded him; I have adventured to commit unto you these first fruites of my untainted fame" (n.p.).

71. Crooke, *Mikrokosmographia*, 197.

72. Crooke, *Mikrokosmographia*, 197.

73. Crooke, *Mikrokosmographia*, 197.

74. Crooke, *Mikrokosmographia*, 223.

75. Crooke, *Mikrokosmographia*, 249–50.

76. William Harvey, *Anatomical exercitations, Concerning the generation Of Living Creatures: To which are added Particular Discourses, of Births, and of Conceptions; &c.* (London: James Young, 1653), sig. A6r, Wellcome Library, EPB/B/27836/1.

77. McCarthy, "Speculations and Multiple Dedications," 45. As Kari Boyd McBride notes, critical work on the prefaces "has constituted a significant subspecies of Lanyer criticism." For work that deals directly with Lanyer's prefaces, see McBride, "Sacred Celebration: The Patronage Poems," in *Aemilia Lanyer: Gender, Genre, and the Canon*, ed. Marshall Grossman (Lexington: University Press of Kentucky, 1998), 61; Barroll, "Looking for Patrons"; Marie H. Loughlin, "'Fast Ti'd unto Them in a Golden Chaine': Typology, Apocalypse, and Woman's Genealogy in Aemilia Lanyer's *Salve Deus Rex Judaeorum*," *Renaissance Quarterly* 53, no. 1 (2000): 133–79; Jessica L. Malay, "Positioning Patronage: Lanyer's *Salve Deus Rex Judæorum* and the Countess of Cumberland in Time and Place," *The Seventeenth Century* 28, no. 3 (2013): 251–74; Su Fang Ng, "Aemilia Lanyer and the Politics of Praise," *ELH* 67, no. 2 (2000): 433–51; Rogers, "The Passion of a Female Literary Tradition"; Julianne Sandberg, "Book, Body, and Bread: Reading Aemilia Lanyer's Eucharist," *Philological Quarterly* 96, no. 1 (2017): 1–25; Ina Schabert, "The Lady's Supper: Aemilia Lanyer's *Salve Deus Rex Judaeorum* as a Female Celebration of the Eucharist," in *Performances of the Sacred in Late Medieval and Early Modern England*, ed. Tobias Döring and Susanne Rupp (Amsterdam: Rodopi, 2005), 155–67; Schnell, "Breaking 'the Rule of *Cortezia*'"; and Helen Wilcox, "Creating Paratextual Communities: Reading Amelia Lanyer and Thomas Coryate," in *Literary Community-Making: The Dialogicality of English Texts from the Seventeenth Century to the Present*, ed. Roger D. Sell (Amsterdam: John Benjamins Publishing Company, 2012), 17–35.

78. McCarthy, "Speculations and Multiple Dedications," 52.

79. A foundational (and oft-cited) claim for critical work on Lanyer's *Salve Deus* has been Barbara Lewalski's argument that the prefaces construct a "contemporary community of learned and virtuous women" who become the "spiritual heirs" to Lanyer's biblical and historical examples throughout the poem (Lewalski, "Of God and Good Women," 212). For some time, scholarship on Lanyer endorsed Lewalski's reading of the prefaces and Lanyer's relationship to her female patrons, but critical work has gradually shifted to challenge and complicate the "community of good women" narrative. Particularly noteworthy challenges to Lewalski's argument have been Barbara Bowen's "Invention of White Womanhood" ("how much does Lanyer's community of

women depend on a shared whiteness?" [279]); Ann Baynes Coiro's "Writing in Service" ("to what extent would a woman be *breaking* company with other women by publishing?" [360]); Jonathan Goldberg's chapter on Lanyer in *Desiring Women Writing* ("elements of power that are conventionally treated simply as an attribute of gender need to be rethought" [20]); and Erica Longfellow's expert complication of the female-female erotics of *Salve Deus* within the mystical marriage tradition (Longfellow, *Women and Religious Writing in Early Modern England* [Cambridge: Cambridge University Press, 2004], especially Chapter 2).

80. In Susanne Woods's edition of Lanyer's *Salve Deus*, she provides details on all extant copies of the text, including which prefatory materials are included or excluded in each volume. On the various political and historical contexts that may have contributed to these prefatory differences, see Woods, "Textual Introduction," xlvii–li; Lewalski, *Writing Women*; McCarthy, "Speculations and Multiple Dedications," 46–50; and Barroll, "Looking for Patrons."

81. Critical work on Lanyer's prefaces has tended to overemphasize biographical analysis, often using the prefaces to study and speculate on Lanyer's social networks. Scholars are divided, though, on whether the prefaces function primarily for patronage or devotional aims. Some situate the "patronage poems" as part of Lanyer's consciously gendered authorial self-fashioning, a prefatory framework that presages the feminist leanings of her Passion narrative. See, for instance, Lewalski, *Writing Women*, 323–41; Beilin, *Redeeming Eve*; and McBride, "Sacred Celebration." Lisa Schnell, Ann Baynes Coiro, and Mary Ellen Lamb also position the dedications as part of a broader feminist project, but one infused with "a profound ambivalence" toward any attempt at female solidarity. See Schnell, "Breaking 'the Rule of *Cortezia*,'" 371; Coiro, "Writing in Service"; and Lamb, "Patronage and Class." Other scholars instead foreground the prefaces' religious iconography and read them within the context of a broader female devotional tradition. See Theresa M. DiPasquale, *Refiguring the Sacred Feminine: The Poems of John Donne, Aemilia Lanyer, and John Milton* (Pittsburgh: Duquesne University Press, 2008), especially pages 162–81; Achsah Guibbory, "The Gospel According to Aemilia: Women and the Sacred," in *Aemilia Lanyer: Gender, Genre, and the Canon*, ed. Marshall Grossman (Lexington: University Press of Kentucky, 1998), 191–211; Femke Molekamp, "Reading Christ the Book in Aemilia Lanyer's *Salve Deus Rex Judaeorum* (1611): Iconography and the Cultures of Reading," *Studies in Philology* 109, no. 3 (2012): 311–32; Micheline White, "A Woman with Saint Peter's Keys?: Aemilia Lanyer's *Salve Deus Rex Judaeorum* (1611) and the Priestly Gifts of Women," *Criticism* 45, no. 3 (2003): 323–41; and Elizabeth Hodgson, *Grief and Women Writers in the English Renaissance* (Cambridge: Cambridge University Press, 2014), especially Chapter 2. These critical conversations on Lanyer's prefaces are foundational for my arguments here, but my reading more closely aligns with Lorna Hutson and Yaakov Mascetti, both of whom distance their claims from biographical contexts and focus on Lanyer's prefatory material as a theorization of a particular kind of reading experience. See Hutson, "Why the Lady's Eyes Are Nothing Like the Sun," in *Women, Texts and Histories 1575–1760*, ed. Clare Brant and Diana Purkiss (London: Routledge, 1992): 13–38; and Mascetti, "'Here I Have Prepar'd my Paschal Lambe': Reading and Seeing the Eucharistic Presence in Aemilia Lanyer's *Salve Deus Rex Judaeorum*," *Partial Answers: Journal of Literature and the History of Ideas* 9, no. 1 (January 2011): 1–15.

82. Molekamp, "Reading Christ the Book," 312.

83. Lanyer's use of the verb "transfigure" here (from the Latin *transfigūrāre*, "to change the shape of") is especially noteworthy in relation to the incarnational experiments I am tracing throughout this chapter.

84. Femke Molekamp, *Women and the Bible in Early Modern England: Religious Reading and Writing* (Oxford: Oxford University Press, 2013), 1.

85. Trettien, *Cut/Copy/Paste*, 36. Also see Peter Stallybrass on how the codex as an "emergent technology" adopted early into Christianity ushered in practices of "discontinuous reading" (Stallybrass, "Books and Scrolls: Navigating the Bible," in *Books and Readers in Early Modern England*, ed. Jennifer Andersen and Elizabeth Sauer [Philadelphia: University of Pennsylvania Press, 2002], 42–79).

86. Trettien, *Cut/Copy/Paste*, 93; 30.

87. Trettien, *Cut/Copy/Paste*, 40. For more on Little Gidding and early modern text technologies of cutting and pasting, see Adam Smyth, *Material Texts in Early Modern England* (Cambridge: Cambridge University Press, 2018), especially pages 17–54.

88. Bowen, "Invention of White Womanhood," 275.

89. Lochrie, *Margery Kempe*, 168.

90. Richard Rambuss, *Closet Devotions* (Durham, NC: Duke University Press, 1998), 2. On queer temporal drag, see Elizabeth Freeman, *Time Binds: Queer Temporalities, Queer Histories* (Durham, NC: Duke University Press, 2010).

91. Miller, "(M)other Tongues," 153.

92. For more on the theme of maternal lineage in *Salve Deus*, see Lewalski, *Writing Women*; Lewalski, "Of God and Good Women"; Pearson, "Women Writers and Women Readers"; and White, "Saint Peter's Keys." On Mary Sidney Herbert as Lanyer's authorial foremother, see Rogers, "The Passion of a Female Literary Tradition."

93. Rayna Kalas, *Frame, Glass, Verse: The Technology of Poetic Invention in the English Renaissance* (Ithaca, NY: Cornell University Press, 2018), 8.

94. Kalas, *Frame, Glass, Verse*, 8.

95. Kalas, *Frame, Glass, Verse*, 83; 56.

96. Kalas, *Frame, Glass, Verse*, 56.

97. In the background here is Kimberly Johnson's excellent work on the impact of Reformation debates surrounding the Eucharist (Christ's flesh as absent or present), particularly on the formulation of lyric poetry. "In the period following the religious Reformation of the sixteenth century," Johnson argues, "the lyric poem becomes a primary cultural site for investigating that capacity of language to manifest presence." See Johnson, *Made Flesh: Sacrament and Poetics in Post-Reformation England* (Philadelphia: University of Pennsylvania Press, 2014), 6.

98. Thomas Gibson, *The anatomy of humane bodies epitomized wherein all parts of man's body, with their actions and uses, are succinctly described, according to the newest doctrine of the most accurate and learned modern anatomists* (London: M. Flesher, 1682), 145, original emphasis, Wellcome Collection, EPB/B/62485.

99. Gibson, *The anatomy of humane bodies*, 145.

100. Gibson, *The anatomy of humane bodies*, 147.

101. Crooke, *Mikrokosmographia*, 228.

102. Crooke, *Mikrokosmographia*, 228.

103. On the figure of Cleopatra within Lanyer's *Salve Deus*, see Bowen, "Invention of White Womanhood," 280–83; Kim F. Hall, *Things of Darkness: Economies of Race and Gender in Early Modern England* (Ithaca, NY: Cornell University Press, 1995), especially pages 182–84; and Sanchez, "Ain't I a Ladie?," 24–28.

104. Bowen, "Invention of White Womanhood," 276.

105. Bowen, "Invention of White Womanhood," 276.

106. Bowen, "Invention of White Womanhood," 279.

107. Bowen, "Invention of White Womanhood," 275.

108. See Stallybrass, "Books and Scrolls" on the biblical codex as "indexical"—a "technology that uses bookmarks like prosthetic fingers to take the reader easily from place to place" (43).

109. On excess and dilation, see Parker, *Literary Fat Ladies*, 13–15.

110. Lanyer uses the term "apologie" in the early modern connotation of *apologia*, or defense. For a reading of Eve's apology within the *apologia* rhetorical tradition, see Lyn Bennett, *Women Writing of Divinest Things: Rhetoric and The Poetry of Pembroke, Wroth and Lanyer* (Pittsburgh: Duquesne University Press, 2004), especially Part 3.

111. Guibbory, "Gospel According to Aemilia," 200.

112. The section on Eve is frequently anthologized and hailed as the most radical of Lanyer's proto-feminist overtures. For representative readings of "Eves Apologie," see Beilin, *Redeeming Eve*, 177–207; Elizabeth Hodgson, "A 'Paraditian Creature': Eve and Her Unsuspecting Garden in Seventeenth-Century Literature," in *Biblical Women in Early Modern Literary Culture 1550–1700*, ed. Victoria Brownlee and Laura Gallagher (Manchester: Manchester University Press, 2015), 41–58; Mueller, "Feminist Poetics"; McGrath, "'Let Us Have Our Libertie Againe'"; and B. R. Siegfried, "An Apology for Knowledge: Gender and the Hermeneutics of Incarnation in the Works of Aemilia Lanyer and Sor Juana Inés de La Cruz," *Early Modern Literary Studies* 6, no. 3 (2001): n.p.

113. John Garrison, "Aemilia Lanyer's *Salve Deus Rex Judaeorum* and the Production of Possibility," *Studies in Philology* 109, no. 3 (2012): 305; 310.

114. Genesis 3:16, *King James Bible Online*, 2022, https://www.kingjamesbibleonline.org/, accessed December 20, 2022.

115. Pettus, *Volatiles from the history of Adam and Eve*, 135–36.

116. Bowen, "Invention of White Womanhood," 287.

117. "prove, *v.*1a" (*OED*).

118. Mary Astell, *A Serious Proposal to the Ladies*, ed. Patricia Springborg (Guelph: Broadview Press, 2002), 83.

119. Thomas Geminus, *Compendiosa totius anatomiae delineatio, aere exarata* (London: John Hereford, 1545), Wellcome Collection, EPB/D/2731.

120. In one extant copy of Thomas Geminus's text, held by the New York Academy of Medicine (pictured in Figure 24), an unknown reader has hand-colored the naked figures and inscribed a Genesis verse along the top of the plate: "The Eyes of them both were opened, and they knew that they were naked" (Genesis 3:7). See Geminus, *Compendiosa totius anatomiae delineation, aere exarata* (London: J. Kingston, 1559), New York Academic of Medicine, VAULT 1.

121. Adriaan van de Spieghel, *De formato foetu liber singularis aeneis figuris exornatus* (Padua, 1626), Tab. IIII, National Library of Medicine, WZ 250 S755dh 1627. Spieghel's plates were first printed in 1623, long after Vesalius's *Fabrica* was published, but the plates are much older, originally prepared for a large anatomical treatise by Giulio Casserio, who worked with Vesalius in Padua. On this publication history, see Glenn Harcourt, "Andreas Vesalius and the Anatomy of Antique Sculpture," *Representations* 17 (1987): 28–61.

122. I am grateful to Jessica Rosenberg for drawing my attention to the vibrant lexicon of early modern terms related to "handbook."

123. Sir Philip Sidney, *Astrophil and Stella*, in *Sir Philip Sidney: The Major Works*, ed. Katherine Duncan-Jones (Oxford: Oxford University Press, 2008), Sonnet 1.

124. Edmund Spenser, *The Faerie Queene* (London: William Ponsonby, 1596), sig. A1^v; and John Donne, "Elegy 19, To his mistress going to bed," *John Donne: The Complete English Poems*, ed. A. J. Smith (New York: Penguin Books, 1996), 124; and "labour | labor, *v.*" (*OED*).

125. Maus, *Inwardness and Theater*, 186; and Elizabeth D. Harvey, *Ventriloquized Voices: Feminist Theory and English Renaissance Texts* (London: Routledge, 1992).

126. Maus, *Inwardness and Theater*, 190.

127. Maus, *Inwardness and Theater*, 190.

128. Harvey, *Ventriloquized Voices*, 79.

129. Maus, *Inwardness and Theater*, 183.

130. Clarke, "Aemelia Lanyer."

CHAPTER 5

1. For more on the recovery of Pulter's manuscript, see Margaret J. M. Ezell, "The Laughing Tortoise: Speculations on Manuscript Sources and Women's Book History," *English Literary Renaissance* 38, no. 2 (2008): 331–55; and Leah Knight and Wendy Wall, "The Manuscript," in *The Pulter Project: Poet in the Making*, ed. Leah Knight and Wendy Wall (2018), http://pulterproject.northwestern.edu.

2. Lady Hester Pulter, *Poems, Emblems, and The Unfortunate Florinda*, ed. Alice Eardley (Toronto: Iter, 2014); and Knight and Wall, eds., *The Pulter Project*.

3. Robyn Adams and Lisa Jardine, "The Return of the Hooke Folio," *Notes & Records of the Royal Society* 60 (2006): 235.

4. Howard Gest, "Homage to Robert Hooke (1635–1703): New Insights from the Recently Discovered Hooke Folio," *Perspectives in Biology and Medicine* 52, no. 3 (2009): 398.

5. Adams and Jardine, "The Return of the Hooke Folio," 238. For more on the recovery of the Hooke Folio, see Louisiane Ferlier, "Lost and Found," *The Royal Society Blog*, January 12, 2021, https://royalsociety.org/blog/2021/01/lost-and-found/, accessed April 2, 2023; and Mark Henderson, "'Britain's Leonardo Rescued on Safety Net," *The Sunday Times*, October 8, 2007.

6. As Adams and Jardine report, when the original bookbinder of the Royal Society meeting records "found the 400 pages of [Hooke's] minutes missing, he left a suitable number of stubs empty in case they should ever reappear. Binding began again as usual with the new secretarial era in December 1683. Nothing more vividly captures the way the Hooke folio completes the early Royal Society records than the thick wadge of empty stubs—waiting for their documents—comprising most of volume MS 559 and the beginning of MS 560 of the draft minutes" ("The Return of the Hooke Folio," 239).

7. Alice Eardley, "'Shut Up in a Countrey Grange': The Provenance of Lady Hester Pulter's Poetry and Prose and Women's Literary History," *Huntington Library Quarterly* 80, no. 2 (2017): 349.

8. Ellen Rooney, "Form and Contentment," *Modern Language Quarterly* 61, no. 1 (2000): 18.

9. Alexander Read, *The Manuall of the Anatomy or Dissection of the Body of Man, Which Usually Are Shewed in the Publike Anatomicall Exercises. Methodically Digested into 6 Books* (London: J.H., 1638), Wellcome Collection, EPB/A/5355.

10. For more on the intersections of natural history, colonialism, and global trade economies, see Alix Cooper, *Inventing the Indigenous: Local Knowledge and Natural History in Early Modern Europe* (Cambridge: Cambridge University Press, 2007); Richard Drayton, *Nature's Government: Science, Imperial Britain and the 'Improvement' of the World* (New Haven, CT: Yale University Press, 2000); Kay Dian Kriz, "Curiosities, Commodities, and Transplanted Bodies in Hans Sloane's 'Natural History of Jamaica,'" *The William and Mary Quarterly* 57, no. 1 (2000):

35–78; and Londa Schiebinger, *Plants and Empire: Colonial Bioprospecting in the Atlantic World* (Cambridge, MA: Harvard University Press, 2007).

11. Nehemiah Grew, *The anatomy of plants. With an idea of a Philosophical History of Plants. And several other lectures, Read before the royal society* (London: W. Rawlins, 1682), 9; and Al Coppola, "'Without the Help of Glasses': The Anthropocentric Spectacle of Nehemiah Grew's Botany," *The Eighteenth Century* 54, no. 2 (Summer 2013): 270.

12. Unless otherwise noted, all references to Pulter's poems follow the titles and poem and line numbers assigned in Knight and Wall, *The Pulter Project*. Poems are cited by poem and line number in the body of the text.

13. William B. Ashworth Jr., "Natural History and the Emblematic World View," in *Grasping the World: The Idea of the Museum*, ed. Donald Preziosi and Claire Farago (Aldershot: Ashgate, 2004), 144–59.

14. Robert Hooke, *Micrographia: or some Physiological Descriptions of minute bodies made by magnifying glasses. With Observations and Inquiries thereupon* (London: Jo. Martyn and Ja. Allestry, 1665), British Library, 39.e.1. All references to *Micrographia* are from this edition and will be cited by page number in the body of the text.

15. Ashworth, "Emblematic World View," 154; 144.

16. On Pulter's reference to the "naked sunburnt female slave," see Frances E. Dolan, "A Female Slave," (Curation, Poem 35), in *The Pulter Project*.

17. One seventeenth-century compendium on insects, compiled from the writings of several natural historians, including Edward Wotton, Conrad Gesner, and Thomas Moffet (the very same Moffet who dedicated a treatise on silkworms to Mary Sidney Herbert), featured a beehive on its frontispiece and described bees as "patterns and precedents of political and economical virtues" (*The theater of insects, or, Lesser living creatures* [London, 1658]). For another example of insects featured emblematically, see Samuel Purchas, *A theatre of politicall flying-insects* (London: R. I., 1657). Also of note here is Francis Bacon's metaphorical use of ants, spiders, and bees to describe different kinds of scientific approaches. In *Novum Organum*, Bacon writes: "Those who have treated of the sciences have been either empirics or dogmatic. The former like ants only heap up and use their store, the latter like spiders spin out their own webs. The bee, a mean between both, extracts matter from the flowers of the garden and the field, but works and fashions it by its own efforts. The true labor of philosophy resembles hers, for it neither relies entirely or principally on the powers of the mind, not yet lays up in the memory the matter afforded by the experiments of natural history and mechanics in its raw state, but changes and works it in the understanding." See Francis Bacon, *The New Organon*, ed. Lisa Jardine and Michael Silverthorne (Cambridge: Cambridge University Press, 2000), 79.

18. Francis Bacon, "The Argument of the Several Parts," in *Of the advancement and proficience of learning; or, The partitions of sciences* (London: Leon Lichfield, 1640), 24, Early English Books Online Text Creation Partnership, 2011, https://quod.lib.umich.edu/e/eebo2/A72146.0001.001, accessed April 2, 2023. Historians of science have traced the first recorded experiments with a convex lens as a tool of magnification to the eleventh-century Arabian scholar Ibn al Haitham. Haitham's text, *Opticae thesaurus alhazeni arabis* was translated into Latin in the twelfth century, and became foundational for late-medieval and early modern advancements in microscopy both in England and on the continent. For more on this history, see S. Bradbury, *The Evolution of the Microscope* (Oxford: Pergamon Press, 1967); Marian Fournier, *The Fabric of Life: Microscopy in the Seventeenth Century* (Baltimore: Johns Hopkins University Press, 1996); Edward G. Ruestow, *The Microscope in the Dutch Republic: The Shaping of Discovery* (Cambridge: Cambridge University Press, 1996); and Catherine Wilson, *The Invisible World: Early Modern Philosophy and the Invention of the Microscope* (Princeton, NJ: Princeton University Press, 1995).

19. David Bardell, "The First Record of Microscopic Observations," *Bioscience* 33, no. 1 (1983): 36–38.

20. Although Hooke rails against the violence of dissecting and harming Nature's creatures, he frequently dissected his specimens for closer study, as I will discuss later in this chapter. See, for instance, his description of dissecting the gray drone-fly (184–85).

21. Frances E. Dolan, ed., "The Pismire," by Hester Pulter (Poem 35, Amplified Edition), in *The Pulter Project.*

22. Frances E. Dolan, "Hester Pulter's Dunghill Poetics," *Journal for Early Modern Cultural Studies* 20, no. 2 (2020): 16–42.

23. Dolan, "Dunghill Poetics," 22. Also see David Cressy on Robert Boyle and other Royal Society members' interest in saltpetre, a substance found in dunghills (Cressy, "Salt Petre, State Security and Vexation in Early Modern England," *Past and Present* 212 [2011]: 73–111); and Vera Keller on the garden as an experimental and philosophical space in the transatlantic colonial culture of seventeenth-century England (Keller, "A 'Wild Swing to Phantsy': The Philosophical Gardener and Emergent Experimental Philosophy in the Seventeenth-Century Atlantic World," *Isis* 112, no. 3 [2021]: 507–30).

24. "The Pismire" is not an emblem poem, but it follows the structure of many of Pulter's emblem poems and, as Frances E. Dolan notes, it "reflects on its own status as emblem in line 28" (Dolan, ed., "The Pismire"). For more on Pulter's observational work in the poems, see Lara Dodds, "Hester Pulter Observes the Eclipse: Or, the Poetics of the Astronomical Event," *Journal for Early Modern Cultural Studies* 20, no. 2 (2020): 144–68.

25. Lorraine Daston and Elizabeth Lunbeck, "Observation Observed," in *Histories of Scientific Observation*, ed. Lorraine Daston and Elizabeth Lunbeck (Chicago: University of Chicago Press, 2011), 1.

26. Bacon, *The New Organon*, 179.

27. Bacon, *The New Organon*, 179.

28. Bacon, *Advancement and proficience of learning*, 26.

29. Bacon, *Advancement and proficience of learning*, 86; 26

30. Bacon, *Advancement and proficience of learning*, 26–28.

31. Bacon's *Sylva Sylvarum*, an extensive collection of case studies testing and modeling Bacon's scientific methods, was published posthumously in 1627 bound together with *The New Atlantis*, Bacon's unfinished utopian novel that portrays an idealized scientific society based on his arguments for observation and experiment. On the implications of this coupled publication, see David Colclough, "'The Materialls for the Building': Reuniting Francis Bacon's *Sylva Sylvarum* and *New Atlantis*," *Intellectual History Review* 20, no. 2 (2010): 181–200.

32. Bacon, *Advancement and proficience of learning*, 24–25. Bacon's primary critiques against syllogism (or, deductive reasoning based on a set of linked assertions or assumptions) pertain to its focus on "words" rather than "things." Syllogism, he argues, is "barren of operations and remote from practice; and in relation to the active part of sciences, altogether incompetent" (*Advancement and proficience of learning*, 25). For more on Bacon's complex argument against words, see Debapriya Sarkar, *Possible Knowledge: The Literary Forms of Early Modern Science* (Philadelphia: University of Pennsylvania Press, 2023), especially pages 10–14; and Chapter 3.

33. Thomas Sprat, *The history of the Royal-Society of london, For the Improving of natural knowledge* (London: T. R., 1667), 90, Early English Books Online Text Creation Partnership, 2011, https://quod.lib.umich.edu/e/eebo/A61158.0001.001, accessed May 8, 2023.

34. Sprat, *History of the Royal-Society*, 262.

35. On Pulter's use of Pliny, see Frances E. Dolan, "The State of Renaissance Studies II: Hester Pulter's Renaissance," *English Literary Renaissance* 50, no. 1 (2019): 32–39; and Ezell, "Laughing Tortoise."

36. Geoffrey Whitney, *A Choice of Emblems, and Other Devices* (Leiden: Christopher Plantyn, 1568), 51.

37. Susan Isaac describes Topsell's book as "a vast collection of would-be observational data," hinting at its relevance to a history of observational methods. Later editions of Topsell's bestiary were published with Thomas Moffat's *Theater of insects*, and there is room to consider how the publication history of Topsell's text testifies to its gradual incorporation into a more officially scientific archive. See Susan Isaac, "The Familiar and the Fantastic: *The historie of foure-footed beastes* by Edward Topsell, 1607," *Royal College of Surgeons of England* (blog), March 16, 2018, https://www.rcseng.ac.uk/library-and-publications/library/blog/the-familiar-and-the-fantastic/, accessed April 11, 2023; Katherine Acheson, "Gesner, Topsell, and the Purposes of Pictures in Early Modern Natural Histories," in *Printed Images in Early Modern Britain: Essays in Interpretation*, ed. Michael Hunter (Farnham: Ashgate, 2010), 127–44; and Ana Maria Pavel, "The Influence of Natural History: Marine Animals and Strange Creatures in Edward Topsell's The Historie of Serpents," *The Annals of Ovidius University Constanța, Philology Series* 32, no. 2 (2021): 37–51.

38. See, for instance, George Wither, *A collection of emblemes, ancient and modern quickened with metricall illustrations* (London: A.M., 1635), 19.

39. Rachel Dunn, "Breaking a Tradition: Hester Pulter and the English Emblem Book," *The Seventeenth Century* 30, no. 1 (2015): 62.

40. Sarah Hutton, "Hester Pulter (c. 1596–1678): A Woman Poet and the New Astronomy," *Etudes Epistémè* 14 (2007): para. 4.

41. Hutton, "A Woman Poet and the New Astronomy," para. 4.

42. Robert M. Schuler, "Three Renaissance Scientific Poems," *Studies in Philology, Texts and Studies* 75, no. 5 (1978): 1.

43. In addition to Hutton's essay, early work on Pulter's science includes Jayne Archer's study of Pulter's "imaginative and intellectual engagements with the experimental, esoteric and emblematic traditions in alchemy." See Jayne Archer, "A 'Perfect Circle'? Alchemy in the Poetry of Hester Pulter," *Literature Compass* 2 (2005): 2. My work is deeply indebted to these pioneering studies, even when I take issue with the frameworks through which they read Pulter's writing.

44. Dolan, "Dunghill Poetics," 17.

45. Ruth Connolly, "Hester Pulter's Childbirth Poetics," *Women's Writing* 26, no. 3 (2019): 282–303; and Amanda Zoch, "Rewriting the Lying-In: Hester Pulter, Katherine Philips, and the Felt Mortality of Pregnancy," *Early Modern Women* 15, no. 1 (2020): 3–25.

46. Dodds, "Hester Pulter Observes the Eclipse"; and Victoria E. Burke, "Playing Football with the Stars: Hester Pulter Rethinks the Metaphysical Astronomy Poem," *Journal for Early Modern Cultural Studies* 20, no. 2 (2020): 169–91.

47. Liza Blake, "Hester Pulter's Particle Physics and the Poetics of Involution," *Journal for Early Modern Cultural Studies* 20, no. 2 (2020): 72.

48. Blake, "Poetics of Involution," 87–88.

49. Howard Marchitello, *The Machine in the Text: Science and Literature in the Age of Shakespeare and Galileo* (Oxford: Oxford University Press, 2011), 3.

50. Howard Marchitello, "Science Studies and English Renaissance Literature," *Literature Compass* 3, no. 3 (2006): 341.

51. Elizabeth Spiller, "Shakespeare and the Making of Early Modern Science: Resituating Prospero's Art," *South Central Review* 26, no. 1/2 (2009): 25.

52. Wendy Beth Hyman, "Seeing the Invisible under the Microscope: Natural Philosophy and John Donne's Flea, *Philological Quarterly* 98, no. 1/2 (2019): 163.

53. Hyman, "Seeing the Invisible," 162.

54. Allan Megill, "Introduction: Four Senses of Objectivity," *Rethinking Objectivity* (Durham, NC: Duke University Press, 1994), 1.

55. Lorraine Daston, "Baconian Facts, Academic Civility, and the Prehistory of Objectivity," in *Rethinking Objectivity*, ed. Allan Megill (Durham, NC: Duke University Press, 1994), 47.

56. Megill, "Four Senses of Objectivity," 301.

57. On Pulter's "naked emblems" within the context of early modern English emblem books, see Dunn, "Breaking a Tradition," especially pages 56–60.

58. Donna Haraway, "Situated Knowledges: The Science Question in Feminism and the Privilege of Partial Perspective," *Feminist Studies* 14, no. 3 (1988): 582.

59. Haraway, "Situated Knowledges," 592.

60. The full title of this poem in the manuscript is "The Garden, or the Contention of flowers. To my Deare Daughter M^{ris} ~~Anne Pulter~~, at her desire written."

61. This line (and the entire poem) includes significant revisions in Pulter's hand. The original line seems to read: "her virgintal Leaves all field mort to bee." One hand then revises "virgintal" to "virgent," and then the same hand inserts "virgin" in the inter-linear space above "virgent," crossing out the "-al" at the end of "virgintal." The same hand strikes out "field mort" and inserts "withered" and "Fillimott" in the inter-linear spaces above and below the original word.

62. Lotte Mulligan, "Self-Scrutiny and the Study of Nature: Robert Hooke's Diary as Natural History," *Journal of British Studies* 35, no. 3 (1996): 312.

63. Felicity Henderson, "Unpublished Material from the Memorandum Book of Robert Hooke, Guildhall Library MS 1758," *Notes & Records of the Royal Society* 61 (2007): 129–75. Entries from this edition cited by date in the body of the text.

64. Henderson, "Memorandum Book of Robert Hooke."

65. Mulligan, "Robert Hooke's Diary as Natural History," 321.

66. Mulligan, "Robert Hooke's Diary as Natural History," 314–15.

67. Robert Hooke, *The Diary of Robert Hooke, M.A. F.R.S. 1672–1680*, ed. Henry W. Robinson and Walter Adams (London: Taylor and Francis, 1635). Entries from this edition are cited by date in the body of the text.

68. Robert Hooke, "Hooke Folio," Manuscript (London, c.1678-1683), HF_591, The Royal Society, MS/847. For an open-access, transcribed, and digitized copy of the Hooke Folio, see Robyn Adams, Jan Broadway, Jenni Thomas, and Anna Winterbottom, eds., *The Hooke Folio Online*, Centre for Editing Lives and Letters, 2007, http://www.livesandletters.ac.uk/cell/Hooke/Hooke.html, accessed May 15, 2023.

69. Hooke, "Hooke Folio," HF_595.

70. For more on climate- and geographical-based theories of race in seventeenth-century England, see Jean E. Feerick, *Strangers in Blood: Relocating Race in the Renaissance* (Toronto: University of Toronto Press, 2010); Sujata Iyengar, *Shades of Difference: Mythologies of Skin Color in Early Modern England* (Philadelphia: University of Pennsylvania Press, 2005); and Cristina Malcolmson, *Studies of Skin Color in the Early Royal Society: Boyle, Cavendish, Swift* (London: Routledge, 2016).

71. David Carroll Simon, *Light Without Heat: The Observational Mood from Bacon to Milton* (Ithaca, NY: Cornell University Press, 2018), 7. For Simon, "many of the genres we now call

'literary' remain hospitable to—indeed, generative of—observational moods," which are also particularly ubiquitous within "early modernity's scientific imagination" (214).

72. Kristin M. Girten, "Mingling with Matter: Tactile Microscopy and the Philosophic Mind in Brobdingnag and Beyond," *The Eighteenth Century* 54, no. 4 (2013): 502.

73. For some representative arguments that situate *Micrographia* within a broader history of vision and ocular supremacy, see Mary Baine Campbell, *Wonder and Science: Imagining Worlds in Early Modern Europe* (Ithaca, NY: Cornell University Press, 2002), especially Chapter 6; S. Bradbury and G. L'E. Turner, eds., *Historical Aspects of Microscopy* (Cambridge: Royal Microscopical Society, 1967); Felicity Henderson, "Robert Hooke and the Visual World of the Early Royal Society," *Perspectives on Science* 27, no. 3 (2019): 395–434; and Erin Webster, *The Curious Eye: Optics and Imaginative Literature in Seventeenth-Century England* (Oxford: Oxford University Press, 2020), especially Chapter 3. For a particularly interesting, materially minded version of this argument, see Meghan C. Doherty on Hooke's "engagement with mid-seventeenth-century visual culture" and, particularly, the art of engraving (Doherty, "Discovering the 'True Form': Hooke's *Micrographia* and the Visual Vocabulary of Engraved Portraits," *Notes and Records* 66, no. 3 [2012]: 211–34).

74. Girten, "Mingling with Matter," 505.

75. On "bookwork," see Whitney Trettien, *Cut/Copy/Paste: Fragments from the History of Bookwork* (Minneapolis: University of Minnesota Press, 2021).

76. Doherty, "Discovering the 'True Form,'" 211–12.

77. Here, in addition to Kristin M. Girten's concept of "tactile micrography," I think with Ian Lawson's argument on Hooke's material practices: "for Hooke, microscopy encompassed practices that manipulated a loosely connected system of materials extending from the ants he kept in a jar on his desk to the sunlight admitted by his window." See Lawson, "Crafting the Microworld: How Robert Hooke Constructed Knowledge about Small Things," *Notes and Records* 70, no. 1 (March 2016): 26.

78. With the term "micro-graphic," I synthesize the multiple connotations of "graphic" available in early modern parlance: first in terms of the "graphic arts" (painting, etching, engraving, and printing); second, "producing by words the effect of a picture; vividly descriptive"; and, third, "of or pertaining to writing; consisting of letters." See "graphic, *adj.* and *n.*" and "graphical, *adj.*" (*OED*).

79. Mary Thomas Crane, "Surface, Depth, and the Spatial Imaginary: A Cognitive Reading of the Political Unconscious," *Representations* 108, no. 1 (2009): 86.

80. For a detailed discussion of the different hands included in Pulter's manuscript, see Sarah C. E. Ross, "Women and Religious Verse in English Manuscript Culture, c. 1600–1688: Lady Anne Southwell, Lady Hester Pulter, and Katherine Austen" (PhD diss., Oxford University, 2000); and Wendy Wall, "What Else Is In the Manuscript?: Or, Where Did Pulter's Poems Live?" (Exploration), in *The Pulter Project*.

81. Carolyn Dinshaw, *Getting Medieval: Sexualities and Communities, Pre- and Postmodern* (Durham, NC: Duke University Press, 1999), 21.

82. Ben Burton and Elizabeth Scott-Baumann, eds., *The Work of Form: Poetics & Materiality in Early Modern Culture* (Oxford: Oxford University Press, 2014).

83. Dianne Mitchell, "Lyric Backwardness," in *The Oxford Handbook of Early Modern Women's Writing in English, 1540–1700*, ed. Elizabeth Scott-Baumann, Danielle Clarke, and Sarah C. E. Ross (Oxford: Oxford University Press, 2022), 190.

84. Elizabeth Zeman Kolkovich, "In Defense of Indulgence: Hester Pulter's Maternal Elegies," *Journal for Early Modern Cultural Studies* 20, no. 2 (2020): 43.

85. William Sherman coins the term "manicule" to describe "the small pointing hand" we frequently find in margins of medieval and early modern manuscripts and printed books. In his study of this mark, he comes across more than fifteen different terms for the symbol, including "pointing hand," "mutton fist," "bishop's fist," and "maniple." See William Sherman, *Used Books: Marking Readers in Renaissance England* (Philadelphia: University of Pennsylvania Press, 2007), 29–34.

86. All the pronouns in these inserted lines have been altered from "her" to "his" or vice versa, suggesting that the scribe mistook the originals and Pulter went through and changed them. It is worth noting that the pronoun flexibility here—and the revisions that call our attention to that flexibility—aligns the gender of the "hert" of this poem with the "hunted hart" of Emblems 22 (Poem 87) and 25 (Poem 90).

87. For additional examples of Pulter adding lines to poems in the manuscript, see Poem 2 (she does not cut, but adds two new cramped lines in the margins) and Poem 6 ("Universall Dissolution," where she makes several cuts and additions throughout the poem).

88. Perhaps we see here Pulter enacting a form of the "disknowledge" Katherine Eggert tracks through late English Renaissance literature, "the conscious and deliberate setting aside of one compelling mode of understanding the world—one discipline, one theory—in favor of another." See Katherine Eggert, *Disknowledge: Literature, Alchemy, and the End of Humanism in Renaissance England* (Philadelphia: University of Pennsylvania Press, 2015), 3.

89. Kolkovich, "In Defense of Indulgence," 65; and Archer, "Alchemy in the Poetry of Hester Pulter." Liza Blake's point about the "insistent repetition of key words and concepts across [Pulter's] poetic corpus" is also helpful grounding here (Blake, "Poetics of Involution," 71).

90. Leah Knight and Wendy Wall, eds., "Why must I thus forever be confined," by Hester Pulter (Poem 57, Elemental Edition); and Victoria E. Burke, ed., "Why must I thus forever be confined," by Hester Pulter (Poem 57, Amplified Edition), in *The Pulter Project*. Sarah C. E. Ross and Elizabeth Scott-Baumann include only the first hundred unbroken lines in their anthology of women poets of the English Civil Wars, with the assigned title taken from the first line ("Why must I thus forever be confined"). See Sarah C. E. Ross and Elizabeth Scott-Baumann, eds., *Women Poets of the English Civil War* (Manchester: Manchester University Press, 2018), 133–36.

91. Sarah C. E. Ross, "Hester Pulter's Devotional Complaints: 'Then Will I Hallelujahs Ever Sing,'" *Journal for Early Modern Cultural Studies* 20, no. 2 (2020): 105. Also see Ross's discussion of Pulter's "sense of solitary imprisonment at Broadfield" in *Women, Poetry, and Politics in Seventeenth-Century Britain* (Oxford: Oxford University Press, 2015), 140–41.

92. Eardley, "'Shut Up in a Countrey Grange,'" 347; and Maureen E. Mulvihill, "The Book of Hester: Editing an Uncanonical Text," *Rare Book Hub* (*Rare Book Monthly* series), November 2016, https://www.rarebookhub.com/articles/2102, accessed May 10, 2023. For additional autobiographical readings, see Victoria E. Burke, "Playing Football with the Stars," 169–71; Burke, "Why must I thus forever be confined," in *The Pulter Project*; and Ross and Scott-Baumann, eds., *Women Poets of the English Civil War*, 136. As Knight and Wall note in their critical introduction to *The Pulter Project*, we now have good evidence that "the isolation in the countryside her poems lament was not unremitting." See Knight and Wall, "The Poet," in *The Pulter Project*; and also Dolan, "Hester Pulter's Renaissance," 32–33.

93. Ezell, "Laughing Tortoise."

94. Dodds, "Hester Pulter Observes the Eclipse," 162.

95. I number the poems here according to the *Pulter Project* elemental editions, but Poem 58 does not currently exist on the *Pulter Project* website. Instead, Leah Knight and Wendy Wall make their editorial decision clear with a note: "We do not consider these lines to constitute a separate

poem. Instead, we see these six lines as the conclusion to Poem 57, 'Why Must I Thus Forever Be Confined.' We have therefore inserted these lines into the preceding poem in this edition."

96. Liza Blake also attends closely to Pulter's punctuation marks (or lack thereof) in "The Hope." See Liza Blake, ed., "The Hope," by Hester Pulter (Poem 65, Amplified Editions B–D), in *The Pulter Project*.

97. Claire M. L. Bourne, "Dramatic Pilcrows," *The Papers of the Bibliographical Society of America* 108, no. 4 (December 2014): 413–52 (452). John Lennard dates the first appearance of parentheses in English print to 1494, but argues that conventions regarding the typographical feature remain in flux for some time. See John Lennard, *But I Digress: The Exploitation of Parentheses in English Printed Verse* (Oxford: Clarendon Press, 1992); and also Jonathan P. Lamb on Philip Sidney's parentheses as "a unique stylistic habit" in "Parentheses and Privacy in Philip Sidney's *Arcadia*," *Studies in Philology* 107, no. 3 (2010): 320.

98. As Jenny C. Mann observes, "a defining characteristic of the figure is that its removal leaves a 'whole,' 'full,' or 'entire' sense behind" (Mann, *Outlaw Rhetoric: Figuring Vernacular Eloquence in Shakespeare's England* [Ithaca, NY: Cornell University Press, 2012], 100).

99. Cited in Mann, *Outlaw Rhetoric*, 100.

100. Mann, Outlaw Rhetoric, 100.

101. George Puttenham, *The Arte of English Poesie*, ed. Edward Arber (London: Alex Murray & Sons, 1869), 180.

102. "sigh, *n.*" (*OED*).

103. Domenico Laurenza, "Art and Anatomy in Renaissance Italy: Images from a Scientific Revolution," *The Metropolitan Museum of Art Bulletin* 69, no. 3 (2012): 5.

NEW HISTORIES OF LITERATURE AND SCIENCE (A CODA)

1. Jane Barker, "Poems on several occasions. in three parts," Magdalen Library, University of Oxford, Magdalen MS 343, line 66.

2. Bathsua Makin, *An essay To Revive the Antient Education of Gentlewomen, in Religion, Manners, Arts & Tongues* (London: J. D., 1673), 35, Early English Books Online Text Creation Partnership, 2011, https://quod.lib.umich.edu/e/eebo2/A51611.0001.001, accessed May 11, 2023.

3. Makin, *Antient Education of Gentlewomen*, 37.

4. Makin, *Antient Education of Gentlewomen*, 43–44.

5. Margo Hendricks, "Feminist Historiography," in *A Companion to Early Modern Women's Writing*, ed. Anita Pacheco (Oxford: Blackwell Publishing, 2002), 367.

6. See Surekha Davies, *Renaissance Ethnography and the Invention of the Human: New Worlds, Maps and Monsters* (Cambridge: Cambridge University Press, 2016); Pablo F. Gómez, *The Experiential Caribbean: Creating Knowledge and Healing in the Early Modern Atlantic* (Chapel Hill: University of North Carolina Press, 2017); Evelyn Fox Keller, *A Feeling for the Organism: The Life and Work of Barbara McClintock* (New York: Henry Holt, 1983); Katharine Park, *Secrets of Women: Gender, Generation, and the Origins of Human Dissection* (Baltimore: MIT Press, 2006); and Banu Subramaniam, *Ghost Stories for Darwin: The Science of Variation and the Politics of Diversity* (Urbana: University of Illinois Press, 2014).

7. On Hutchinson and physics, see Jonathan Goldberg, *The Seeds of Things: Theorizing Sexuality and Materiality in Renaissance Representations* (New York: Fordham University Press, 2009), especially Chapter 4; Jessie Hock, *The Erotics of Materialism: Lucretius and Early Modern Poetics* (Philadelphia: University of Pennsylvania Press, 2021), especially Chapter 4; and Charlotte

Newcombe, "Understanding Lucretius: Atomism and Form in Lucy Hutchinson's Translation of *De rerum natura*," in *Early Modern Literatures and Sciences: Forms of Knowledge in Women's Writing*, ed. Liza Blake and Whitney Sperrazza (forthcoming).

8. On this point, see the powerful documentary *Picture a Scientist*, directed by Sharon Shattuck and Ian Cheney (Uprising and Wonder Collaborative, 2020).

9. Eve Kosofsky Sedgwick, *Touching Feeling: Affect, Pedagogy, Performativity* (Durham, NC: Duke University Press, 2003), 8.

10. N. Katherine Hayles, "Deciphering the Rules of Unruly Disciplines: A Modest Proposal for Literature and Science," in *Literature and Science*, ed. Donald Bruce and Anthony Purdy (Amsterdam: Rodopi, 1994), 31.

11. Rayna Kalas, *Frame, Glass, Verse: The Technology of Poetic Invention in the English Renaissance* (Ithaca, NY: Cornell University Press, 2018), 16.

12. Margaret Cavendish, *Poems, or several fancies In verse: with the Animal Parliament, in prose* (London: A. Maxwell, 1668), sig. Aa4^{r}, British Library, C.118.g.4.

INDEX

Note: Page numbers in italics refer to figures.

ACKNOWLEDGMENTS

Above all, my gratitude goes to Penelope Anderson and Ellen MacKay. Penelope, for treating everything I bring to her attention with unmatched thoughtfulness and care, and for teaching me how to read women's writing on its own terms. Ellen, for providing a clarity of vision that has consistently enriched the inquiries at the heart of this project, and for modeling what it truly means to be a galvanizing mentor in all things. These women have provided fierce support not only of this project but of my person at every stage of my career thus far, and their mentorship has far exceeded anything I hoped for or expected in my professional life.

This book has taken many forms over the years, and I will never be able to account for all the intellectual and social relations that have brought it into being. I have had energizing exchanges about this project with people across many locations—Illinois; Indiana; Kansas; New York; Texas; Washington, DC; and London—and for all the influences this argument has absorbed along the way, I am incredibly grateful. What follows is, I am sure, an incomplete accounting of those influences.

At Indiana University, my graduate work was nurtured by an outstanding group of scholars and teachers. Scott Herring taught me how to think structurally about a scholarly monograph in his queer theory seminar, and Nico Meli's interdisciplinary history of medicine seminar shaped the thinking that anchors this book's methodological investments. Special thanks are also due to Joan Linton, Judith Anderson, Linda Charnes, Shane Vogel, and Patty Ingham—all of whom persistently encouraged me, supported me, and challenged my thinking during my graduate education. I also benefited enormously from the strong Renaissance Studies program at IU, which provided the opportunity to think with formidable scholars across what can sometimes be trenchant disciplinary divides. I am grateful, especially, to Constance Furey, Sarah Van der Laan, Alison Calhoun, Bill Newman, Sonia Velázquez, and Hall Bjørnstad for constantly reminding me that England was not even close to the center of the Renaissance world.

In the early days of shaping this book out of my dissertation research, I had the immense gift of joining the interdisciplinary community at the Hall Center for the Humanities as a Postdoctoral Fellow. I am so grateful for the flexibility of that research position, which simply allowed me time to write and think. I found a lifelong mentor and friend in Jonathan Lamb, who was always willing to talk Bruno Latour, pedagogy, and job market insights in the Hall Center hallways. He has reflected pieces of this book back to me at several key moments and, in doing so, has helped me sharpen its contours. He is also exceptionally good at reminding me, simply, "say what you mean," and continues to prove a generous reader and interlocutor for my work. Special thanks are also due to Geraldo Sousa, David Bergeron, Misty Schieberle, Laura Mielke, and Megan Kaminski for warmly enfolding a Postdoctoral Fellow into the English Department community at KU. And to the Hall Center staff during my time in residence—Sarah Bishop, Eliott Reeder, Andrew Hodgson, Kathy Porsch, and Dan Consolver—thank you for making every day a delight.

I wrote much of this book in the company of imaginative scholars and wonderful colleagues at the Rochester Institute of Technology. I am particularly indebted to Anne Royston and Danielle Pafunda—always willing to respond to emails about turns of phrases and ready to jump in the car for apple picking and cider donut eating at a moment's notice (that em dash is for you, Anne). Lisa Hermsen, Sean Grass, Robert Glick, and Jess Lieberman have my deepest gratitude for their extraordinary mentorship during some particularly challenging years of global pandemic research and teaching. I also thank Amit Ray, Trent Hergenrader, Corinna Schlombs, Ren Van Meenen, Silvia Benso, Elena Sommers, Rose Wilson, Steve Galbraith, Amelia Hugill-Fontanel, and Rebekah Walker for their collegiality and the terrific support they all provided during my junior faculty years. My students at RIT, many of whom were women, transgender, nonbinary, and gender-fluid individuals navigating the sometimes murky waters of STEM fields, taught me more than anyone else about the real-world stakes of this project and the importance of establishing longer, deeper histories of shared humanities and STEM thinking. As I transition to Texas A&M, I have already found immense support for this project and my work. Sincerest gratitude to all of my new colleagues, and especially to Margaret Ezell, Kevin O'Sullivan, Dorothy Todd, Kathy Torabi, Emily Johansen, Sara DiCaglio, and Sarah Potvin for warmly welcoming me into this new, outstanding intellectual community.

This book has been richly informed by conversations about poetry, women's writing, the history of science, and material texts with Liza Blake, Dianne Mitchell, Jessica Rosenberg, Katherine Hunt, Hillary Nunn, Maggie Simon, Danielle

Clarke, Debapriya Sarkar, Joshua Calhoun, Pavneet Aulakh, Dyani Taff, Wendy Wall, Jen Boyle, Matthew Harrison, Brad Pasanek, Wendy Beth Hyman, and Joey Gamble. Very special thanks is due to Michelle Dowd, Megan Heffernan, and Melissa Sanchez, who generously read every page of an early manuscript draft. At a shadowy corner table in the RSA conference hotel in San Juan, these three scholars provided feedback that made this book's argument and organization what it is today. Megan introduced me to Margaret Cavendish when I was a master's student at the University of Chicago and, therefore, it is truly her guidance that seeded this long-term project. I have also benefited from conversations about and feedback on my argument at professional conferences for the Shakespeare Association of America, Renaissance Society of America, Society for the Study of Early Modern Women and Gender, and Modern Language Association, as well as works-in-progress series at the Hall Center for the Humanities, University of Kansas Medical Center, and Indiana University's College Arts and Humanities Institute. For all those communities and the people who shape them, I am immensely grateful.

Financial support from the National Endowment for the Humanities, Folger Shakespeare Library, Wellcome Library, and American Philosophical Society allowed me to complete the archival research necessary for this project. Additional funding from both the Glasscock Humanities Center and Department of English at Texas A&M helped support the book's production costs. Thank you to the people at those institutions who continue to value research, thinking, and writing as endeavors worth funding. Thank you to the research librarians, staff members, and scholars at the British Library, Folger Shakespeare Library, Clendening History of Medicine Library, Newberry Library, New York Academy of Medicine, and Wellcome Library. At the Wellcome, in particular, Stefania Signorello spent hours with me studying anatomical fugitive sheets and identifying material evidence on how these three-dimensional figures were constructed. My vision for this project was enhanced by daily walks from the British Library to the Wellcome Library during the summers of 2022 and 2023. Traversing Euston Road and the short physical distance between those two institutions—one a mecca for literary scholars and the other a monument to histories of medicine and science—became a kind of embodied experience of the work I was trying to do with this book's argument. And the Virginia Woolf bust in the gardens at Tavistock Square gave me life to keep working every day.

In the shift from research archives to printed book, my deepest gratitude to Jenny Tann, Henry S. Turner, Mary Thomas Crane, Walter Biggins, and Lily Palladino for supporting the vision of *Anatomical Forms*. Jenny has offered exceptional guidance along the way and has been an ideal partner in the publication process. I am so appreciative of her expertise and enthusiasm for the work. Many thanks, also, to the anonymous readers at the University of Pennsylvania Press for their tremendously useful comments on the manuscript. And to Molly Yarn, indexer extraordinaire, for her keen insights as we packaged the book for publication.

With full acknowledgment that this is the most unfinished expression of gratitude I offer, thank you to my family, both the family I was born into and the family I have made over the years. My immediate and extended family members have never failed to support me even when, at many times throughout this process, my work has taken me far from home and from the ability to share in family time. I am especially grateful to Sara Sheldon, Daniel Sperrazza, and Tyler Sperrazza, who have provided unwavering love and support in all things. My mom brought us with her to the halls of Canisius College in her days as adjunct faculty, and sitting in the back of those Criminal Justice night classes, I learned that you could bring your whole self to your profession. My dad read *The Hobbit* aloud to us at bedtime, and hearing the joy in his voice as he introduced this old book to us, I learned that reading could be a source of great pleasure throughout one's life. My brother models a focused work ethic and a generosity of spirit that I strive always to emulate. He is a true intellectual companion as well as my best friend. Thank you to Megan Conte, Allison Leventhal, Erin Yazgan, Lorenzo Rubio, James Bishop, and Amanda Zoch—the dearest friends and the people who constantly remind me to step away from my computer and turn my face to the sun. And, most especially, thank you to Matthew Seidel, who was with me for every word of this project and whose support is imprinted on every page.

www.ingramcontent.com/pod-product-compliance
Lightning Source LLC
LaVergne TN
LVHW091525200926
848483LV00002B/308